Advances in Experimental Medicine and Biology

Volume 1445

Advances in Experimental Medicine and Biology provides a platform for scientific contributions in the main disciplines of the biomedicine and the life sciences. This series publishes thematic volumes on contemporary research in the areas of microbiology, immunology, neurosciences, biochemistry, biomedical engineering, genetics, physiology, and cancer research. Covering emerging topics and techniques in basic and clinical science, it brings together clinicians and researchers from various fields.

Advances in Experimental Medicine and Biology has been publishing exceptional works in the field for over 40 years, and is indexed in SCOPUS, Medline (PubMed), EMBASE, BIOSIS, Reaxys, EMBiology, the Chemical Abstracts Service (CAS), and Pathway Studio.

2022 CiteScore: 6.2

Xiaoyan Qiu Editor •
Jing Huang Deputy Editor •
Xiaojun Xu Deputy Editor

Non B cell-Derived Immunoglobulins

The Structure, Characteristics and the Implication on Clinical Medicine

Springer

Editor
Xiaoyan Qiu
Department of Immunology
School of Basic Medical Sciences,
Peking University
Beijing, Beijing, China

Deputy Editor Xiaojun Xu
Department of Immunology
School of Basic Medical Sciences,
Peking University
Beijing, Beijing, China

Deputy Editor Jing Huang
Department of Immunology
School of Basic Medical Sciences,
Peking University
Beijing, Beijing, China

ISSN 0065-2598 ISSN 2214-8019 (electronic)
Advances in Experimental Medicine and Biology
ISBN 978-981-97-0510-8 ISBN 978-981-97-0511-5 (eBook)
https://doi.org/10.1007/978-981-97-0511-5

This Springer imprint is published by the registered company Springer Nature Singapore Pte Ltd.
The registered company address is: 152 Beach Road, #21-01/04 Gateway East, Singapore 189721, Singapore

Foreword

The earliest studies of adaptive immunity focused on the analysis of serum immunoglobulin (Ig) produced in response to challenges from pathogens. Ig was first discovered as an antibody, and it has long been regarded as an immune defense molecule; its non-antibody activity has been neglected. The origin and generation mechanism of Ig have been a mystery, and a number of hypotheses regarding the mechanism of Ig production, including the template hypothesis, natural selection hypothesis, and clonal selection hypothesis, were proposed and accepted at different periods in immunology history. It was discovered only in the 1960s that B cells, not other immune cells, were found to be able to produce high-level antibodies. Susumu Tonegawa revealed that Ig gene recombination is a necessary condition for the generation of Ig in B cells at the end of the 1970s; thereby, the concept that Ig is only produced by B cells has been established and widely accepted. Prof. Qiu from Peking University revealed that Ig can also be widely produced by non-B cells; these Igs display not only antibody activities but also cellular biological activities under physiological conditions. Moreover, it has been documented that these Igs maintain the basic life activities of cells by promoting cell proliferation and adhesion. In addition, Ig produced by non B cells directly participates in the pathological process of malignant tumours and immune-related diseases, such as kidney disease; importantly, Igs were proven for the first time to be produced by malignant tumours, and this type of Ig has a strong tumour cell growth-promoting effect. Prof. Qiu's finding has been supported by more results from different labs around the world. The initial discovery of the existence of non B-derived Igs has evolved into a comprehensive study on non B-Ig, not only for its role in basic immunology but also for its potential clinical applications.

The book covers non B-Ig-related basic immunology and the fundamental principles relating to clinical immunology. It is designed for immunologists, clinicians, and postgraduate students who have basic immunological knowledge. With the rapid advancement in immunology, this book, I believe, will help us to broaden our understanding of immunology and the pathogenesis of various immune-related diseases.

Beijing, China Youhui Zhang

Acknowledgements

This book is designated as the comprehensive collection of research work related to non B immunoglobulins, which were initially discovered over 30 years ago. Much about the origin, structure, physicochemical characteristics as well as functions of non B-Ig has been unveiled. I am grateful to Prof. Youhui Zhang (Institute of Oncology, Chinese Academy of Medical Sciences) for his encouragement and comments since he served as a reviewer of my Ph.D. thesis; the completion of the non B-Ig-related research would not have been possible without his unwavering support. I am grateful to Prof. Peter J Delves (University College London) for his patience, commentary, and discussions as the manuscripts took shape. His professional experience is invaluable to the entire project. I would also like to extend my gratitude to all the authors of this book, including young clinicians and immunologists. Their significant contributions to the writing of the book, despite their demanding clinical and research commitments, have been crucial, especially as the whole project took longer than anticipated. They also provided professional suggestions for clarity and accuracy.

Many individuals have supported our research in various ways over the past 20 years. The realisation of this book was made possible only when all initial hypotheses were validated. I am grateful to Prof. Guizhen Yang (Norman Bethune Health Science Centre of Jilin University, my Ph.D. supervisor) and Prof. Peixian Tang (Academy of Military Medical Sciences, my post-doctoral supervisor) for their steadfast support. My gratitude also goes to Prof. Lieping Chen (Yale University) for his support, particularly his assistance in revising my manuscripts, which were first published in Cancer Res, 2003. I must also express my sincere appreciation to Prof. Wei He, Prof. Xuetao Cao, and Prof. Bo Huang (Institute of Basic Medicine, Chinese Academy of Medical Sciences), Prof. Zhigang Tian (Institute of Immunology, University of Science and Technology of China), Prof. Xiyun Yan (Institute of Biophysics, Chinese Academy of Sciences) for their affirmation and robust support of this study; Prof. Ning Fu (Southern Medical University) for her selfless help and encouragement over the past 20 years; Prof. Yang Ke, Weigang Fang, Dalong Ma, Yuxin Yin, Xian Wang, You Wan, Baoxue Yang, and Wei Kong (all of Peking University) for their long-term support. My gratitude is extended to my collaborator Prof. C. Cameron Yin (the University of Texas MD Anderson Cancer Centre, USA), Prof. Youli Zu (Weill Medical College, Cornell University, USA), Prof. Yupei Zhao (Union Hospital, Chinese Academy of Medical Sciences), Prof. Tao Xu (Peking University

People's Hospital), and Prof. Yue Wang (Peking University Third Hospital). Lastly, my deepest thanks to Prof. Erdan Dong (Peking University), Prof. Wei Hong (NSFC), and NSFC for their continuous financial support for more than 20 years.

Xiaoyan Qiu
Jing Huang
Xiaojun Xu

Contents

Editor and Contributors

About the Editor

Dr. Xiaoyan Qiu Prof. Xiaoyan Qiu is the director of the Department of Immunology at the School of Basic Medical Sciences, Peking University, and also the director of the Key Laboratory of Medical Immunology, Ministry of Health, China. Her research focuses on non-B cell-derived immunoglobulins and cancer immunotherapy. In 1989, she was the first to discover that immunoglobulins can be produced by non-B cells, including normal cells, and are highly expressed in malignant tumor cells. These non-B-Igs play important roles in physiological and pathological conditions, such as antibody functions, immune regulation, maintenance of cellular growth and survival, as well as mediation of inflammation and tumor occurrence. She has published more than 80 articles in international peer-reviewed journals, such as Cancer Research, Leukemia, and Cellular & Molecular Immunology. Her discovery on non-B Igs has breakthrough implications for basic and clinical immunology.

Dr. Jing Huang is an Associate Professor of Immunology at Peking University. Dr. Jing obtained her Ph.D. degree in Immunology at Peking University in 2008. From 2013 to 2014, she was a visiting scholar in the Department of Cancer Immunology & AIDS at Dana-Farber Cancer Institute. Her research interests include cancer immunology and immunotherapy. In recent years, her research has focused on emphasizing non-B cell-derived immunoglobulins that regulate tumor immunity. Dr. Jing has published more than 30 peer-reviewed papers in international journals including Cancer Research, Cellular & Molecular Immunology, and Brain Behavior and Immunity. She also serves as a reviewer for several peer-reviewed immunology and cancer journals.

Dr. Xiaojun Xu graduated from University College London (UCL) with a Ph.D. degree. He then worked at King's College School of Medicine and Dentistry (London) and Royal London Hospital before joining the Department of Immunology at Peking University as an associate professor in 2003. Dr. Xu has been engaged in teaching immunology for a long time and has a deep understanding of immunology theory. His research interests include autoimmune diseases, primary immunodeficiency diseases, and immunotherapy.

Contributors

Ming Chu Department of Immunology, School of Basic Medical Sciences, Peking University, Beijing, China

Ming Cui Department of General Surgery, Peking Union Medical College Hospital, Chinese Academy of Medical Sciences, Peking Union Medical College, Beijing, China

Hui Dai Department of Immunology, School of Basic Medical Sciences, NHC Key Laboratory of Medical Immunology, Peking University, Beijing, China

Peter J. Delves Division of Infection and Immunity, Department of Immunology, UCL (University College London), London, UK

Yuqing Deng Department of Immunology, Hebei Medical University, Shijiazhuang, China

Zhenling Deng Department of Nephrology, Peking University Third Hospital, Beijing, China

Tianrui Fan Shanghai Discovery and Development Center, Abiosciences, Shanghai, China

Ning Fu Department of Immunology, Southern Medical University, Guangzhou, China

Erya Gao People's Medical Publishing House Co., LTD, Beijing, China

Zihan Geng Institute of Chinese Materia Medica, China Academy of Chinese Medical Sciences, Beijing, China

Xiaoting Gong Department of Clinical Laboratory, Tongde Hospital of Zhejiang Province, Hangzhou, Zhejiang, China

Peng Hao Telebrain Medical Technology Co., Ltd., Beijing, China

Fanlei Hu Department of Rheumatology and Immunology, Peking University People's Hospital & Beijing Key Laboratory for Rheumatism Mechanism and Immune Diagnosis, Beijing, China

Jing Huang Department of Immunology, School of Basic Medical Sciences, Peking University, Beijing, China

Xinmei Huang Department of Respiratory and Critical Care Medicine, Nanjing Drum Tower Hospital, The Affiliated Hospital of Nanjing University Medical School, Nanjing, China

Dongyang Jiang Department of Cardiology, Pan-vascular Research Institute, Shanghai Tenth People's Hospital, Tongji University School of Medicine, Shanghai, China

Guohui Li Department of Pharmacology, College of Medicine, University of Arizona, Tucson, AZ, USA

Qinyuan Liao Department of Immunology, Guilin Medical University, Guilin, Guangxi, China

Wei Liu Department of Immunology, College of Basic Medicine, Hebei Medical University, Shijiazhuang, China

Yang Liu Neuro-Oncology Branch, Center for Cancer Research, National Cancer Institute, Bethesda, MD, USA

Junfan Ma Department of Clinical Research, Sinocelltech Group Limited, Beijing, China

Teng Ma Cancer Research Center, Beijing Chest Hospital, Capital Medical University, Beijing, China

Caipeng Qin Department of Urology, Peking University People's Hospital, Beijing, China

Xiaoyan Qiu Department of Immunology, School of Basic Medical Sciences, Peking University, Beijing, China

Wenwei Shao Academy of Medical Engineering and Translational Medicine, Tianjin University, Tianjin, China

Zhengzuo Sheng Department of Thoracic Surgery, Fu Xing Hospital, Capital Medical University, Beijing, China

Zhan Shi Department of Ultrasound in Medicine, The Second Affiliated Hospital of Zhejiang University School of Medicine, Hangzhou, China

Jingshu Tang State Key Laboratory of Bioactive Substances and Functions of Natural Medicines, Institute of Materia Medica, Chinese Academy of Medical Sciences & Peking Union Medical College, Beijing, China

Chong Wang Department of Analytical Development, Cell Therapy Technology Product Engine, Oncology Therapeutic Area Unit, Takeda Pharmaceutical Company, Cambridge, MA, USA

Qianqian Wang School of Food and Drug, Shenzhen Polytechnic University, Shenzhen, China

Yue Wang Department of Nephrology, Peking University Third Hospital, Beijing, China

Zihan Wang Department of Nephrology, Peking University Third Hospital, Beijing, China

Lina Wu Central Laboratory, Key Laboratory of Carcinogenesis and Translational Research (Ministry of Education/Beijing), Peking University Cancer Hospital & Institute, Beijing, China

Miaoran Xia Department of Immunology, School of Basic Medical Sciences, Capital Medical University, Beijing, China

Lin Xiao Department of Clinical Laboratory, National Cancer Center/National Clinical Research Center for Cancer/Cancer Hospital, Chinese Academy of Medical Sciences and Peking Union Medical College, Beijing, China

Tao Xu Department of Urology, Peking University People's Hospital, Beijing, China

Xiaojun Xu Department of Immunology, School of Basic Medical Sciences, Peking University, Beijing, China

Huige Yan Department of Orthopaedic Surgery, Sir Run Run Shaw Hospital, Zhejiang University School of Medicine, Hangzhou, China

C. Cameron Yin Department of Hematopathology, The University of Texas MD Anderson Cancer Center, Houston, TX, USA

Sha Yin Xi'an Children's Hospital, Affiliated Children's Hospital of Xi'an Jiaotong University, Xi'an, Shaanxi, China

Meng Yu Shanghai R&D Department, Shanghai Baiying Biotechnology Co., Ltd., Shanghai, China

Chi Zhang Department of Immunology, School of Basic Medical Sciences, Peking University, Beijing, China

Jingxuan Zhang Platform Biology, Suzhou Abogen Biosciences, Suzhou, Jiangsu, China

Li Zhang Division of HIV/AIDS and Sex-Transmitted Virus Vaccines, Institute for Biological Product Control, National Institutes for Food and Drug Control (NIFDC) and WHO Collaborating Center for Standardization and Evaluation of Biologicals, Beijing, China

Liang Zhang Department of Biomedical Sciences, City University of Hong Kong, Hong Kong, SAR, China

Meng Zhang Department of Cardiology, Aerospace Center Hospital, Beijing, China

Shuai Zhang Strategic Marketing Department of Central Marketing, GeneralElectric Healthcare Co., Shanghai, China

Youhui Zhang Department of Immunology, Cancer Institute, Chinese Academy of Medical Science and Peking Union Medical College, Beijing, China

Jie Zheng Hematologic Disease Laboratory, Department of Stem Cell Transplantation, Beijing Children's Hospital, Capital Medical University, Beijing, China

Wenjing Zhou Institute of Thoracic Oncology, West China Hospital, Sichuan University, Chengdu, China

Xiaohui Zhu Department of Obstetrics and Gynecology, Center for Reproductive Medicine, Peking University Third Hospital, Beijing, China

Zhu Zhu Department of Immunology, School of Basic Medical Sciences, Peking University, Beijing, China

Illustrations and Illustration Coordinator

Yan Zhang Department of Immunology, School of Basic Medical Sciences, Peking University, Beijing, China

Part I

Non B Cell-Derived Immunoglobulins: From Gene to Structure and Function

1 Non B Cell-Derived Immunoglobulin, A Brighter Horizon for the Future

Ming Chu, Ning Fu, Liang Zhang, Meng Yu, Youhui Zhang, and Xiaoyan Qiu

Abstract

The canonical theory of immunology stating that "Immunoglobulin (Ig) is produced by B lymphocytes and exerts antibody activity" has been established since the 1970s. However, the discovery of non B cell-derived Igs (non B-Igs), which can exert multiple biological activities in addition to their antibody activities, necessitates a reevaluation of the classic concept of Ig. This has been documented with a number of characteristics related to their structure, modification, genetic regulation as well as the functions associated with clinical conditions, particularly multiple cancers. The discovery of non B-Ig provides us with a new perspective to better understand not only basic immunology, but also various Ig-related clinical manifestations including autoimmune diseases, chronic inflammation, and anaphylaxis. Notably, non B-Ig can directly promote the occurrence of malignant tumours.

Keywords

Non B cell-derived immunoglobulins (non B-Igs)

M. Chu (✉) · X. Qiu
Department of Immunology, School of Basic Medical Sciences, Peking University, Beijing, China
e-mail: famous@bjmu.edu.cn; qiuxy@bjmu.edu.cn

N. Fu
Department of Immunology, Southern Medical University, Guangzhou, China
e-mail: zfzn@smu.edu.cn

L. Zhang
Department of BIomedical Sciences, City University of Hong Kong, Hong Kong, SAR, China
e-mail: liangzhang.28@cityu.edu.hk

M. Yu
Shanghai R&D Department, Shanghai Baiying Biotechnology Co., Ltd., Shanghai, China
e-mail: 1711110028@pku.edu.cn

Y. Zhang
Department of Immunology, Cancer Institute, Chinese Academy of Medical Science and Peking Union Medical College, Beijing, China

1.1 The Historical Background and Limitations of the Classic Theory About the Formation, Structure, and Function of Ig

1.1.1 The Broadly Accepted Classic Concept That Ig Only Exhibits Antibody Activity Needs to Be Reevaluated from a Fresh Perspective

The discovery of Ig began with the identification of the antibody. In 1890, von Behring and Kitasato first identified an "antitoxin" produced as a result of injecting diphtheria toxin into animals (Fig. 1.1) (von Behring et al. 1991). Over

X. Qiu et al. (eds.), *Non B cell-Derived Immunoglobulins*, Advances in Experimental Medicine and Biology 1445, https://doi.org/10.1007/978-981-97-0511-5_1

the subsequent decade, antibactericidal substances were frequently reported in animal serums. Preiffer and Bordet noted the lysis of *Vibrio cholerae* in the intestines of immunised guinea pigs and dubbed the substance responsible as bacteriolysin (Pfeiffer 1895). Gruber and colleagues found that serums from animals immunised with bacteria could induce bacterial agglutination, and the substance causing this agglutination was termed agglutinin (Scheibner et al. 1977). Subsequently, antitoxin, bacteriolysin, and agglutinin were collectively referred to as "antibody (Ab)", and the substance that could induce the production of Ab became known as an antigen (Ag).

From these early observations, the concept of antigen and antibody became firmly established. However, the nature of the antibody was unclear until 1937 when Tiselius and, separately, Kabat and coworkers used serum electrophoresis to discover that the level of gammaglobulin in serum increased significantly following Ag immunisation (Fig. 1.1). Furthermore, the gammaglobulin demonstrated clear antibody activity, leading to the proposal that gamma globulin was indeed the antibody (Tiselius 1937). In 1968, a special committee of the World Health Organisation (WHO), and in 1972, the International Society for Immunology, defined proteins with either antibody activity or antibody structure as immunoglobulin (Ig) (Fig. 1.1). Although a few exceptions, such as Ig produced by multiple myeloma, have been found to have an antibody structure but not the immune defence activity, Ig has not been identified to have other activities. Thus, the concept of "Ig function equates to antibody" has been generally accepted. Yet, it has never been explained why vertebrates produce large amounts of Ig at nearly constant levels daily without antigen stimulation conditions, and whether the Ig might serve some non-antibody function, apart from immune defence. Recently, we have found that Igs, including IgM, IgG, and IgA, are abundantly present in T cell-deficient nude mice. However, the nude mice are unable to produce specific IgG or IgA antibodies against specific antigens when immunised with thymus-dependent (TD) antigens. Further analysis showed that IgG in the circulation of nude mice often combined with other molecules, such as apolipoprotein and cytokines. This discovery suggests that Ig not only exhibits antibody activity according to the classical concept but also performs various biological activities that have yet to be identified.

1.1.2 Several Influential Hypotheses Had Emerged Regarding the Mechanism of Antibody Production

In the annals of history, several significant doctrines about the source of antibodies have emerged. Before the 1960s, immunologists were primarily focused on the biochemical process of

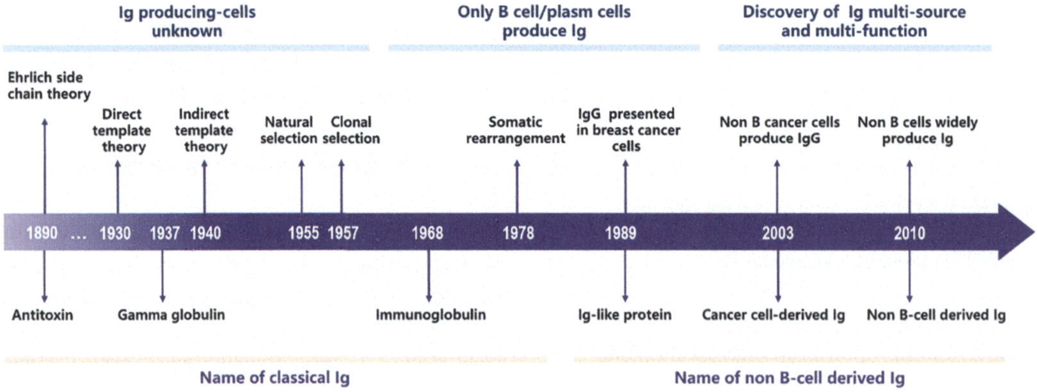

Fig. 1.1 A historical briefing of immunoglobulin

antibody production, rather than the actual cell types producing the antibodies. In 1897, Paul Ehrlich proposed the "side chain theory", suggesting that antitoxin molecules on the cell surface (i.e., cell receptor) recognised the toxin entering the body. After the invasion of the toxin into cells, the cells stimulated by the toxin produced more anti-toxin molecules from the cell surface and released them into the bloodstream (Ehrlich 1900). However, due to the inability to explain the issue of antibody diversity, this hypothesis had not been widely accepted. In 1930, Haurowitz put forward the "direct template hypothesis" that antigen molecules serve as templates that determine the structure of antibodies (Breinl et al. 1930). Nevertheless, the "direct template hypothesis" couldn't account for immune memory. Accordingly, Pauling and colleagues modified the "direct template hypothesis" and proposed an indirect template hypothesis, wherein rather than being guided directly by the antigen to synthesise antibodies, the antibody-producing cells produced a corresponding structure using the antigen as a template, and integrated the corresponding structure into their genomes and transmitted the genetic information to their daughter cells (Pauling 1940). In 1955, Jerne proposed the "natural selection hypothesis", which posited that under normal conditions, the body synthesised various specific antibodies in advance, known as natural antibodies (Fig. 1.1). Upon the entry of an antigen into the body, it selectively binds to the corresponding antibodies, which then stimulates the antibody-producing cells to produce more such antibodies (Jerne 1955). In 1957, Mac Farlane Burnet further proposed the "clonal selection hypothesis", wherein the body produced a variety of clones of immune cells, each expressing a specific receptor generated by genetic somatic mutation (Fig. 1.1). When a particular receptor was selected and bound by its specific Ag, the process induced clonal proliferation and antibody production (Mackay 2008). Following the formation of specific cell clones, some would divide to form memory cells. In addition, the immune cells recognising antigens during the embryonic phase would either be inhibited or undergo apoptosis, thus establishing immunological tolerance to the corresponding antigens. This theory explained most immunological phenomena exquisitely.

1.1.3 B Cells and Plasma Cells Have Been Found to Secrete Antibodies

In the 1960s, immunologists began to identify Ig-producing cells. As the function of Ig was first discovered through its antibody activity, immunologists principally focused on immune cells, but not on other types of cells, in their search for Ig-producing cells. Gowans discovered that a graft versus host reaction occurred following the transplantation of small lymphocytes (Gowans 1962). Additionally, after irradiation, there was a significant reduction in both the number of small lymphocytes and antibody production. These results implied that antibody-producing cells were part of the small lymphocyte population. Subsequent findings using immunofluorescence staining with anti-isotype serum showed that Ig could be located on the membrane of bone marrow-derived lymphocytes (B cells), but not the thymus-derived lymphocytes (T cells). Following on from this, Miller and his colleagues found that bone marrow-derived B cells could differentiate into plasma cells to produce secreted antibody (Miller 1961). As a result, the concept of antibodies/Ig being generated by B cells/plasma cells became widely accepted. Logically, the "immunocompetent cells" in Burnet's "clonal selection theory" were replaced by "B cells". However, whether non B cells could also produce Ig was largely overlooked.

1.1.4 The Concept That "Only B Cells/Plasma Cells Can Produce Ig" Was Further Reinforced by a Misinterpretation of the Findings by Susumu Tonegawa

For a long period, immunochemists recognised that Ig demonstrates an incalculable diversity within an individual, with the diversity primarily located in the first functional domain at its N-terminus (known as the variable region). The mechanism behind this Ig diversity remained a mystery. Australian immunologist Frank Macfarlane Burnet once proposed that the genes encoding the Ig variable region consisted of different gene fragments, with their diversity stemming from the random rearrangement of these gene fragments. In 1978, Suzumu Tonegawa confirmed Burnet's hypothesis for the first time (Fig. 1.1). Tonegawa discovered that Ig genes encoding the Ig variable region did indeed include numerous variable (V), diversity (D), and joining (J) fragments within the genome, which could be randomly rearranged to form diverse VHDJH (for Ig heavy chain) or VLJL (for light chain) genes. For this significant discovery, Tonegawa Susumu was awarded the Nobel Prize in Physiology or Medicine in 1987 (Brack et al. 1978; Tonegawa et al. 1977, 1978; Tonegawa 1976). In his study, Tonegawa utilised a mouse myeloma cell known to produce Ig as a model and used the mRNA of the Igκ produced by the myeloma cell itself as a probe. He demonstrated through a Southern blot that there was Ig gene rearrangement in the genome of the myeloma cell, while no band of Ig gene rearrangement was observed in the mouse embryonic tissue, liver tissue or spermatid used as a negative control. This result is consistent with the notion of "B cells/plasma cells generate Ig" mentioned above, which further reinforced the concept of "only B cells/plasma cells generate Ig"—a concept that has been documented in immunological textbooks to this day. It is important to note that this is entirely a historical misunderstanding, limited by the insufficient understanding of scientists at the time regarding Ig diversity and Ig gene rearrangement mechanisms. It has since been revealed that different cell types can exhibit different VHDJH or VLJL rearrangement patterns. It should be highlighted that the probe used to analyse the VLJL rearrangement on the genome in the mouse myeloma cells by Tonegawa was the Igκ chain mRNA from the myeloma cells themselves, not from other tissues. As a result, it is reasonable that only the myeloma cells themselves, but not the mouse embryos, liver tissues, and spermatogenic cells, showed the Igκ rearranged band. In fact, Zheng et al. confirmed that epithelial cancer cells and normal tissue cells can also express Ig, and both VHDJH and VLJL expressed by non B cells displayed unique rearrangement characteristics (Huang et al. 2008; Zheng et al. 2009). This implies that the classical concept "only B lineage can produce Ig" needs to be corrected.

1.2 The Discovery and Identification of Non B Cell-Derived Igs

A growing body of evidence demonstrates that, in addition to B cells' capability of producing Ig (B-Ig), non B cells broadly produce Ig (non B-Ig). Moreover, non B-Ig presents features distinct from traditional B-Ig in variable region sequence, physicochemical properties, and function.

1.2.1 Non B Cell-Derived Ig Was Initially Discovered in Epithelial Tumour Cells

In 1989, Qiu first discovered that IgG staining was clearly present in breast cancer cells, but not in epithelial cells of normal breast and benign breast tumour tissues (such as fibroadenoma) (Fig. 1.1). Subsequently, Qiu and colleagues found that IgG is also present in a range of other malignant tumour cells. Importantly, Ig gene rearrangement, a specific event once thought to solely occur in B cells, was also found in these non B lineage cancer cells; this finding was first published in Chinese Journals in 1996. In 2003, for the first time in the field, we reported that epithelial cancer cells could synthesise and secrete IgG, and the IgG displayed growth factor-like activity and promoted survival and growth of cancer cells (Fig. 1.1) (Qiu et al. 2003). Subsequently, many researchers independently confirmed the discovery. Inspired by our original observations, we hypothesised that the reason why some tumour cells can produce Ig is genomic instability. However, a series of subsequent experiments indicate that under physiological conditions nearly all non B lineage cells can produce Ig (non B-Ig), and non B-Ig not only has different origins from that of classic Ig but also exhibits different characteristics in the structure of the variable region, physicochemical properties and function. Our key findings have been: (1) Non B-Ig, including IgM, IgG, IgA, and IgD, is widely expressed in different lineage cells including epidermis, liver epithelium, neurons, spermatogenic cells, monocytes, and cardiomyocytes (Zhang et al. 2010; Huang et al. 2008; Jiang et al. 2015; Shao et al. 2016). (2) Non B-Ig primarily distributes in local tissues and exhibits hydrophobicity. (3) Non B-Ig includes both 4 peptide chain structures and free light chains (Wang et al. 2022). (4) The regulation of non B-Ig in non B cells presents unique characteristics, such as non B-Ig gene transcription depending on the Octamer Binding Transcription Factor 1 (Oct-1) instead of the Oct-2 on which B-Ig relies (Fig. 1.1) (Zhu et al. 2010). (5) Unlike the B-Ig that exhibits diversity in its variable region, the non B-Ig shows a conservative V(D)J or VJ sequence pattern among the same lineage-origin cells (Zheng et al. 2009). (6) Non B-Ig not only displays the immune defence function as antibody does, but also promotes cell survival, proliferation and migration, and is involved in cancer initiation and metastasis as an extracellular matrix protein (Wang et al. 2022; Liu et al. 2015; Liao et al. 2015; Sheng et al. 2016; Adamovic et al. 2009; Chen et al. 2007, 2010, Babbage et al. 2006; Qiu et al. 2013). (7) Cancer cells can highly express IgG, especially cancer stem cells (CSCs) can produce the IgG with a unique sialylation modification (sialylated IgG), which can promote tumour stemness and inhibit the proliferation of T cells which play the role as immune checkpoints, therein the tumour cell-secreted IgG can be used as a new target for tumour targeted therapy and immunotherapy (Zheng et al. 2007; Tang et al. 2018; Yoshimi et al. 2002).

1.2.2 A New Horizon of Immunoglobulin Study

From the brief historical review of Ig study, we are fully aware of the fact that the concept of Ig source and function has been firmly established, i.e., "Ig is only produced by B cell lineage, Ig only serves as antibody activity". Undoubtedly, the discovery of non B-Ig is a great challenge to the traditional understanding about Ig. Our strenuous study for more than 30 years has unambiguously confirmed that B cell is not the sole source of Ig and Ig has many functions except for the antibody activity. Here, we propose a new concept of "Ig multi-source and multi-function". We believe that the discovery of non B-Ig has an epoch-making significance, which open a new door to reveal the pathogenesis of pathogenic Ig-mediated immune-related diseases.

In basic research, it will lead us to redefine and understand the origin, structure, function, and genetic properties of Ig. The discovery of non B-Ig led us to understand that the classic concept of Ig source and function has hindered the revelation of Ig's multi-source and multi-function. As early as 30 years ago, several groups had found that IgG was significantly present in neurons of mouse and rabbit fetal or adult brain (Yoshimi et al. 2002; Chun et al. 1991). One reasonable explanation for this finding would be neuronal endocytosis of the Ig. However, in 2008, Huang et al. found that the IgG was not transported via endocytosis by the neurons but was produced by the neurons themselves (Huang et al. 2008). In addition, it was found that strong IgG-positive stainings in mouse spermatogenic cells, but no reasonable explanation for its origin has been proposed. Similarly, in 2009, Huang et al. demonstrated that abundance of IgG and

IgM was produced by spermatogenic cells (Huang et al. 2009). More interestingly, it is known that large amounts of IgG deposits in the tumour microenvironment (TME), which was commonly considered as anti-tumour antibodies produced by B cells. However, Qiu et al. have found that a variety of non B cell-origin cancer cells, especially cancer stem cells, can produce and secrete IgG, resulting in direct promotion of cancer initiation and metastasis. It can be seen how the long-held belief of the classic concepts has hindered the in-depth understanding of the more important non-antibody function of Ig. In fact, the classic concept that "Ig is produced by the B cell lineage, has a four-polypeptide chain structure, is hydrophilic and exerts antibody activity" applies to the B-Ig and circulating Ig only, but not the novel non B-Ig.

In the clinic, the findings will push us to re-understand the pathogenesis of many Ig-mediated diseases. Autoimmune diseases, including systemic lupus erythematosus (SLE), rheumatoid arthritis (RA), myasthenia gravis, and autoimmune hepatitis, have been defined as B cell-derived autoantibody-mediated conditions. However, the autoantigens recognised by the autoantibodies, such as DNA, histone, and cytoskeleton proteins, are usually located in cell nuclei rather than on the cell surface of the target cells. We have found that IgG, IgM, and IgA located in both the cytoplasm and cell nucleus of normal cells can bind to RNA-binding protein in cell nuclear, as well as the cytoplasmic intermediate filaments (unpublished data). In addition, it is known that Ig mediates 90% of renal inflammatory diseases, such as IgA-type renal disease caused by IgA deposition in the mesangial area, and the membranous nephropathy due to IgG4 deposition between podocytes and basement membrane. To date, the Igs are commonly considered to be produced by B cells, the Ig pass through the blood vessels and deposit in the mesangial area or between the podocyte and basement membrane, respectively. However, abolishing the B cells with anti-CD20 antibody did not benefit patients with IgA nephropathy (Deng et al. 2018). Recently, Deng et al. found that mesangial cells could produce and secrete IgA, and the podocyte could produce and secrete IgG4 (Deng et al. 2018; Jing et al. 2018). These findings suggest that local mesangial cells or podocyte undergoing pathological stress may produce the pathological Igs of either the IgA or IgG4 class.

In addition, systemic amyloidosis and Ig light chain deposition disease are two major incurable diseases in clinical practice, both are caused by the deposition of hydrophobic, monoclonal or oligoclonal free Ig light chain in the extracellular matrix of the heart, kidney and/or liver tissue, resulting in irreversible tissue damage. Because some cases were found to be accompanied by multiple myeloma, it has been considered that its incidence is related to multiple myeloma in the field. However, evidence of multiple myeloma cannot be found in most patients. In fact, Wang et al. have proved that non B cells widely produce free Igκ light chains that showed high homology to the pathogenic free Ig light chains described above, which also exhibit hydrophobic and extracellular deposition characteristics, and are largely released into the extracellular spaces under cell stress conditions to promote the inflammatory response. This finding suggests that the pathogenic free Ig light chain that leads to systemic amyloidosis and Ig light chain deposition disease is not a particular product of the B cell lineage, and may open new avenues for understanding the pathogenesis, improving diagnosis and treatment of these two types of diseases.

In summary, due to limitations of classical concepts, the source of these pathogenic Igs has always been considered to be produced by B cells, but B cell removal therapy has only partially alleviated the condition of some patients. This further suggests that the B-Ig is not the main culprit in all patients with autoimmune disease or Ig-mediated kidney inflammation. The discovery of non B-Ig will certainly prompt understanding of the pathogenesis of Ig-mediated disease, and bring about more effective treatments.

References

Adamovic T et al (2009) Microalterations of inherently unstable genomic regions in rat mammary carcinomas as revealed by long oligonucleotide array-based comparative genomic hybridization. Cancer Res 69(12):5159–5167

Babbage G et al (2006) Immunoglobulin heavy chain locus events and expression of activation-induced cytidine deaminase in epithelial breast cancer cell lines. Cancer Res 66(8):3996–4000

Brack C et al (1978) A complete immunoglobulin gene is created by somatic recombination. Cell 15(1):1–14

Breinl F et al (1930) Chemische Untersuchung des Präzipitates aus Hämoglobin und Anti-Hämoglobin-Serum und Bemerkungen über die Natur der Antikörper. Hoppe-Seyler's Zeitschrift für physiologische Chemie 192:45–57

Chen Z et al (2007) Immunoglobulin G expression in carcinomas and cancer cell lines. FASEB J 21(11):2931–2938

Chen Z et al (2010) Immunoglobulin G is present in a wide variety of soft tissue tumors and correlates well with proliferation markers and tumor grades. Cancer 116(8):1953–1963

Chun JJ et al (1991) The recombination activating gene-1 (RAG1) transcript is present in the murine central nervous system. Cell 64(1):189–200

Deng H et al (2018) Expression of immunoglobulin a in human mesangial cells and its effects on cell apoptosis and adhesion. Mol Med Rep 17(4):5272–5282

Ehrlich P (1900) Croonian lecture. On immunity with special reference to cell life. Proc R Soc Lond 66:424–448

Gowans JL (1962) The fate of parental strain small lymphocytes in F1 hybrid rats. Ann N Y Acad Sci 99:432–455

Huang J et al (2008) Expression of immunoglobulin gene with classical V-(D)-J rearrangement in mouse brain neurons. Int J Biochem Cell Biol 40(8):1604–1615

Huang J et al (2009) Expression of immunoglobulin gene with classical V-(D)-J rearrangement in mouse testis and epididymis. J Histochem Cytochem 57(4):339–349

Jerne NK (1955) The natural-selection theory of antibody formation. Proc Natl Acad Sci USA 41(11):849–857

Jiang D et al (2015) IgG and IgA with potential microbial-binding activity are expressed by normal human skin epidermal cells. Int J Mol Sci 16(2):2574–2590

Jing Z et al (2018) Expression of immunoglobulin G in human podocytes, and its role in cell viability and adhesion. Int J Mol Med 41(6):3296–3306

Liao Q et al (2015) Aberrant high expression of immunoglobulin G in epithelial stem/progenitor-like cells contributes to tumor initiation and metastasis. Oncotarget 6(37):40081–40094

Liu Y et al (2015) Binding of the monoclonal antibody RP215 to immunoglobulin G in metastatic lung adenocarcinomas is correlated with poor prognosis. Histopathology 67(5):645–653

Mackay IR (2008) Autoimmunity since the 1957 clonal selection theory: a little acorn to a large oak. Immunol Cell Biol 86(1):67–71

Miller JF (1961) Immunological function of the thymus. Lancet 2(7205):748–749

Pauling L (1940) A theory of the structure and process of formation of antibodies. J Am Chem Soc 62:2643–2657

Pfeiffer R (1895) Weitere Mittheilungen über die specifischen Antikörper der Cholera. Z Hyg Infekt 20:198–219

Qiu X et al (2003) Human epithelial cancers secrete immunoglobulin g with unidentified specificity to promote growth and survival of tumor cells. Cancer Res 63:6488–6495

Qiu X et al (2013) Immunoglobulin gamma heavy chain gene with somatic hypermutation is frequently expressed in acute myeloid leukemia. Leukemia 27(1):92–99

Scheibner E et al (1977) Theorie der agglutination theory of agglutination. Zentralbl Veterinarmed B 24(10):821–829

Shao W et al (2016) Identification of liver epithelial cell-derived Ig expression in μ chain-deficient mice. Sci Rep 6:23669

Sheng Z et al (2016) IgG is involved in the migration and invasion of clear cell renal cell carcinoma. J Clin Pathol 69(6):497–504

Tang J et al (2018) Lung squamous cell carcinoma cells express non-canonically glycosylated IgG that activates integrin-FAK signaling. Cancer Lett 430:148–159

Tiselius A (1937) Electrophoresis of serum globulin: electrophoretic analysis of normal and immune sera. Biochem J 31(9):1464–1477

Tonegawa S (1976) Reiteration frequency of immunoglobulin light chain genes: further evidence for somatic generation of antibody diversity. Proc Natl Acad Sci USA 73(1):203–207

Tonegawa S et al (1977) Cloning of an immunoglobulin variable region gene from mouse embryo. Proc Natl Acad Sci USA 74(8):3518–3522

Tonegawa S et al (1978) Sequence of a mouse germline gene for a variable region of an immunoglobulin light chain. Proc Natl Acad Sci USA 75(3):1485–1489

von Behring E et al (1991) The mechanism of diphtheria immunity and tetanus immunity in animals. 1890. Mol Immunol 28(12):1319–1320

Wang Q et al (2022) A widely expressed free immunoglobulin κ chain with a unique Vκ4-1/Jκ3 pattern promotes colon cancer invasion and metastasis by activating the integrin β1/FAK pathway. Cancer Lett 540:215720

Yoshimi K et al (2002) IgG-immunostaining in the intact rabbit brain: variable but significant staining of hippocampal and cerebellar neurons with anti-IgG. Brain Res 956(1):53–66

Zhang S et al (2010) Immunoglobulin gene locus events in epithelial cells of lactating mouse mammary glands. Cell Mol Life Sci 67(6):985–994

Zheng H et al (2007) Immunoglobulin alpha heavy chain derived from human epithelial cancer cells promotes the access of S phase and growth of cancer cells. Cell Biol Int 31(1):82–87

Zheng J et al (2009) Immunoglobulin gene transcripts have distinct VHDJH recombination characteristics in human epithelial cancer cells. J Biol Chem 284(20):13610–13619

Zhu X et al (2010) Distinct regulatory mechanism of immunoglobulin gene transcription in epithelial cancer cells. Cell Mol Immunol 7(4):279–286

2 The Expression of Non B Cell-Derived Immunoglobulins

Jie Zheng, Guohui Li, Wei Liu, Yuqing Deng, and XiaoJun Xu

Abstract

Although V(D)J recombination and immunoglobulin (Ig) production are traditionally recognised to occur only in B lymphocytes and plasma cells, the expression of Igs in non-lymphoid cells, which we call non B cell-derived Igs (non B Igs), has been documented by growing studies. It has been demonstrated that non B-Igs can be widely expressed in most cell types, including, but not limited to, epithelial cells, cardiomyocytes, hematopoietic stem/progenitor cells, myeloid cells, and cells from immune-privileged sites, such as neurons and spermatogenic cells. In particular, malignant tumour cells express high level of IgG. Moreover, different from B-Igs that mainly localised on the B cell membrane and in the serum and perform immune defence function mainly, non B-Igs have been found to distribute more widely and play critical roles in immune defence, maintaining cell proliferation and survival, and promoting progression. The findings of non B-Igs may provide a wealthier breakthrough point for more therapeutic strategies for a wide range of immune-related diseases.

Keywords

Immunoglobulin · Expression · Cancer · Non B cell · Tumour progression · Tumour metastasis · Immune regulation

J. Zheng (✉)
Hematologic Disease Laboratory, Department of Stem Cell Transplantation, Beijing Children's Hospital, Capital Medical University, Beijing, China
e-mail: zhengjie@bch.com.cn

G. Li
Department of Pharmacology, College of Medicine, University of Arizona, Tucson, AZ, USA
e-mail: guohuili@arizona.edu

W. Liu · Y. Deng
Department of Immunology, Hebei Medical University, Shijiazhuang, China
e-mail: liuweihebmu@hebmu.edu.cn; dengyuqing@hebmu.edu.cn

X. Xu
Department of Immunology, School of Basic Medical Sciences, Peking University, Beijing, China
e-mail: xiaojunxu@bjmu.edu.cn

Immunoglobulins (Igs), per the canonical definition, are a group of highly specialised recognition proteins expressed either on the B lymphocyte cell membrane or in secretion. These can recognise various antigens from pathogens such as bacteria, viruses, and other disease-causing organisms. The membrane-bound form of Ig serves as a receptor for antigens known as the B cell receptor (BCR), whilst the secreted form, produced by terminally differentiated B cells also

X. Qiu et al. (eds.), *Non B cell-Derived Immunoglobulins*, Advances in Experimental Medicine and Biology 1445, https://doi.org/10.1007/978-981-97-0511-5_2

called plasma cells, is the antibody (Ab). Both BCR and Ab from the same B cell possess the same antigen specificity. The B cell has traditionally been acknowledged as the sole producer of Igs. However, this long-held belief has recently been challenged by burgeoning evidence demonstrating that Igs can be generated by numerous lineages of cells rather than sole B cells, thus the term 'non B-Igs' was coined to distinguish these Igs from those classically derived from B cells. In addition to their origin source, non B-Igs differentiate from those of B cell-derived ones in multifaceted characteristics, as presented in detail in the following corresponding sections.

2.1 Non B-Igs Expression in Non B Cancers

The pathway leading to the study of non B-Igs expression was first forged by Dr. Qiu in the late 1980s, and the ground-breaking findings were subsequently published in three key journals in China (Qiu and Yang 1996a, b, 1997). In these studies, IgG-positive signal staining was observed not only in the plasma of cancer cells from many types of cancers, including breast cancer, colon cancer, liver cancer, and the corresponding tumour cell lines, but also in highly proliferative cells with elevated tumorigenesis risk (Qiu and Yang 1997). Western blot (WB) analysis showed the proteins reacting with anti-human IgG illustrated the same antigenicity and molecular weight as those of IgG. Subsequently, Ig expression in non B cancer cells was further confirmed by the results of reverse transcription-polymerase chain reaction (RT-PCR) and Southern blot analyses which revealed Ig gene recombination occurred in non B cells (Qiu and Yang 1996a, b, 1997). Based on these observations, Dr. Qiu hypothesized that the IgG-like signal in the cancer cells was not a staining accident or a false positive signal, but that these Igs were expressed therein. Henceforth, this hypothesis is termed the 'non B-Igs hypothesis'. This revolutionary hypothesis has been supported by more emerging evidence, which will be introduced chronologically in the following parts.

It was reported by Kimoto in 1998 that the gene transcripts for the heavy chain constant regions (C region) of some Igs and the T-cell receptor (TCR) were present in five carcinoma cell lines (Kimoto 1998), which supports the non B-Igs hypothesis. Qiu published her comprehensive findings that tumour cells express Igs with compelling evidence in 2003 (Qiu et al. 2003). In that study, not only the expressions but also the functions of Igs in human cancers of epithelial origin, including carcinomas of the breast, colon, liver, lung, and established epithelial cancer lines, were investigated. Most notably, the major concern about the non B-Igs hypothesis arguing that the Ig detected in cancer tissue is contamination from tumour-infiltrating B cells, was directly addressed by strict and elegant designs using multiple techniques.

First, in that study, the Ig-like protein was confirmed to be IgG by NH2-terminal sequence analysis which showed the same 14 amino acid residues of the purified protein as the classical IgG protein. Second, the expression of the Ig gene in cancer samples was determined by in situ hybridisation (ISH) and Northern blot analysis. Furthermore, Igγ chain mRNA was also detected in all 5 lung cancer tissues and 3 of the 5 adjacent lung tissues. Third, RT-PCR and subsequent sequencing showed that the IgG heavy chains (including V, D, J, and CH1 regions) were expressed in the cancer cells isolated by laser capture microdissection (LCM). The third complementarity determining regions (CDR3) segments, which include the VDJ recombination sequence, were detected in HeLa S3 (cervical cancer) (GenBank No. AY05372), HeLa MR (cervical cancer), A549 (lung adenocarcinoma) (GenBank No. AY035373), and HT-29 (colon cancer) (GenBank No. AY035374). Furthermore, HeLa S3, and HT-29 demonstrated the monoclonality of the VDJ recombination, although occasionally it contained single nucleotide deletions, insertions or mutations, whereas HeLa MR and A549 did not. Interestingly, some identical CDR3 sequences were shown in both HT-29 and HeLa MR, which indicated that there is an identically specific function of the CDR3 between HT-29 and HeLa MR. Additionally, human cancer cells were found to actively express the recombination activating genes 1 and 2 (RAG1 and RAG2), the enzymes required for the formation of VDJ recombination, which further suggested that human epithelial cancer cells are capable of processing VDJ recombination. Moreover, for the first time, it was discovered that the IgG serves as a growth factor-like molecule, promoting the survival and proliferation of cancer cells.

Motivated by the discoveries of non B-Igs, a number of groups have reported accumulated evidence of Ig protein and corresponding transcriptional expression in many human epithelial cancers (Fig. 2.1), numerous cancer cell lines (listed in parentheses after the corresponding cancer) and their cultured media, e.g. oral epithelial cancer (WSU-HN6 and CAL27) (Qiu et al. 2008; Wang and Gan 2021), salivary cystic carcinoma (Qiu et al. 2008; Peng et al. 2017; Lv et al. 2017), salivary gland cancer (SACC-83) (Lv et al. 2017), breast cancer (MCF-7, ZR75, MDA-MB-231, SKBR3, T47D, Bcap37, and BT474) (Qiu and Yang 1996a, b, 1997; Kimoto 1998; Qiu et al. 2003; Babbage et al. 2006; Zhang et al. 2010; Ma et al. 2013; Yang et al. 2013; Liao et al. 2015; Wang et al. 2020; Li et al. 2012; Lee 2009; Zheng et al. 2007a; Kang et al. 2012; Welinder et al. 2013), liver cancer (HepG-2, Hep3B, BCL-7402, and Hep-2) (Qiu et al. 2003; Chen et al. 2007; Lei et al. 2014; Lee et al. 2009a; Hu et al. 2012), lung cancer (SK-MES-1 and A549) (Qiu et al. 2003; Lee 2009; Chen et al. 2007; Liu et al. 2015; Wang et al. 2019), nasopharyngeal cancer (CNE1 and HNE2) (Zheng et al. 2007a, b; Hu et al. 2011), laryngeal carcinoma (HEp2) (Kimoto 1998; Wang et al. 2013), oesophageal carcinoma (Zhang et al. 2012), gastric cancer (Li et al. 2012, 2016), bladder cancer (T24, 5637, BIU87, and EJ) (Liang et al. 2013; Sheng et al. 2016a), colon cancer (SW116, HT-29, SW480, LOVO, and HCT116) (Kimoto 1998; Qiu et al. 2003; Geng et al. 2019; Wu et al. 2014; Niu et al. 2012; Deng et al. 2006; Jiang et al. 2019; Hu et al. 2012), prostate cancer (PC3, Du145, and LNCaP) (Chen et al. 2007; Liu et al. 2012; Qin et al. 2020; Xu et al. 2016), renal cancer (CAKI-I, 786-O, 293, and ACHN) (Lv et al. 2017; Liang et al. 2013; Hu et al. 2012; Sheng et al. 2016b), cervical cancer (Hela, C-33A, CA33, and ME-180) (Qiu et al. 2003; Li et al. 2004, 2012; Chen et al. 2007; Lee et al. 2009a; Zheng et al. 2007b; Hu et al. 2012; Wang et al. 2014), ovarian cancer (CaOV3, SK-OV-3, and OC-3-VGH) (Qiu et al. 2003; Ji et al. 2015), thyroid cancer (Qiu et al. 2012), parathyroid cancer (Cui et al. 2020a), pancreatic cancer (HC48, SW1990, MIA PaCa-2, PANC-1, AsPC-1, T3M4, CFPAC-1, HPAF, and BxPC-3) (Kimoto 1998; Hu et al. 2012; Li et al. 2011; Wan et al. 2015; Cui et al. 2020b; Chen et al. 2019); many soft tissue sarcomas, e.g. osteosarcoma, liposarcoma, rhabdomyosarcoma, etc. (A673, U-2OS, and HT1080) (Chen et al. 2010, 2011), myeloid tumours, e.g. acute myelogenous leukaemia (AML) (HEL, NB4, HL-60, OCI-AML3, THP-1, U937, and KG-1) (Huang et al. 2014; Qiu et al. 2013; Wang et al. 2015; Wu et al. 2020), and benign prostatic hyperplasia (Liu et al. 2012) (details are shown in Table 2.1). Despite the variety of tumour types in derivation, Igs from various tumour tissues or cell lines show similarities in both isotype diversity and molecular structures with B lymphocyte-produced Igs. These characterisations of widespread expression profile and comparatively conserved structure between tumours suggest critical roles of Igs in tumorigenesis, as has been revealed in survival and growth processes in several types of tumours by previous study, which will be discussed in other chapters of this book (details in Chap. 5) (Cui et al. 2021; Pan et al. 2013). Their functions related to other biological processes such as development, metastasis, as well as in other types of tumours, await further studies.

2.1.1 Epithelial Cancers Produce Non B-Igs

Epithelium constitutes the largest tissue within the human body, encompassing all glandular epithelium, mucosal epithelium, and skin epidermis. At present, it has been found that all cancer cells

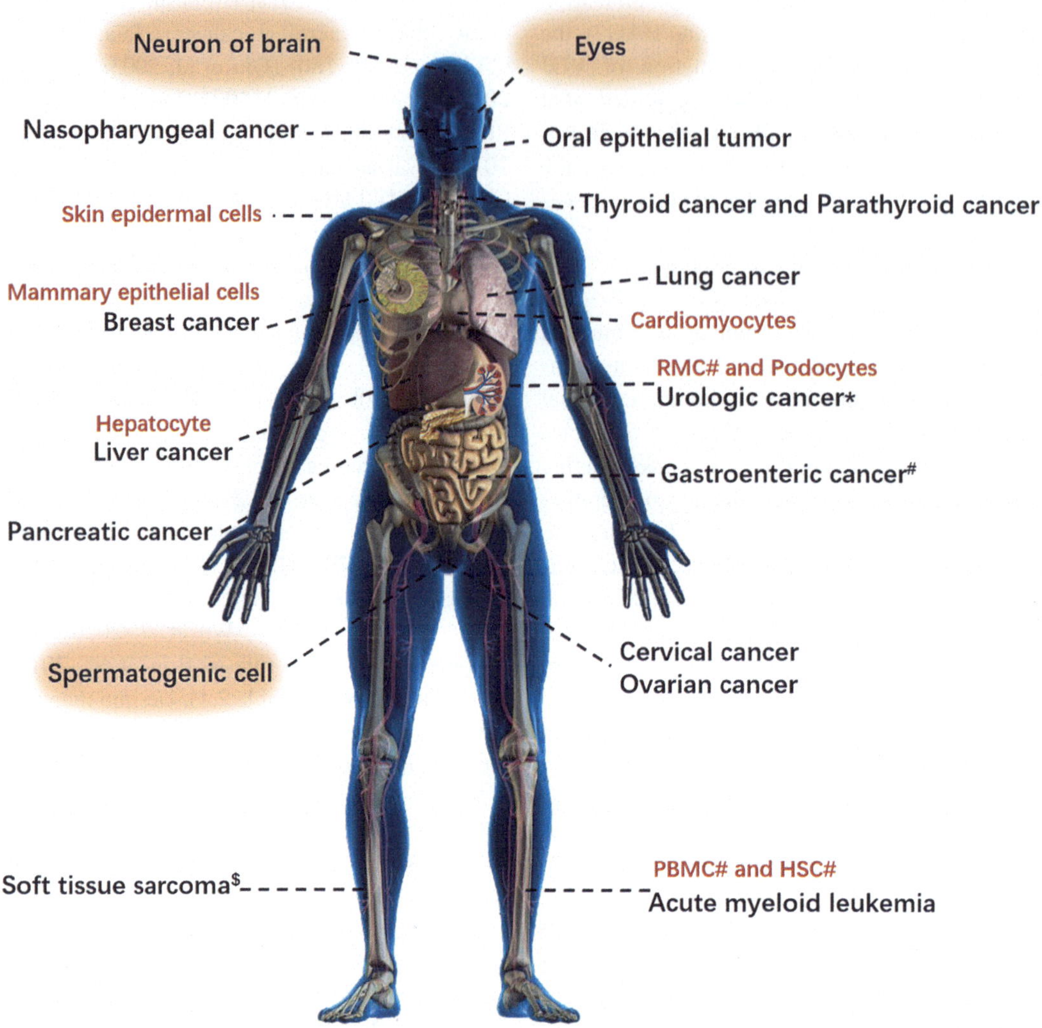

Fig. 2.1 Expression profile of non B-Igs. Red characters refer to the human normal cells. Black characters refer to the cancer cells. Highlight characters (orange) refer to the cells from immune privileged sites. *Urologic cancer includes prostate cancer, renal cancer and bladder cancer. #Gastroenteric cancer includes oesophageal cancer and renal cell carcinoma. $Soft tissue sarcoma includes histiocytoma, leiomyoma, malignant fibrous histiocytoma, fibrosarcoma, leiomyosarcoma, and rhabdomyosarcoma. #*RMC* renal mesangial cells, *PBMC* peripheral blood monocytes, *HSC* human stem cell

of epithelial origin express Igs, especially IgG, at a high level.

2.1.1.1 Breast Cancer Cells Produce Non B-Igs

Breast cancer represents the most prevalent cancer among women, accounting for 24.5% of total cancer cases and 15.5% of cancer deaths worldwide (Jemal et al. 2009). As such, there is a pressing need to investigate the pathological mechanism of breast cancer and its therapies. Current evidence indicates that breast cancer cells express high level of IgG, which can promote tumour metastasis and serve as a potential therapeutic target for breast cancer. In 1996, Qiu et al. discovered that IgG staining was evident in breast cancer cells (Qiu and Yang 1996a); this phenomenon was subsequently identified in other

Table 2.1 Cancer-derived Ig expression in different types of malignancies (Cui et al. 2021)

Malignancy types	Detected Ig isotypes	Positive ratio in patients cancer cells (%)	Positive cell lines	Biological functions	Clinical significance	References
Breast cancer	IgG	81	MCF-7, MDA-MB-231, SKBR3, T47D, ZR75–1, MDA-435	Promotes growth, promotes tumour immune escape	Histological subtype, metastasis, clinical stage	Qiu and Yang (1996a), Kimoto (1998), Babbage et al. (2006), Ma et al. (2013), Yang et al. (2013), Wang et al. (2020), Li et al. (2012) and Lee (2009)
	IgA	97	SKBR3, MCF-7, MDA-MB-231, Bcap37	NA	Lymph node metastasis	Kimoto (1998), Babbage et al. (2006), Zheng et al. (2007a), Kang et al. (2012) and Welinder et al. (2013)
Lung cancer	IgG	66	A549, NCI-H520, SK-MES-1, Calu-6, H441	Promotes growth, promotes invasion and migration	Local invasion, tumour differentiation, lymph node metastasis, prognosis	Qiu et al. (2003), Lee (2009), Chen et al. (2007) and Liu et al. (2015)
Colon cancer	IgG	55	HT-29, LoVo, SW480, HCT116, SW1116	Promotes growth, promotes migration and invasion	Tumour differentiation, TNM stage, lymph node metastasis, inflammatory infiltration	Kimoto (1998), Qiu et al. (2003), Geng et al. (2019), Wu et al. (2014), Niu et al. (2012), Deng et al. (2006) and Jiang et al. (2019)
Pancreatic cancer	IgG	87	HC48, SW1990, AsPC-1, BxPC-3, MIA PaCa-2, PANC-1, T3M4, CFPAC-1, HPAF	Inhibits the cytotoxic activity of NK cells, inhibits apoptosis, promotes growth, promotes migration and invasion, induces inflammation	Tumour differentiation, chemoresistance, metastasis, prognosis	Kimoto (1998), Li et al. (2011), Wan et al. (2015), Cui et al. (2020b) and Chen et al. (2019)

(continued)

Table 2.1 (continued)

Malignancy types	Detected Ig isotypes	Positive ratio in patients cancer cells (%)	Positive cell lines	Biological functions	Clinical significance	References
Liver cancer	IgG	NA	BCL-7402, HepG2, Hep3B, Hep-2	Promotes growth, promotes migration, inhibits apoptosis	NA	Qiu et al. (2003), Chen et al. (2007), Lei et al. (2014) and Lee et al. (2009a)
Gastric cancer	IgG	44	MGC-803, MKN28, AGS, BGC-823, SGC-7901	Promotes growth, promotes migration and invasion	Prognosis	Li et al. (2012, 2016)
Oesophageal cancer	IgG, Igκ, Igλ	77	Eca109, SHEEC	NA	Tumour differentiation, correlation with Ki67	Zhang et al. (2012)
Cervical cancer	IgG, IgA	NA	HeLa, C-33A, CA33, ME-180	Promotes growth, inhibits NK cell effector function, induces inflammation	NA	Qiu et al. (2003), Li et al. (2004, 2012), Chen et al. (2007), Lee et al. (2009a), Zheng et al. (2007b) and Wang et al. (2014)
	Igκ	78	HeLa	Promotes malignant transformation	NA	Qiu et al. (2003), Chen et al. (2007) and Li et al. (2004)
Ovarian cancer	IgG	NA	CaOV3, SK-OV-3, OC-3-VGH	Promotes growth, promotes migration and invasion	NA	Qiu et al. (2003) and Ji et al. (2015)
Prostate cancer	IgG, Igκ	92	PC3, DU145, LNCaP	Promotes growth, promotes invasion and migration, inhibits apoptosis	Tumour differentiation	Chen et al. (2007), Liu et al. (2012), Qin et al. (2020), Xu et al. (2016) and Pan et al. (2013)
Bladder cancer	IgG	75–91	T24, BIU-87, 5637, EJ	Promotes growth, promotes migration and invasion, inhibits apoptosis	Tumour differentiation, recurrence	Liang et al. (2013) and Sheng et al. (2016a)

(continued)

Table 2.1 (continued)

Malignancy types	Detected Ig isotypes	Positive ratio in patients cancer cells (%)	Positive cell lines	Biological functions	Clinical significance	References
Renal cancer	IgG	90	786-O, ACHN, Caki-1, 293	Promotes growth, promotes migration and invasion, inhibits apoptosis	Tumour differentiation, clinical stage, prognosis	Lv et al. (2017) and Sheng et al. (2016b)
Nasopharyngeal cancer	IgA, Igκ	NA	CNE1, HNE2	Promotes growth, promotes malignant transformation	NA	Zheng et al. (2007a, b) and Hu et al. (2011)
Laryngeal cancer	IgM	63	HEp2	NA	Lymph node metastasis, clinical stage, prognosis	Kimoto (1998) and Wang et al. (2013)
Oral cancer	IgG, IgA	86	WSU-HN6, CAL27	Promotes growth, migration and invasion, inhibits apoptosis	NA	Qiu et al. (2008) and Wang and Gan (2021)
Salivary gland cancer	IgG	52–60	SACC-83	Promotes growth, mediates motility, regulates the EMT	Nerve invasion, metastasis, prognosis	Lv et al. (2017)
Thyroid cancer	IgG	80	NA	NA	Differential diagnosis, lymph node metastasis	Qiu et al. (2012)
Parathyroid cancer	IgG	78	NA	NA	Differential diagnosis, recurrence, prognosis	Cui et al. (2020a)
Soft tissue tumour	IgG	97	A673, U-2OS, HT1080	NA	Tumour differentiation	Chen et al. (2010, 2011)
Acute myeloid leukaemia	IgG	79	HEL, NB4, HL-60, OCI-AML3, THP-1	Promotes growth, inhibits apoptosis	Tumour differentiation, prognosis	Qiu et al. (2013) and Wu et al. (2020)
	IgM	50	THP-1, OCI-AML3, HL-60, U937, HEL, KG-1, NB4	Promotes growth	NA	Huang et al. (2014)
	Igκ	94	HEL, HL-60, KG-1, NB4, OCI-AML3, THP-1	Promotes migration	NA	Wang et al. (2015)

The presence of NA in the table is due to the absence of further studies. Cited with minor modification
NA not available

epithelial cancer cells, such as colon cancer and lung cancer cells (Qiu et al. 2003). Qiu et al. further demonstrated that Ig gene rearrangement and IgG expression occurred in non B cell-derived breast cancer cells through a series of methods, such as RT-PCR, Southern blot, ISH, immunohistochemistry (IHC), and WB. In 2003, Qiu et al. was the first to report in Cancer

Research that epithelial tumour cells, including breast cancer cells, could produce IgG, and that this IgG possesses growth factor-like activity (Qiu et al. 2003). Since then, several groups have reported that breast cancer cells can express Igs. In 2006, Babbage et al. discovered the existence of V(D)J rearrangements of the Ig gene in 4 out of 6 breast cancer cell lines, and the expression of IgM VH as pre- and post-switched transcripts in cell line ZR75 and MDA-MB-231 (Babbage et al. 2006); they also found a significantly increased IgH gene copy number in rat breast cancer induced by a chemical carcinogen (Babbage et al. 2006). In another study conducted by Yang et al. (2013), the expression of IgG in 186 breast cancers, 20 benign breast lesions, and 30 normal breast tissues was investigated. Compared to medullary carcinoma, carcinoma in situ, and benign lesions, both infiltrating ductal carcinoma and infiltrating lobular carcinoma displayed significantly higher levels of expression for the heavy chain of Igγ and light chain of Igκ. Moreover, a higher expression of IgG correlated with more malignant histological subtypes and advanced disease stages, suggesting a positive relationship between non B-IgG expression and breast cancer histological subtypes, and AJCC stages of the cancer, which may have clinical prognostic value. However, it should be noted that the expression of Igγ and Igκ is not proportionate to that of other classic breast cancer markers, such as estrogen receptor (ER), progesterone receptor (PR), HER2/neu, p53, PCNA, and nm23 (a tumour metastasis inhibitory gene) (Yang et al. 2013). Whilst the underlying mechanisms causing this divergence warrant further investigation, cancer-derived Igs (cIgs) have shown potential for clinical application.

It's worth mentioning that in recent years, a monoclonal antibody (mAb) RP215, which can specifically recognize IgG expressed in non B cancer cells, has been used to detect non B-IgG expressed in epithelial tumour cells. It has been identified that this non B-IgG, recognized by RP215, features a unique glycosylation modification which carries a unique N-glycosylation modification at Asn162 site of the CH1 domain, and the glycan end of the non B-IgG has a high sialylation modification. Therefore, the non B-IgG is referred to as sialylated IgG (SIA-IgG). SIA-IgG is primarily expressed in epithelial tumour cells, particularly in stem/progenitor cells. It relies on sialic acid modification to promote malignant biological behaviours, such as carcinogenesis and metastasis (Lee et al. 2008, 2012).

In fact, as early as 2015, Liao et al. reported that the high expression of IgG, recognized by RP215, might be characteristic of primary breast cancer stem/progenitor cells (Liao et al. 2015). In that study, it was discovered that the IgG recognized by RP215 was significantly overexpressed in the basal-like cell layer around the breast cancer nest, or in some invasive single or multiple cancer cell clusters. Interestingly, these IgG-high-expressing cells are co-located with cancer stem cell (CSCs) markers. Through tissue microarray (TMA) analysis of 45 pairs of breast cancer and its lymph nodes, stronger RP215 staining was observed in metastatic cancer cells than in situ breast cancer. The knockout of IgG in MDA-MB-231 eliminated both the ability of cancer cells to anchor to the extracellular matrix and the characteristics of CSC, such as self-renewal, proliferation, migration, and invasion (refer to Chap. 14 of this book for details). Mechanistic investigation found that SIA-IgG potentiated tumour growth via effector T cell inhibition (Wang et al. 2020). These results clearly demonstrate that SIA-IgG can serve as not only a new biomarker to predict metastasis and poor prognosis, but also a possible therapeutic target in breast cancer.

2.1.1.2 Lung Cancer Cells Produce Non B-Igs

Lung cancer, particularly non-small cell lung cancer (NSCLC), currently has the highest incidence rate of all malignant tumours (Jemal et al. 2009). It's been proven that lung cancer cells express high level of IgG, especially SIA-IgG, which can significantly promote the initiation and progression of NSCLC. SIA-IgG has been highlighted as a potential new target for NSCLC treatment. As early as 2003, Qiu et al. reported that both NSCLC and small cell lung cancer (SCLC) cells can express IgG (Qiu et al. 2003). Subsequently, increasing evidence revealed that lung cancer cell lines express IgG, Igμ, Iγ Cγ sterile transcript, RAG1, RAG2, and activation-induced cytidine deaminase(AID), and lung cancer cells from 11 patients express both IgG and Igκ (Chen et al. 2007). Importantly, a strong correlation was noted between IgG expression and poor differentiation, local invasion, metastasis, and poor prognosis in NSCLC patients, suggesting it might be used as a new prognostic biomarker for lung cancer (Wang et al. 2019).

In recent years, NSCLC, but not SCLC, has been shown to primarily express SIA-IgG as described above (Liu et al. 2015). Liu et al. have identified the frequency of SIA-IgG expression in 140 lung adenocarcinoma (ADC) patient samples. Results showed high SIA-IgG expression in 90% of lung squamous cell carcinoma, 50% of basal cell-like cancer cells in cancer nests, and in invasive cancer cells that migrated into the tumour stroma. Furthermore, SIA-IgG staining was significantly higher in metastatic tumours compared to primary tumours. Moreover, SIA-IgG staining strongly correlated with differentiation, local invasion, metastasis, and clinical outcome of patients with lung ADC. Functional studies unveiled that SIA-IgG played pivotal roles in multiple processes such as cell migration, invasion, survival, and in vivo growth (Liu et al. 2015; Tang et al. 2018). Mechanistic studies found that SIA-IgG can combine with integrin on the surface of cancer cells to promote the activation of Focal Adhesion Kinase (FAK) and Steroid Receptor Coactivator (Src) (FAK-Src) pathway, thereby promoting tumour formation and metastasis. Recently, it was discovered that SIA-IgG was expressed in cancer stem cells, and the SIA-IgG depends on its sialic acid modification to promote biological behaviours of CSC, such as carcinogenesis and metastasis. Additionally, the SIA-IgG depends on its sialic acid modification to inhibit T cell proliferation and mediate immune escape (Wang et al. 2020).

2.1.1.3 Colorectal Cancer Cells Produce Non B-Igs

The occurrence of IgG in carcinoma cells of colonic origin, but not in their normal epithelial cell counterparts, was first reported by Qiu and Yang (1996a). Then, Kimoto et al. detected the transcripts of Ig heavy chain and TCR in 5 epithelial-derived cancer cell lines, including SW116 (human colon carcinoma) using nested RT-PCR in 1998 (Kimoto 1998). Following this, elevated levels of IgG and IgA were reported in cell lines and various types of epithelial cancer cells, including colorectal cancer (CRC) (Qiu et al. 2003). Furthermore, Niu et al. revealed the expression of IgG genes with V(D)J recombination in 80 cases of CRC, 4 colon cancer cell lines, and a tumour-bearing immune-deficient mouse model (Niu et al. 2012). Interestingly, these Ig heavy chains showed conservative VDJ rearrangement characteristics as revealed by sequencing (Niu et al. 2012; Deng et al. 2006). In addition to IgG, other isotypes of Ig have also been detected in colon cancers. For instance, IgM has been found to be expressed in CRC cells (HT-29 and SW480) (Hu et al. 2012). Moreover, both mRNA and protein of IgA and Igκ were detected in CRC cells (Hu et al. 2003; Geng et al. 2007). These results demonstrated that Ig expression is ubiquitous in colon cancers and may be crucial for cancer cells. Indeed, further analysis revealed that IgG expression was associated with tumour differentiation, pTNM stage, lymph node involvement, and inflammatory infiltration (Niu et al. 2012), and positively correlated with the expressions of Cyclin D1, NF-κB, and PCNA in HT-29. Conversely, blockage of IgG led to an increase in cancer cell apoptosis (Niu et al. 2012; Deng et al. 2006), and inhibited the abilities of cancer cells for anchor-independent colony formation and invasion (Deng et al. 2006). These findings suggest that IgG is involved in the development and growth of colorectal cancer cells, and might play a self-protecting role during cancer development. Therefore, blocking IgG could be a potential therapy for treating CRC (refer to Chap. 11 of this book for details).

2.1.1.4 Ovarian Cancer and Cervical Cancer Cells Produce Non B-Igs

Indeed, the monoclonal antibody RP215, which can specifically recognize the SIA-IgG from cancer cells, was first obtained using the extract of the human OC-3-VGH ovarian cancer cell line as an antigen. In the 1980s, in an attempt to obtain monoclonal antibodies recognizing cancer cells specifically, Lee et al. extracted the protein of ovarian cancer cell lines as an antigen and prepared hybridoma in immunised mice. Of these, RP215 is a monoclonal antibody identified from 3000 hybridomas that only recognizes cancer cells but not normal cells. When initially identifying the specificity of RP215, Lee et al. found that the affinity chromatography component of RP215 included an Ig heavy chain, but there were also some other membrane proteins present. The exact antigen recognized by RP215 remained unknown for a long time, being temporarily called CA215, and it was speculated that CA215 might be a group of glycoproteins. Only recently was it determined that RP215 recognizes only SIA-IgG produced by non B cancer cells, including ovarian cancer cells. RP215 has been investigated for use in cancer treatment. A single dose (10 mg/kg) injection of RP215 reduced the tumour size (in the case of OC-3-VGH ovarian cancer cells) by 50% compared to the control within 2 weeks, with little, if any, effect on body weight or apparent toxicity (Lee et al. 2009b). Thus, RP215 might be utilised as an anti-cancer drug, offering promise for cancer treatment.

Likewise, cervical cancer cells have also been reported to express IgG (Chen et al. 2007; Wang et al. 2014; Liu et al. 2017). It was discovered that IgG expression in cervical cancer cell lines such as Hela cells and SiHa cells was significantly elevated by lipopolysaccharide (LPS) exposure (Wang et al. 2014), and that IgG expression was markedly reduced by RNA interference (RNAi) using small hairpin RNA (shRNA) targeted at the heavy constant of Igγ1 (IGHG1). This attenuated the LPS-induced proinflammatory cytokine production in HeLa cells (Wang et al. 2014), consistently inhibited the phosphorylation levels of NF-κB and MAPK (ERK, JNK, and p38), and in turn suppressed NF-κB transcriptional activities including nuclear translocation and the NF-κB responsive element activity. Moreover, it was found that IgG was recruited to Toll-like receptor 4 (TLR4) in the cytoplasm after LPS stimulation, and that the LPS-initiated signalling pathways were inhibited by the reduction of IgG expression via downregulating TLR4 expression. Similar results were observed in vivo in both a mouse model of endotoxaemia and human cervical cancer tissue using IHC (Wang et al. 2014). These findings demonstrate that IgG binds to TLR4 and enhances its expression to increase LPS-induced proinflammatory cytokine production in cervical cancer cells. Besides IgG, other isotypes of Ig such as IgM have also been found to be expressed in both primary cervical cancer tissue and human cervical cancer cell lines, such as Hela, Hela S3, and Hela MR, using IHC performed on tissue microarrays and RT-PCR (Hu et al. 2012). The cIgM was not only expressed on the cell surface in the form of BCR-like complex with CD79 but was also spontaneously secreted into supernatant. Interestingly, similar to B1 cell-derived natural IgM, these cIgMs displayed significant natural antibody activity, which was regulated by the TLR9-MyD88 pathway (Hu et al. 2012).

2.1.1.5 Pancreatic Cancer Cells Produce Non B-Igs

Pancreatic cancer (PC) is one of the most lethal malignant diseases with poorly understood pathogenesis (Amedei et al. 2014), although some risk factors have been identified (Lotze 2011; DiMagno et al. 1999; Matsubara et al. 2012; Nath et al. 2005). Dysfunction of Ig expression and an autoimmune mechanism have been implicated as being associated with the pathogenesis of pancreatitis, pancreatic cancer, and diabetes (Silva et al. 2011; Wilkin 1990). However, the pathogenesis has not been fully documented.

The expression of IgG in PC and human PC cell lines (MIA PaCa-2, PANC-1, AsPC-1, and BxPC-3) was examined by Gu et al. using IHC, immunofluorescence (IF), ISH, and RT-PCR in 2015 (Wan et al. 2015). IgG, its mRNA, and the enzymes of RAG1, RAG2, and AID were present in human PC cells, but the IgG-specific receptors (CD16, CD32, CD64, and FcRn) were not found in PC cells and adjacent islets. Interestingly, compared with PC patients without diabetes, the expression of IgG in the islet endocrine cells from PC patients with diabetes increased significantly. The quantity of IgG-positive islet cells in poorly differentiated cancers was also higher than those in well-differentiated cancers. This suggested that islet cell-produced IgG might be related to the onset of PC-associated diabetes and the growth and differentiation of PC cells (Wan et al. 2015). Indeed, the blockage of IgG by either siRNA or anti-IgG antibodies resulted in increased apoptosis in cultured PC and pancreatic ductal adenocarcinoma (PDAC) cells (Wan et al. 2015; Cui et al. 2020b). A mechanistic study revealed that PDAC cell-derived IgG can enhance PDAC cell debris-induced interleukin 1β (IL-1β) release by M2 macrophages, and the increased IL-1β promoted both the epithelial-mesenchymal transition and the metastasis of PDAC cells (Chen et al. 2019). High cIgG expression correlated to poor tumour differentiation, metastasis, poor disease-free survival (DFS), and overall survival (OS) in a clinical follow-up census, which could be used as an independent prognostic factor for DFS and OS to improve the accuracy of an established prognosis prediction model for 1-year OS and 2-year OS (Cui et al. 2020b). These findings demonstrate that the cIgG was synthesised by pancreatic cancer cells rather than being taken up from surrounding tissues or the circulation, and showed potential clinical applications in both prognosis and treatment. These findings also open new avenues for understanding the pathogenesis of PC.

2.1.1.6 Oral Epithelial Carcinoma Cells Produce Non B-Igs

Oral tumours can originate from glandular epithelium and squamous epithelium respectively. Similar to other tumours of epithelial origin, oral epithelial tumours also express Igs, especially IgG, which is involved in the progression of oral cancer. Zhu et al. have examined expression of both IgG and IgA in 18 cases of pleomorphic adenoma, 22 cases of benign or malignant tumour of oral epithelium, 25 cases of normal biopsies of oral squamous epithelium, and 24 cases of normal biopsies of salivary gland epithelium, and found that both mRNAs and proteins of IgG and IgA were expressed in pleomorphic adenomas (especially in cytoplasm-rich tumour cells), oral epithelial carcinoma (cytoplasm, cellular membrane or glandular lumen-like structures), and even a few in normal secretory and ductal epithelial cells (Qiu et al. 2008). In recent years, SIA-IgG was found in salivary adenoid cystic carcinoma (SACC); moreover, SIA-IgG level positively correlated with short survival time, epithelial-mesenchymal transition and nerve invasion (Peng et al. 2017; Lv et al. 2017). The knockdown of IgG in SACC cells promoted the epithelial-mesenchymal transition (Peng et al. 2017), decreased cellular proliferation and invasion, and increased the expression levels of E-cadherin (Lv et al. 2017; Peng et al. 2017), alpha-smooth muscle actin (Lv et al. 2017), and matrix metalloproteinase 9 (Peng et al. 2017). As early as 2009, Qiu et al. found that transcripts of IgG heavy chain with a unique VH5–51/D3–16/JH4 rearrangement displayed in oral squamous cell carcinoma (Zheng et al. 2009). Recently, Wang et al. revealed that the cIgG expressed by oral squamous cell carcinoma was involved in multiple critical processes including cell proliferation, migration, and invasion. The aforementioned processes were markedly inhibited, and cisplatin-induced cell apoptosis was significantly promoted after cIgG was knocked down. Mechanistic investigation unravelled a cooperative interaction between the cIgG and the cisplatin: the cIgG expression could be potentiated by the cisplatin which also upregulated the phosphorylation of AKT and protein kinase 1 (PDK1), and the knockdown of cIgG blocked the cisplatin-induced AKT, PDK1, and Src phosphorylation, thereby attenuated the phosphorylation of AKT and PDK1. Logically, further investigation found that the cIgG and cisplatin cooperated in regulating AKT and PDK1 phosphorylation by converging on PTP-BAS, a negative regulator of Src activation, with the knockdown of cIgG increasing PTP-BAS (at both mRNA and protein levels), whereas cisplatin decreasing its protein expression. Consistently, the knockdown of PTP-BAS raised phosphorylation of Src, PDK1, and AKT, and counteracted the cisplatin-induced cell proliferation inhibition which could be promoted by IgG knockdown. Conversely, the Src overexpression enhanced the AKT and PDK1 phosphorylation (Wang and Gan 2021). These findings suggest that IgG has growth factor-like activity and is involved in tumorigenesis as well as promotes the invasion and drug resistance of cancer cells. This SIA-IgG might serve as a useful biomarker to predict the prognosis of the disease and as an effective target molecule for oral tumour therapy.

2.1.1.7 Bladder Cancer and Renal Cell Carcinoma (RCC) Cells Produce Non B-Igs

Bladder cancer, accounting for 90% of all primary bladder tumours, is the fourth most common cancer in men (Bos et al. 2014). It has been identified that bladder cancer and renal cell carcinoma cells can also express IgG (Liang et al. 2013; Sheng et al. 2016a). Sheng et al. first found the IgG staining in bladder cancer cells (Sheng et al. 2016a). The expression of IgG level was thought to be correlated with histologically high-grade cancer, large tumour size, and high recurrence rate. Moreover, the IgG transcripts showed that several sets of VHDJH rearrangements were frequently shared among bladder cancer cell lines, and even among 5637 and BIU87 cell lines. As expected, knockdown of IgG significantly reduced the proliferation, migration, and invasion of 5637 and BIU87 cells (Sheng et al. 2016a), indicating the involvement of IgG in progression in bladder cancer cells (refer to Chap. 8 of this book for details).

The renal cell carcinoma (RCC) is the most common type of kidney cancer in adults (Jonasch et al. 2014). Sheng et al. demonstrated the SIA-IgG expression in RCC cells but not adjacent normal kidney tissues (Sheng et al. 2016b), especially the renal tubular epithelial cells. High SIA-IgG expression significantly correlated with poor differentiation and advanced stage of RCC. Coherently, knockdown of IgG in RCC cell lines significantly suppressed cancer cell proliferation, migration, and invasion (refer to Chap. 8 of this book for details). That meant IgG could affect the biological behaviours of RCC cells and could serve as a novel target for RCC therapy.

2.1.1.8 Prostate Cancer Cells Produce Non B-Igs

Prostate cancer is the second leading cause of death in male patients worldwide (Rawla 2019). Benign prostatic hyperplasia (BPH) is a condition that occurs when the prostate gland enlarges. IgG was discovered in secretory epithelial cells and in secretory granules within the ductular lumen in BPH tissues in a rabbit prostate, but the IgG was thought to be produced by B lymphocytes and transported into the prostate tissue (Ablin et al. 1971).

Using a formalin-fixed paraffin-embedded tissue microarray (including 68 tissue samples obtained from 59 adenocarcinomas and 9 normal prostate tissues) and 50 tissue samples (including 20 BPHs and 30 prostatic adenocarcinomas), Liu et al. found that IgG expression occurred in 35% of the BPH tissues compared to 92.1% of the prostate cancer tissues, but not in any normal prostate tissues (Liu et al. 2012). Subsequently, the expression of IgG and Igκ in the cytoplasm of prostate cancer cells was confirmed (Chen et al. 2007; Jiang et al. 2014). Moreover, the IgG expression level correlated positively with not only Gleason score and histological grade but also cancer cell survival. Specifically, the decrease in IgG expression resulting from blocking by an anti-human IgG monoclonal antibody reduced cell proliferation and increased apoptosis in cultured prostate cancer cells (Liu et al. 2012). Recently, Qin et al. reported that SIA-IgG expressed in prostate cancer cells was associated with clinicopathological characteristics such as the ability to anchor to the extracellular matrix and cell–cell adhesion. Furthermore, prostate cancer cells expressing IgG transcripts demonstrated unique VHDJH rearrangement patterns. Mechanistic investigations found that androgen deprivation therapy (ADT) increased SIA-IgG level by upregulating the expression of SOX2, a transcription factor of cancer stem cells, thereby promoting the progression of prostate cancer. Additionally, the findings revealed that SIA-IgG activated mitogen-activated protein kinase (MAPK)/extracellular-signal-regulated kinase (ERK) and AKT so as to play a role in the maintenance of stemness and the cancer progression, which was further confirmed by in vivo experiments demonstrating that the growth of prostate cancer cells was significantly attenuated after SIA-IgG was knocked down using either siRNA or RP215 (Qin et al. 2020).

2.1.1.9 Papillary Thyroid Cancer Cells Produce Non B-Igs

Papillary thyroid cancer is the most common endocrine malignancy (Ian et al. 2002). The cellular deposits of IgG were detected in 80% of 41 papillary thyroid cancer cases by Lucas et al. in 1996, which was assumed to be the IgG deposition from the serum rather than local production (Lucas et al. 1996). The IgG was found to be co-localized with C3d, C4d, and C5, indicating the existence of immune complexes (ICs) (Lucas et al. 1996; Yamakawa et al. 1994). Therein it was proposed that the co-expression of IgG and C3d, C4d, and C5 represented a tumour-specific host immune response induced by initiation of the classical complement pathway by virtue of binding of IgG to tumour antigens (Lucas et al. 1996). Conversely, after the expression of the heavy chain of IgG and light chain of Igκ was thoroughly examined in a papillary thyroid cancer tissue microarray by Qiu et al. using IHC, ISH, and LCM-assisted RT-PCR (Qiu et al. 2012), it was conclusively determined that papillary thyroid cancer cells could indeed produce IgG heavy chain and Igκ light chain Igκ, as evidenced by the detection of both proteins and mRNA transcripts. These cIgGs might contribute to the ICs in papillary thyroid cancer, and the ICs also consisted of host-synthesised IgGs which recognised both papillary thyroid cancer surface antigens and anti-idiotypic cIgGs. The ICs might prevent host IgG from binding to its antigen thereby protecting the cancer cells from antibody-dependent cell-mediated cytotoxicity and complement-dependent cytotoxicity, resulting in promotion of tumour growth, progression, and metastasis (Qiu et al. 2012). This could provide new insights for understanding the mechanism of tumorigenesis and developing novel tumour therapies.

2.1.2 Soft Tissue Sarcoma Cells Produce Non B-Igs

Soft tissue sarcoma is a rare type of cancer that originates in the tissues connecting, supporting, or surrounding other body structures. These tissues include muscles, fats, blood vessels, nerves, tendons, and the linings of joints. Despite its rarity, to date, more than 50 subtypes of soft tissue sarcomas have been reported (Bourcier et al. 2019). Even though soft tissue sarcomas differ markedly from epithelial neoplasia in both clinical behaviours and biological features (Ray-coquard et al. 2018), they also have been found expressing IgG, as evidenced in paraffin-embedded tissue sections from a total of 80 samples using IHC. This demonstrated expression of the heavy chain of IgG and light chain of Igκ in fibrous histiocytoma, leiomyoma, malignant fibrous histiocytoma, fibrosarcoma, leiomyosarcoma, and rhabdomyosarcoma, but not in fibroma (Chen et al. 2010). Given that Ig expression level strongly correlates with proliferation markers including PCNA, Ki67, and cyclin D1, it is reasonable to suggest that Ig may serve as a useful biomarker for cell proliferation in sarcomas (Chen et al. 2010).

The expression of IgG heavy chain and Igκ in sarcoma cell lines A673, U-2OS, and HT1080 was also ascertained by Chen et al. (2011). The findings showed that these cIgs were mainly present in the cytoplasm and on the cell membrane, albeit in lesser quantities (Chen et al. 2011). Further analysis (by immuno-electron microscopy) revealed that IgG localized on the rough endoplasmic reticulum (RER) and the cell membrane. In addition, RAG1, RAG2, AID enzymes, Iκ-Cκ germline transcript, and V(D)J recombination were also found in the three sarcoma cell lines. Considering the following facts: (1) RER is the site where proteins are synthesised, (2) RAG1, RAG2, and AID are necessary enzymes for Ig synthesis, and (3) I-C germline transcription is used as an indicator of chromatin accessibility, it is conceivable that IgG was synthesised locally by these sarcoma cell lines.

2.1.3 Myeloblasts of Acute Myeloid Leukaemia (AML) Produce Non B-Igs

It has been recognized that myeloid cells and lymphocytes originate from haematopoietic precursor cells undergoing a continuous process in the bone marrow (Laurenti et al. 2018; Bassler et al. 2019), and that myeloid cells do not produce Igs (Lefranc and Lefranc 2020). However, acute myeloid leukaemia (AML) cells were found expressing and secreting IgG, IgM, and Igκ (Huang et al. 2014; Qiu et al. 2013; Wang et al. 2015). IgG was frequently expressed in AML cells and primary myeloblasts, playing a role in the survival of AML cells (Qiu et al. 2013). IgM was expressed in myeloblasts, monocytes, and neutrophils with a similar frequency, and was crucial for cell proliferation (Huang et al. 2014). Igκ was frequently expressed in myeloblasts, mature monocytes, and neutrophils, but it was either not expressed or expressed only rarely in mature neutrophils or monocytes, playing a critical role in cell migration (Wang et al. 2015) (refer to Chap. 15 of this book for details).

2.2 Non B-Ig Expression in Normal Cells

The discovery that non-B cells can produce Ig was initially made in cancer cells, not in normal cells. It was once hypothesized that this was due to the heterologous expression of the Ig gene caused by the genomic instability of the cancer cell. However, mounting evidence has shown that normal non-B cells, generally non cancer cells, do express Ig such as IgG, IgM, and IgA (Babbage et al. 2006; Bos et al. 2014). However, the expression level is significantly lower than that of cancer cells and thus has been overlooked for a long time.

The normal cells found expressing non B-Igs to date are summarised in Table 2.2 and will be presented individually in the following sections.

2.2.1 Epithelial Cells Produce Non B-Igs

The epithelium is of critical importance in host defence and immune surveillance, as it is the first cell layer that primarily encounters the majority of microorganisms. Originally, it was thought that the epithelium serves only as a passive barrier against invading pathogens, typically sufficient to restrain commensal microbes. There is a large amount of Ig in the skin, mucous, and glandular cavities, but it has been believed that the Ig secreted by plasma cells in the dermis or lamina propria of the mucous reaches the skin, mucous surface, and glandular cavity through the epithelial basement membrane and epithelial cells. In reality, numerous pieces of evidence have proven that epithelial cells not only serve as a physical barrier but also spontaneously secrete Ig. On one hand, Ig plays the role of antibody activity to protect the skin and mucosa; on the other hand, the Ig can promote the proliferation and survival of epithelial cells.

2.2.1.1 Mouse Epithelial Cells Produce Non B-Igs

The IgG in mice colostrum and milk was believed to derive from circulating blood or be produced by the local plasma cells. IgG in mice mammary gland epithelial cells was thought to be transferred from the circulating blood through the neonatal fragment crystallisable (Fc) region of Ig receptors (FcRn) expressed on the membrane of mammary gland epithelial cells (Cianga et al. 2003). In support of this, significantly increased IgG was detected in the sera and milk from the transgenic mice that overexpressed the bovine FcRn in their lactating mammary glands (Lu et al. 2007). However, the fact that no selective accumulation of endogenous mouse IgG or injected bovine IgG was found in the milk of the transgenic female mice raises questions about whether FcRn mediates the transfer of IgG from plasma to milk (Lu et al. 2007).

In 2010, Zhang et al. reported the IgG expression in the epithelial cells of mammary glands from pregnant and lactating mice (Zhang et al. 2010). To further prove that breast epithelium can produce Ig during lactation, they sorted cytokera-

Table 2.2 Non B-Igs expression in various normal cell types

Cell types	Species	Detected Ig isotypes	Detected enzymes, gene transcripts, rearrangement related to Ig synthesis	Biological functions	References
Epithelial cells of mammary glands	BALB/c, KM, and C57BL/6 mice	IgG	AID	NA	Zhang et al. (2010)
Epithelial cells of liver and small intestine	μMT mice	IgG1, IgG2a, IgG2b, IgM, IgA, IgD, Igκ, and Igλ	*Ighg*, *Ighm*, *Igha* and *Ighd* (except for *Ighe*), *Igk*, and *Igl*	IgM binds various bacteria and recognise single-stranded DNA (ssDNA), double-stranded DNA (dsDNA), lipopolysaccharide (LPS), and insulin with low affinity, anti-infection	Cianga et al. (2003) and Lu et al. (2007)
Epithelial cells of pancreas, stomach, lung, kidney, and urethra	μMT mice	IgG1, IgG2a, IgG2b, IgM, IgA, IgD, Igκ, and Igλ	NA	NA	Cianga et al. (2003)
Epithelial cells of skin	Human	IgG and IgA	*IGHG* and *IGHA*	Recognise several pathogens, especially *Staphylococcus aureus*	Shao et al. (2016a)
Epithelial cells of renal tubular	Human	IgG	*IGHG*, *IGK*, *IGL*, *RAG1*, *RAG2*, and *AID*	NA	Shao et al. (2016b) and Jiang et al. (2015)
Hepatocytes	Human, rat, and SCID mice	IgG, IgM, and IgA	*IGHG1*, *Ighg1*, RAG1, RAG2, and AID	IgG plays important roles in hepatocyte proliferation, apoptosis, and cancer growth in vitro	Lei et al. (2014)
Cardiomyocytes	C57BL/6, Balb/c and μMT mice	IgM, Igκ	*Igk-V8*, *Igk-C*, *Igh-6*, *Igj*, *Ighg*, *Igh*, and *Igl-V1*	Igκ combines with cytoskeleton-related protein-plectin to promote the adhesion of intercalated discs and the contraction of cardiomyocytes	Rincheval-Arnold et al. (2002) and Deng et al. (2020, 2021)
CD34⁺ HSC/HPCs, monocytes, macrophages, neutrophils, and hematopoietic stem cells	Human	IgG	*IGHG VHDJH*	NA	Lloyd-Jones et al. (2010)
		IgM	*VHDJH*	NA	Qiu et al. (2013)
		Igκ	NA	Promotes migration	Wang et al. (2015)
		NA	*IGHM*, *IGK*, *IGHE*, *IGHD*, *IGHG1*, and *IGHA1*	NA	Eltzschig et al. (2011)
		IgG, IgM, Igκ, and Igλ	*IGHG1*, *IGHG3*, *IGHM*, *IGK*, and *IGL*	NA	Zhang et al. (2006a) and Kulik et al. (2009)

(continued)

Table 2.2 (continued)

Cell types	Species	Detected Ig isotypes	Detected enzymes, gene transcripts, rearrangement related to Ig synthesis	Biological functions	References
Renal mesangial cells (HRMCs) and podocytes	Human	Igα1, Igα2, Igκ, and Igλ	NA	Play roles in cell adhesion and survival	Zhang et al. (2006b)
		IgG4, Igκ, and Igλ	*RAG1*, *RAG2*, and *AID*	Important for cell viability and adhesion	Haas et al. (2010)

The presence of NA in the table is due to the absence of further studies
NA not available

tin 18-positive mammary gland epithelial cells and identified abundant amounts of Ig heavy chain and light chain transcripts in these epithelial cells of BALB/c, Kunming (KM), and C57BL/6 mice (Zhang et al. 2010). Furthermore, the expression of AID but not RAG1/2 was also detected in these mammary epithelial cells. The finding suggests that the IgG found in murine colostrum is, at least in part, produced by the mammary gland epithelial cells, providing an entry point for investigating the significance of the IgG produced by mammary gland epithelial cells. Subsequently, Shao et al. demonstrated that epithelial cells of mice, especially the μMT mice (B cell-deficient mice), including the pancreas, liver, lung, kidney, stomach, urethra, and the small intestine, can express IgM and Igκ, and the liver cells also express IgG1, IgG2b, and much higher levels of IgG2a, IgA, IgD, and Igλ (Shao et al. 2016a, b). Consistently, Shao et al. reported that rearranged transcripts of heavy chains and light chains at both the mRNA and protein levels were detected in sorted liver epithelial cells of μMT mice by using IHC, WB, the enzyme-linked immunospot (ELISPOT), mass spectrometry (MS), RT-PCR, and DNA sequencing. Surprisingly, a very low concentration [~100 times lower than wild type (WT) mice] of Igs has been found in the bloodstream of μMT mice, which lack mature B cells. Moreover, the level of Igs in non-immune organs from μMT mice was similar to that from WT mice, suggesting that non-B cells can spontaneously express Igs that are retained locally. Class switching was suggested not to occur in liver epithelial cells-derived Igs, as each heavy chain group had its unique variable region sequence. Moreover, mutations in the variable region of the γ and α chains indicate that different classes of Igs have different activities (Lu et al. 2007). Interestingly, sequencing analysis showed that IGKV9–12401/IGKJ201 recombination frequently occurred in both the liver epithelial cells and the intestinal epithelial cells (IECs), but not in the spleen cells (Shao et al. 2016a, b). These results suggest that liver epithelial cells possess the capacity of producing functional Ig rearrangements. Furthermore, IgM expression level was significantly upregulated when liver epithelial cells of mice were stimulated with the cytosine–phosphate–guanosine (CpG), an agonist of TLR9, which mimics bacterial infection (Jiang et al. 2015). Interestingly, the epithelial cell-derived IgM illustrated natural antibody activity, which could bind various bacteria and recognise single-stranded DNA (ssDNA), double-stranded DNA (dsDNA), LPS, and insulin with low affinity. These results highlight the fact that epithelial cells could express IgM, which is polyreactive and could be involved in anti-infection responses (Jiang et al. 2015), indicating that the role of epithelium in immunity needs to be redefined.

The physiological significances of non B-Igs expressed by the epithelial cells remain elusive. Some postulations are presented as follows: In the special lactating physiological state, the IgG derived from both the mammary epithelial cells and the plasma cells may cooperatively support the development and maturation of the newborn's immune system via milk so as to protect the infant's gastrointestinal tract against infections as IgA does (Rincheval Arnold et al. 2002). On the other hand, the similar mutations and the expression preference of certain VκJκ recombinants in mammary epithelial cells from different individuals and different strains imply the important function of Ig in innate immunity or support the proliferation of lactating epithelial cells (Zhang et al. 2010).

2.2.1.2 Human Epithelial Cells Produce Non B-Igs

It was found that IgG was highly expressed in epithelial cancer cells and proliferating cells, whereas IgM was expressed at a similar level in epithelial cancer cells and normal epithelial cells (Kimoto 1998; Wang et al. 2013). Hu et al. first proved that a variety of human epithelial cells can express IgM and its transcripts (Hu et al. 2012). This kind of IgM was further confirmed to have natural antibody activity (Shao et al. 2016a). In 2015, Jiang et al. reported significant staining of IgG and IgA in human skin epidermal cells and HaCaT (an immortal human keratinocyte cell line). Rearranged Igγ and Igα were transcribed in the sorted epidermal cells with a unique VDJ rearrangement pattern in each individual tested (Jiang et al. 2015). Moreover, both IgG and IgA were found to bind several pathogens, especially *Staphylococcus aureus* (Jiang et al. 2015). The expression of non B-Ig in human epithelial cells is further corroborated by two recent studies. Deng et al. identified IgG protein, IGHG, IGK, IGL, RAG1, RAG2, and AID gene expression in human renal tubular epithelial cell lines by using single-cell RNA sequencing, IHC, nested PCR, Sanger sequencing, immunostaining, and WB. Moreover, IgG expression was upregulated by transforming growth factor-β1 (TGF-β1), a key molecule modulating Ig production (Deng et al. 2020, 2021).

Belonging to epithelial cells, hepatocytes contribute to the synthesis of proteins, cholesterol, bile salts, fibrinogen, phospholipids, and glycoproteins. Despite high concentrations of IgGs found in the hepatic and gall bladder bile of both adult humans and rats, these IgGs were thought to be locally produced by B lymphocytes or transported from the blood in the form of immune complexes through FcRn, which is functionally expressed on the apical surface of hepatocytes. This viewpoint, however, was recently challenged by Gu et al. (Lei et al. 2014). In that study, both IgG and its mRNA were found to be present in adult human hepatocytes, rat hepatocytes, human liver cell line L-02, rat hepatocyte cell line BRL-3A, and the human hepatoma cell line HepG-2. These findings suggest that hepatocytes themselves could produce IgG. The detection of IgG protein and gene transcripts in the livers of the severe combined immunodeficient (SCID) mice, which lack B lymphocytes, further confirmed that the IgG found in hepatocytes was not absorbed from circulation but locally synthesised. This was corroborated by the following evidence: (1) the essential enzymes necessary for the synthesis of Igs, including RAG1, RAG2, and AID, were also expressed in hepatocytes; (2) IgG synthesis was significantly reduced after the cultured liver cells were transfected with IgG gene-targeted siRNA, and the downregulation of IgG inhibited the proliferation and metastasis of L-02 and HepG-2 cells. This suggests that hepatocyte-produced IgG plays pivotal roles in hepatocyte proliferation, apoptosis, and cancer growth with profound clinical implications (refer to Chap. 7 of this book for details); (3) IgG expression in hepatocytes was found to be significantly enhanced in the regenerated liver and robustly correlated with Ki-67 expression after partial hepatectomy (PH); and (4) apart from IgG, liver epithelial cells were found to produce various classes of Igs, and hepatocytes were found to express IgM and IgA, but not IgD and IgE. This supports the speculation that hepatocytes are capable of expressing Igs (Lei et al. 2014).

2.2.2 Cardiomyocytes Produce Non B-Igs

Acute myocardial infarction (AMI) is a leading cause of morbidity and mortality (Lloyd-Jones et al. 2010). Reperfusion strategy is the current standard therapy for AMI, but it may lead to paradoxical cardiomyocyte dysfunction, known as ischaemic reperfusion injury (IRI) (Eltzschig et al. 2011). This is often accompanied by IgM deposition in the cardiac tissues (Zhang et al. 2006a), and IgM identified recognizing the soluble cytosolic protein annexin IV and a highly conserved region within non-muscle myosin heavy chain type II A and C in the initiation of reperfusion injury (Kulik et al. 2009; Zhang et al. 2006b). Thought to be produced by B cells from the bloodstream and bound to intracellular self-antigen exposed during myocyte injury, IgM localizes on the membranes, cross-striations, and in the cytoplasm of cardiomyocytes, leading to complement activation and resulting in cardiomyocyte injury (Haas et al. 2010). However, the expression of some Ig transcripts, including *Igk-V8, Igk-C, Igh-6, Igj, Ighg, Igh, and Igl-V1* in murine cardiac tissue and in the cardiac cell line HL-1, was determined by Kang et al. (Kang et al. 2009). Moreover, the transcription of the IgM heavy chain was also found in mouse heart tissue by mining the Gene Expression Omnibus (GEO) database (Zhu et al. 2017). Furthermore, IgM, with a unique VDJ rearrangement, was confirmed to be expressed by the cardiomyocytes from the Balb/c and μMT adult mice, primary cultured cardiac cells separated from newborn mice, and the myocardial cell line (HL-1) through IHC, ELISA, and WB analysis (Zhu et al. 2017). The concentration of IgM increased in the supernatant of cultured μMT mice heart tissue under ischaemic conditions in a time-dependent manner (Zhu et al. 2017). These findings suggest that at least a portion of the IgM deposited in the infarcted area during myocardial infarction is produced by the myocardium itself, and that the expression and secretion of cardiomyocyte-derived IgM could be regulated by ischaemia/hypoxia. In addition, cardiomyocytes were found to express free Igκ, which can bind with cytoskeleton-related protein-plectin to promote the adhesion of intercalated discs and the contraction of cardiomyocytes (Zhu et al. 2017; Qiu et al. 2020). To date, Ig transcripts in the heart tissue of mice have been reported by at least three groups (Zhu et al. 2017; Vasu et al. 2007; Ashton et al. 2006).

2.2.3 Monocytes, Macrophages, Neutrophils, and Haematopoietic Stem Cells Produce Non B-Igs

The discovery that mature myeloid cells (monocytes and neutrophils) (Huang et al. 2014) and myeloblasts from AML (Qiu et al. 2013; Wang et al. 2015) could produce Igs instigated research to investigate whether haematopoietic stem/progenitor cells (HSC/HPCs) could produce Igs. Liu et al. found the IgM heavy chain was expressed in umbilical cord blood (UCB) $CD34^+$ HSC/HPCs after analysing the GEO microarray database (Liu et al. 2016). Subsequently, they confirmed the *IGHM* transcript expression in the UCB $CD34^+$ HSC/HPCs from 4 healthy individuals. Moreover, HSC/HPCs-derived Igs showed a preference for *VHDJH* rearrangement patterns usage, in that among the seven VH families (VH1–VH7), only the VH3 family, such as VH3–15, VH3–21, VH3–23, and VH3–30, was expressed in the HSC/HPCs (Liu et al. 2016).

The IgM gene rearrangement and transcription were detected by Qiu et al. using RT-PCR with primers specific to the *VHDJH* gene in $CD33^+$ monocytes and neutrophils isolated and purified from patients with non-haematopoietic neoplasms and healthy controls, and the IgM expression in myeloid cells was also confirmed by flow cytometry analysis. These findings demonstrate that non-neoplastic monocytes or neutrophils could produce IgM themselves. Further studies showed that IgM was expressed at a higher frequency in monocytes than in neutrophils, suggesting that monocytes might be able to express IgM more frequently than neutrophils under certain physiological and pathological conditions (Huang et al. 2014). Additionally, it was found that Igs (IgG, IgM, Igκ, and Igλ) and rearrangements of both heavy chains (IgG1, IgG3, and IgM) and light chains (both Igκ and Igλ) could be detected in tumour-infiltrating macrophages in humans. Importantly, the diversity of the macrophage-expressed Ig repertoires correlated negatively with tumour volume (Fuchs et al. 2018; Busch et al. 2019). The function of macrophage-secreted Igs in tumours is worthy of further investigation.

2.2.4 Human Renal Mesangial Cells (HRMCs) and Podocytes Produce Non B-Igs

IgA nephropathy, also known as Berger's disease, is a kidney disease stemming from the accumulation of IgA deposits in the kidneys, which causes inflammation damaging kidney tissues (Perše et al. 2019). It is generally considered that the accumulated IgA is an abnormal, glycosylated IgA1. The abnormal IgA1 can be recognised by antiglycan autoantibodies of the IgA1 and/or IgG isotype, resulting in the formation of circulating immune complexes (CIC) (Perše et al. 2019; Tomana et al. 1999; Maillard et al. 2015). The pathogenic CIC deposits in the glomerular mesangium can promote resident mesangial cells to secrete proinflammatory factors and activate complement through the alternative and lectin pathways, eventually resulting in the initiation of glomerular injuries (Maillard et al. 2015). However, the observation that IgA deposition is also seen in renal biopsy of patients without evidence of kidney disease casts doubt on the above postulation; it is reported that the incidence of mesangial IgA deposition in healthy people is around 16% (Suzuki et al. 2003).

In 2018, Deng et al. found the expression of IgA in primary HRMCs (Deng et al. 2018). The IgA heavy chain (α1 and α2) and the light chain (κ and λ) were found distributed in the cytoplasm or on the cell membranes of HRMCs (e.g. C2M12); the IgA heavy chain (α1 and α2) was also secreted into the culture media of HMCs. More interestingly, it was found that the IgA expression in HMCs could be stimulated by AngII, an endogenous proinflammatory factor, whereas the knockdown of Igα1 expression by siRNA inhibited cell adhesion and promoted apoptosis (Deng et al. 2018). These results demonstrate that mesangial cells can express and secrete IgA, which may contribute to the deposition of IgA in patients with IgA nephropathy, and the expression of IgA might be associated with cell functions such as proliferation and adhesion (refer to Chap. 8 of this book for details).

Another terminally differentiated cells of the renal balloon, podocytes (Reiser et al. 2016), were also found to be capable of secreting Igs (Jing et al. 2018). IgG heavy chain (Igγ) IgG4, Ig light chains κ and λ were found to be present in the cytoplasm or on the membrane of podocytes. Moreover, Igγ, γ1, γ3, γ4, κ, and λ were detected in the lysates of podocyte by WB; mass spectrometry result further confirmed the secretion of IgG1 and Igλ by podocytes in the culture supernatant. Furthermore, constant region transcripts of Igγ, γ1, γ3, γ4, κ, and λ were verified by RT-PCR. In addition, the existence of transcriptions of rearrangement-related enzymes, such as RAG1, RAG2, and AID, further suggested that Igs were produced by podocytes, with a similar mechanism to that in B lymphocytes. More intriguingly, siRNA-mediated downregulation of IgG expression reduced podocyte viability and adhesive capabilities (refer to Chap. 8 of this book for details).

2.3 Non B-Ig Expression in Immune-Privileged Sites

Traditionally, antibodies were not believed to pass through the anatomical barriers in the brain, eyes, and reproductive organs under physiological conditions due to their large molecular weight (Forrester et al. 2008). Neither were leukocytes, including lymphocytes. However, growing evidence demonstrates that the synthesis of IgG can be detected in immune-privileged sites, including the eyes, brain, placenta, testicles, and epididymis (please refer to Table 2.3 for a panoramic view), as detailed below.

Table 2.3 Non B-Igs expression in immune-privileged sites

Immune privileged sites	Species	Detected Ig isotypes	Detected enzymes, gene rearrangement related to Ig synthesis	Biological functions	References
Neuron	BALB/c and SCID mice	IgG	*Ighg* and *Igk*	NA	Hedger et al. (1994)
	Mice	NA	*Ighg3*	NA	Yoshimi et al. (2002)
	Mice	NA	*Ighg3 and Ighm*	NA	Naegele et al. (1991)
	SD rats	IgG	NA	Protects neurons against injury induced by complement and microglial activation	Archelos et al. (2000)
	Human	IgG, Igκ, and Igλ	*IGHG1, IGK,* and *IGL, VDJH, RAG1,* and *RAG2*	NA	Wurster et al. (1994)
	Human and SD rats	IgG	*IGHG1 (Ighg1), Iγ-Cγ, IGK (Igk), RAG1 (Rag1),* and *RAG2 (Rag2)*	Improves the survival and reduces apoptosis of dopaminergic neurons induced by 6-hydroxydopamine (6-OHDA) toxicity, and triggers microglia activation	Huang et al. (2008)
Spermatogenic cells	BALB/c and SCID mice	IgG, IgM, and Igκ and Igλ	*Ighg1* constant region and the *Igk* light chain variable region	NA	Hedger et al. (1994)
	Human	IgG	*IGHG1, IGL, RAG1, RAG2,* and *AID*	Blocking IgG inhibits sperm from penetrating zona-free hamster egg	Rani et al. (2021)
Eyes	Human μMT mice	IgG, Igκ, and Igλ	Human *IGHG1, IGK,* and *IGL,* VDJ_H, *RAG1,* and *RAG2,* constant region of human *IGHG1,* murine *IGHG* and *IGK* (except for SCID mice)	NA	Scheurer et al. (2021)

The presence of NA in the table is due to the absence of further studies
NA not available

2.3.1 Neurons of the Central Nervous System (CNS) Produce Non B-Igs

The presence of IgG in the brain has been established in neurons from the brains of rats and cats, as well as in the intact, normal brains of adult rabbits and mice (Dunn et al. 1995; Hedger et al. 1994; Yoshimi et al. 2002; Naegele et al. 1991). Conceivably, this IgG was thought to be transported across the blood–brain barrier (BBB) via breaches occurring either under physiological/pathological conditions, at specific anatomical sites with a more permeable BBB (e.g. the circumventricular organs), or during early developmental stages, which is based on the theory that Ig is only produced by B lymphocytes. It was believed that capillaries present in the brain prior to embryonic day 15 (E15) had not yet formed a mature barrier, so up to this stage, neurons might take up IgGs and concentrate them in their cytoplasm (Archelos et al. 2000; Wurster et al. 1994).

However, IgG expression was found in both adult mouse brain neurons and isolated neonatal mouse neurons, and the IgG immunoreactivity was located in the cytoplasm, axons, and dendrites of the primary culture neurons (Huang et al. 2008). In support of this, it has been documented that Ig and its receptors are widely expressed in both the central and peripheral nervous systems (Rani et al. 2021; Scheurer et al. 2021; Zhang et al. 2011, 2013; Biology et al. 2013). Further investigations showed that this neuron-derived Ig can protect neurons from injury induced by complement and microglial activation (Zhang et al. 2013; Biology et al. 2013). Future studies to reinforce a more comprehensive understanding of the physiological functions and underlying mechanisms are warranted.

2.3.2 Spermatogenic Cells Produce Non B-Igs

As an immunologically privileged site in the body, the testes and seminal plasma possess a generalised immunosuppressive activity. Multiple factors are associated with immunotolerance in the testes and epididymis: (1) the blood–testis barrier and the blood–epididymal barrier (Kaur et al. 2017), (2) immune protection from Sertoli cells expressing immunoregulatory molecules (Allen et al. 1978), and (3) the immunological homeostasis attributed to the regulation of the Fas system under both physiological and pathological conditions (Filippini et al. 2001; Fijak et al. 2006). When IgG and IgA were found in sperm (Allen et al. 1978; Hjort 1996), it was believed that these Igs were transferred into the seminal plasma by binding to an Fc receptor expressed in sperm (Kamada et al. 1991). However, FcRIII immunoreactivity in sperm was not detected by other groups (Bronson et al. 1992; Chiu et al. 2002). Surprisingly, the IgG concentration in normal adult rat testicular fluid was 65% of the level in serum, therefore, compared with other tissues, the permeability of the rat testes to circulating IgG was postulated to be relatively high, and the transendothelial IgG transfer in the testes was not influenced by alterations in testicular activity or hormone secretion (Hedger et al. 1994). However, these viewpoints were challenged by the findings of the expression of rearranged heavy chains of Igγ and Igκ light chains transcripts, and the presence of Ig proteins in adult mouse testicular spermatogenic cells (Huang et al. 2009). Intriguingly, in testis tissue, IgG immunoreactivity was found in both spermatogonia and spermatocytes, whilst IgM immunoreactivity was present in both spermatids and spermatozoa of testis tissue. More interestingly, IgG and IgM were also expressed in SCID mouse testes and epididymis, i.e. the expression of IgG and IgM in spermatogenic cells and epididymal epithelial cells was relatively independent of the immune status or the serum IgG concentration of mice (Huang et al. 2009). It was soon discovered that human sperm can also express IgG, which may facilitate the sperm penetrating zona-free hamster eggs (Yan et al. 2016). The other physiological functions of the non B-Igs in the testes await further investigation.

2.3.3 Eyes Produce Non B-Igs

Eyes were believed to be one of the immune privilege sites (Taylor 2009). However, using IHC, ISH, and RT-PCR on ocular tissues from both humans and mice, several intraocular structures were found to express proteins and mRNA transcripts of IgG heavy chains, Igκ, and Igλ, as well as V(D)J recombinations, RAG1 and RAG2. These intraocular structures included the cornea (the stroma, the epithelium, and the endothelium in humans; the stroma and the endothelium in mice), the iris (the epithelium in humans; the epithelium and the stroma in mice), the ciliary body (the epithelium in humans; the epithelium and the stroma in mice), the sclera and choroid (the lumina of blood vessels and in the connective tissue), and the retina (the ganglion cells, the inner nuclear layer, the inner plexiform layer, and the nerve fibre layer) (Niu et al. 2011). Furthermore, FcRn was detectable in the epithelium and endothelium of the cornea, the epithelium of the ciliary body, the epithelium of the iris, and the vascular endothelium of the intraocular structures. Fcγ receptors (FcγRs) were detected in the pigment epithelium of the iris and processes, and the stroma of the ciliary body (Niu et al. 2011). As we know, FcγR was thought to modulate local immune responses (Heijnen et al. 1997), and FcRn mediate intraocular IgG transport across the BBB (Zhang and Pardridge 2001). The extensive distribution of IgG and its receptors in intraocular structures indicated that locally produced IgG might play a significant role in maintaining the ocular microenvironment, stimulating the growth and regeneration of the ganglion and retinal pigment epithelium (RPE) cells of the retina, regulating the function of neurons, and protecting eyes by eliminating antigens and pathogens, and it might also be involved in the pathogenesis of age-related macular degeneration and some inflammatory diseases (Niu et al. 2011). Further investigation is needed to elucidate the functions of the Ig expressed in eyes.

2.4 Concluding Remarks

Non B-Igs have been found for more than 20 years. In this chapter, we have reviewed the production and the classes/subclasses characteristics of the non B-Igs, the location and the possible functions of non B-Igs under physiological or different pathological conditions. The expression of non B-Igs in tumour tissues and their possible functions in tumorigenesis suggest their potential roles as a prognostic biomarker in clinical application. The expression of non B-Igs in normal or benign tissues might reveal new facets of the immune system. The glycosylated cIgG might be used as an attractive target for antibody-based therapeutic strategies for cancers. Novel technologies and well-designed investigations will both improve our understanding of immunology and provide more treatments for related diseases.

References

Ablin RJ et al (1971) Localization of immunoglobulins in human prostatic tissue. J Immunol 107(2):603–604

Allen GJ et al (1978) Interaction of immunoglobulin fragments with the mammalian sperm acrosome. J Exp Zool 203(2):271–276

Amedei A et al (2014) Pancreatic cancer: role of the immune system in cancer progression and vaccine-based immunotherapy. Hum Vaccin Immunother 10(11):3354–3368

Archelos JJ et al (2000) Pathogenetic role of autoantibodies in neurological diseases. Trends Neurosci 23(7):317–327

Ashton KJ et al (2006) Age-associated shifts in cardiac gene transcription and transcriptional responses to ischemic stress. Exp Gerontol 41(2):189–204

Babbage G et al (2006) Immunoglobulin heavy chain locus events and expression of activation-induced cytidine deaminase in epithelial breast cancer cell lines. Cancer Res 66(8):3996–4000

Bassler K et al (2019) The myeloid cell compartment — cell by cell. Annu Rev Immunol 37:269–293

Biology C et al (2013) Neuron-derived IgG protects dopaminergic neurons from insult by 6-OHDA and activates microglia through the FcγR I and TLR4 pathways. Int J Biochem Cell Biol 45(8):1911–1920

Bos MK et al (2014) Bladder preservation with brachytherapy compared to cystectomy for T1-T3 muscle-invasive bladder cancer : a systematic review. J Contemp Brachytherapy 6(2):191–199

Bourcier K et al (2019) Basic knowledge in soft tissue sarcoma. Cardiovasc Intervent Radiol 42(9):1255–1261

Bronson RA et al (1992) Monoclonal antibodies identify fc gamma receptors on unfertilized human oocytes but not spermatozoa. J Reprod Immunol 21(3):293–307

Busch S et al (2019) Circulating monocytes and tumor-associated macrophages express recombined immunoglobulins in glioblastoma patients. Clin Transl Med 8(1):18

Chen Z et al (2007) Immunoglobulin G expression in carcinomas and cancer cell lines. FASEB J 21(11):2931–2938

Chen Z et al (2010) Immunoglobulin G is present in a wide variety of soft tissue tumors and correlates well with proliferation markers and tumor grades. Cancer 116(8):1953–1963

Chen Z et al (2011) Immunoglobulin G locus events in soft tissue sarcoma cell lines. PLoS One 6(6):e21276

Chen Q et al (2019) Tumour cell-derived debris and IgG synergistically promote metastasis of pancreatic cancer by inducing inflammation via tumour-associated macrophages. Br J Cancer 121(9):786–795

Chiu WWC et al (2002) Use of antisperm antibodies in differential display Western blotting to identify sperm proteins important in fertility. Hum Reprod 17(4):984–989

Cianga P et al (2003) The MHC class I related fc receptor, FcRn, is expressed in the epithelial cells of the human mammary gland. Hum Immunol 64(12):1152–1159

Cui M et al (2020a) Cancer-derived immunoglobulin G: a novel marker for differential diagnosis and relapse prediction in parathyroid carcinoma. Clin Endocrinol 92(5):461–467

Cui M et al (2020b) High expression of cancer-derived glycosylated immunoglobulin G predicts poor prognosis in pancreatic ductal adenocarcinoma. J Cancer 11(8):2213–2221

Cui M et al (2021) Immunoglobulin expression in cancer cells and its critical roles in tumorigenesis. Front Immunol 12:613530

Deng YQ et al (2006) Immunoglobulin expression in colon cancer cell line HT-29 and its biological activities. Chin J Oncol 28(2):88–91

Deng H et al (2018) Expression of immunoglobulin A in human mesangial cells and its effects on cell apoptosis and adhesion. Mol Med Rep 17(4):5272–5282

Deng Z et al (2020) Single-cell RNA sequencing confirms IgG transcription and limited diversity of V_HDJ_H rearrangements in proximal tubular epithelial cells. Sci Rep 10(1):19657

Deng Z et al (2021) Expression of immunoglobulin G in human proximal tubular epithelial cells. Mol Med Rep 23(5):327

DiMagno EP et al (1999) AGA technical review on the epidemiology, diagnosis, and treatment of pancreatic ductal adenocarcinoma. Am Gastroenterol Assoc Gastroenterol 117(6):1464–1484

Dunn JA et al (1995) Transient immunoglobulin-like molecules are present in the subplate zone and cerebral cortex during postnatal development. Cereb Cortex 5(6):494–505

Eltzschig HK et al (2011) Ischemia and reperfusion-from mechanism to translation. Nat Med 17(11):1391–1401

Fijak M et al (2006) The testis in immune privilege. Immunol Rev 213:66–81

Filippini A et al (2001) Immunology and immunopathology of the male genital tract control and impairment of immune privilege in the testis and in semen. Hum Reprod Update 7(5):444–449

Forrester JV et al (2008) Immune privilege or privileged immunity? Mucosal Immunol 1(5):372–381

Fuchs T et al (2018) Expression of combinatorial immunoglobulins in macrophages in the tumor microenvironment. PLoS One 13(9):e0204108

Geng LY et al (2007) Expression of SNC73, a transcript of the immunoglobulin α-1 gene, in human epithelial carcinomas. World J Gastroenterol 13(16):2305–2311

Geng ZH et al (2019) Human colorectal cancer cells frequently express IgG and display unique Ig repertoire. World J Gastrointest Oncol 11(3):195–207

Haas MS et al (2010) Blockade of self-reactive IgM significantly reduces injury in a murine model of acute myocardial infarction. Cardiovasc Res 87(4):618–627

Hedger P et al (1994) Measurement of immunoglobulin G levels in adult rat testicular interstitial fluid and serum. J Androl 15(6):583–590

Heijnen IA et al (1997) Human IgG fc receptors. Int Rev Immunol 16(1–2):29–55

Hjort T (1996) Quantitative determination of IgG and IgA on sperm from infertile patients with and without antisperm antibodies. Am J Reprod Immunol 36(4):211–215

Hu JB et al (2003) Expression of a novel immunoglobulin gene SNC73 in human cancer and non-cancerous tissues. World J Gastroenterol 9(5):1054–1057

Hu D et al (2011) Heterogeneity of aberrant immunoglobulin expression in cancer cells. Cell Mol Immunol 8(6):479–485

Hu F et al (2012) Spontaneous production of immunoglobulin M in human epithelial cancer cells. PLoS One 7(12):e51423

Huang J et al (2008) Expression of immunoglobulin gene with classical V-(D)-J rearrangement in mouse brain neurons. Int J Biochem Cell Biol 40(8):1604–1615

Huang J et al (2009) Expression of immunoglobulin gene with classical V-(D)-J rearrangement in mouse testis and epididymis. J Histochem Cytochem 57(4):339–349

Huang J et al (2014) Rearrangement and expression of the immunoglobulin μ-chain gene in human myeloid cells. Cell Mol Immunol 11(1):94–104

Ian DH et al (2002) Managing patients with papillary thyroid carcinoma: insights gained from the Mayo Clinic's experience of treating 2512 consecutive patients during 1940 through 2000. Trans Am Clin Climatol Assoc 113:241–260

Jemal A et al (2009) Cancer statistics, 2009. CA Cancer J Clin 59(4):1–25

Ji F et al (2015) Prognostic value and characterization of the ovarian cancer-specific antigen CA166-9. Int J Oncol 47(4):1405–1415

Jiang C et al (2014) Immunoglobulin G expression in lung cancer and its effects on metastasis. PLoS One 9(5):e97359

Jiang D et al (2015) IgG and IgA with potential microbial-binding activity are expressed by normal human skin epidermal cells. Int J Mol Sci 16(2):2574–2590

Jiang H et al (2019) Cancer IgG, a potential prognostic marker, promotes colorectal cancer progression. Chin J Cancer Res 31(3):499–510

Jing Z et al (2018) Expression of immunoglobulin G in human podocytes, and its role in cell viability and adhesion. Int J Mol Med 41(6):3296–3306

Jonasch E et al (2014) Renal cell carcinoma. BMJ 349:g4797

Kamada M et al (1991) Archives of andrology identification of IgG and fc-binding proteins in human seminal plasma and sperm. Arch Androl 27(1):1–7

Kang BY et al (2009) LOX-1 dependent overexpression of immunoglobulins in cardiomyocytes in response to angiotensin II. Biochem Biophys Res Commun 379(2):395–399

Kang S et al (2012) In situ identification and localization of IGHA2 in the breast tumor microenvironment by mass spectrometry. J Proteome Res 11(9):4567–4574

Kaur G et al (2017) An overview of a Sertoli cell transplantation model to study testis morphogenesis and the role of the Sertoli cells in immune privilege. Environ Epigenet 3(3):dvx012

Kimoto Y (1998) Expression of heavy-chain constant region of immunoglobulin and T-cell receptor gene

transcripts in human non-hematopoietic tumor cell lines. Genes Chromosom Cancer 22(1):83–86
Kulik L et al (2009) Pathogenic natural antibodies recognizing Annexin IV are required to develop intestinal ischemia-reperfusion injury. J Immunol 182(9):5363–5373
Laurenti E et al (2018) From haematopoietic stem cells to complex differentiation landscapes. Nature 553(7689):418–426
Lee G (2009) Cancer cell-expressed immunoglobulins: CA215 as a pan cancer marker and its diagnostic applications. Cancer Biomark 5(3):137–142
Lee G et al (2008) Molecular identity of a pan cancer marker, CA215. Cancer Biol Ther 7(12):2007–2014
Lee G et al (2009a) Cancer cell expressions of immunoglobulin heavy chains with unique carbohydrate-associated biomarker. Cancer Biomark 5(4):177–188
Lee G et al (2009b) Preclinical assessment of anti-cancer drugs by using RP215 monoclonal antibody. Cancer Biol Ther 8(2):161–166
Lee G et al (2012) Peptide mapping and glycoanalysis of cancer cell-expressed glycoproteins CA215 recognized by RP215 monoclonal antibody. J Carbohydr Chem 31(1):10–30
Lefranc MP, Lefranc G (2020) Immunoglobulins or antibodies: IMGT® bridging genes, structures and functions. Biomedicines 8(9):319
Lei Y et al (2014) Expression and distribution of immunoglobulin G in the normal liver, hepatocarcinoma and postpartial hepatectomy liver. Lab Investig 94(11):1283–1295
Li M et al (2004) Expression of immunoglobulin kappa light chain constant region in abnormal human cervical epithelial cells. Int J Biochem Cell Biol 36(11):2250–2257
Li X et al (2011) The presence of IGHG1 in human pancreatic carcinomas is associated with immune evasion mechanisms. Pancreas 40(5):753–761
Li M et al (2012) Promotion of cell proliferation and inhibition of ADCC by cancerous immunoglobulin expressed in cancer cell lines. Cell Mol Immunol 9(1):54–61
Li L et al (2016) Overexpression of cancer cell-derived immunoglobulin G correlates with poor prognosis in gastric cancer patients. Transl Cancer Res 5(3):285–293
Liang PY et al (2013) Overexpression of immunoglobulin G prompts cell proliferation and inhibits cell apoptosis in human urothelial carcinoma. Tumour Biol 34(3):1783–1791
Liao Q et al (2015) Aberrant high expression of immunoglobulin G in epithelial stem/progenitor-like cells contributes to tumor initiation and metastasis. Oncotarget 6(37):40081–40094
Liu Y et al (2012) IgG gene expression and its possible significance in prostate cancers. Prostate 72(6):690–701
Liu Y et al (2015) Binding of the monoclonal antibody RP215 to immunoglobulin G in metastatic lung adenocarcinomas is correlated with poor prognosis. Histopathology 67(5):645–653
Liu J et al (2016) Immunoglobulin gene expression in umbilical cord blood-derived CD34 + hematopoietic stem/progenitor cells. Gene 575(1):108–117
Liu E et al (2017) Tumor derived IgG suppress the proliferation of T cells in cord blood. J Peking Univ Health Sci 9(5):824–828
Lloyd-Jones D et al (2010) Executive summary: heart disease and stroke statistics-2010 update: a report from the American Heart Association. Circulation 121(7):948–954
Lotze MT (2011) Pancreatic cancer is not noble. J Innate Immun 4(1):4–5
Lu W et al (2007) Over-expression of the bovine FcRn in the mammary gland results in increased IgG levels in both milk and serum of transgenic mice. Immunology 122(3):401–408
Lucas SD et al (1996) Tumor-specific deposition of immunoglobulin g and complement in papillary thyroid carcinoma. Hum Pathol 27(12):1329–1335
Lv WQ, Peng J, Wang HC et al (2017) Expression of cancer cell-derived IgG and extra domain A-containing fibronectin in salivary adenoid cystic carcinoma. Arch Oral Biol 81:15–20
Ma C et al (2013) Immunoglobulin G expression and its potential role in primary and metastatic breast cancers. Curr Mol Med 13(3):429–437
Maillard N et al (2015) Current understanding of the role of complement in IgA nephropathy. J Am Soc Nephrol 26(7):1503–1512
Matsubara S et al (2012) Incidental pancreatic cysts found by magnetic resonance imaging and their relationship with pancreatic cancer. Pancreas 41(8):1241–1246
Naegele JR et al (1991) Expression of a unique 56-kDa polypeptide by neurons in the subplate zone of the developing cerebral cortex. Proc Natl Acad Sci USA 88(2):330–334
Nath SD et al (2005) Fasting serum glucose level and cancer risk in Korean men and women (multiple letters). JAMA J Am Med Assoc 293(18):2210–2211
Niu N et al (2011) Expression and distribution of immunoglobulin G and its receptors in an immune privileged site: the eye. Cell Mol Life Sci 68(14):2481–2492
Niu N et al (2012) IgG expression in human colorectal cancer and its relationship to cancer cell behaviors. PLoS One 7(11):e47362
Pan B et al (2013) Suppression of IGHG1 gene expression by siRNA leads to growth inhibition and apoptosis induction in human prostate cancer cell. Mol Biol Rep 40(1):27–33
Peng J et al (2017) Involvement of non B cell-derived immunoglobulin G in the metastasis and prognosis of salivary adenoid cystic carcinoma. Oncol Lett 14(4):4491–4498
Perše M et al (2019) The role of IgA in the pathogenesis of IgA nephropathy. Int J Mol Sci 20(24):6199
Qin C et al (2020) Cancer-driven IgG promotes the development of prostate cancer though the SOX2–CIgG pathway. Prostate 80(13):1134–1144

Qiu X, Yang G (1996a) Existence of immunoglobulin-like protein in malignant tumor cells. J Norman Bethune Univ Med Sci 22:575–577

Qiu X, Yang GZ (1996b) Characteristic of immunoglobulin like protein in malignant tumor cells and the analyses of immunoglobulin gene. Chin J Immunol 22(5):295

Qiu X, Yang GZ (1997) Analyses of immunoglobulin gene in epithelial cancers. Chin J Cancer Biother 4(2):157–158

Qiu X et al (2003) Human epithelial cancers secrete immunoglobulin G with unidentified specificity to promote growth and survival of tumor cells human epithelial cancers secrete immunoglobulin G with unidentified specificity to promote growth and survival of tumor cells. Cancer Res 63(19):6488–6495

Qiu X et al (2008) Immunoglobulin mRNA and protein expression in human oral epithelial tumor cells. Appl Immunohistochem Mol Morphol 16(3):232–238

Qiu Y et al (2012) Immunoglobulin G expression and its colocalization with complement proteins in papillary thyroid cancer. Mod Pathol 25(1):36–45

Qiu X et al (2013) Immunoglobulin gamma heavy chain gene with somatic hypermutation is frequently expressed in acute myeloid leukemia. Leukemia 27(1):92–99

Qiu X et al (2020) Immunoglobulin kappa light chain produced by cardiomyocytes and participates in maintaining intercalated disc integrity. Res Square 2020:114990

Rani R et al (2021) Neuron-specific spinal cord translatomes reveal a neuropeptide code for mouse dorsal horn excitatory neurons. Sci Rep 11(1):5232

Rawla P (2019) Epidemiology of prostate cancer. World J Oncol 10(2):63–89

Ray-coquard I et al (2018) Options for treating different soft tissue sarcoma subtypes. Future Oncol 14(10s):25–49

Reiser J et al (2016) Podocytes. F1000Res 5:F1000

Rincheval-Arnold A et al (2002) Up-regulation of polymeric immunoglobulin receptor mRNA in mammary epithelial cells by IFN-γ. Mol Cell Endocrinol 194(1–2):95–105

Scheurer L et al (2021) Expression of immunoglobulin constant domain genes in neurons of the mouse central nervous system. Life Sci Alliance 4(11):1–9

Shao W et al (2016a) Epithelial cells are a source of natural IgM that contribute to innate immune responses. Int J Biochem Cell Biol 73:19–29

Shao W et al (2016b) Identification of liver epithelial cell-derived Ig expression in μ chain-deficient mice. Sci Rep 6:1–12

Sheng Z et al (2016a) Involvement of cancer-derived IgG in the proliferation, migration and invasion of bladder cancer cells. Oncol Lett 12(6):5113–5121

Sheng Z et al (2016b) IgG is involved in the migration and invasion of clear cell renal cell carcinoma. J Clin Pathol 69(6):497–504

Silva DG et al (2011) Anti-islet autoantibodies trigger autoimmune diabetes in the presence of an increased frequency of islet-reactive CD4 T cells. Diabetes 60(8):2102–2111

Suzuki K et al (2003) Incidence of latent mesangial IgA deposition in renal allograft donors in Japan. Kidney Int 63(6):2286–2294

Tang J et al (2018) Lung squamous cell carcinoma cells express non-canonically glycosylated IgG that activates integrin-FAK signaling. Cancer Lett 430:148–159

Taylor AW (2009) Ocular immune privilege. Eye (Lond) 23(10):1885–1889

Tomana M et al (1999) Circulating immune complexes in IgA nephropathy consist of IgA1 with galactose-deficient hinge region and antiglycan antibodies. J Clin Invest 104(1):73–81

Vasu VT et al (2007) Genome-wide screening of alpha-tocopherol sensitive genes in heart tissue from alpha-tocopherol transfer protein null mice (ATTP−/−). FEBS Lett 581(8):1572–1578

Wan X et al (2015) Pancreatic expression of immunoglobulin G in human pancreatic cancer and associated diabetes. Pancreas 44(8):1304–1313

Wang LM, Gan YH (2021) Cancer-derived IgG involved in cisplatin resistance through PTP-BAS/Src/PDK1/AKT signaling pathway. Oral Dis 27(3):464–474

Wang H et al (2013) Prognostic significance of immunoglobulin M overexpression in laryngeal squamous cell carcinoma. Acta Otolaryngol 133(10):1080–1087

Wang J et al (2014) Cancer-derived immunoglobulin G promotes LPS-induced proinflammatory cytokine production via binding to TLR4 in cervical cancer cells. Oncotarget 5(20):9727–9743

Wang C et al (2015) IGK with conserved IGKV/IGKJ repertoire is expressed in acute myeloid leukemia and promotes leukemic cell migration. Oncotarget 6(36):39062–39072

Wang G et al (2019) Expression and clinical significance of cancer-derived immunoglobulin G in non-small cell lung cancer by bioinformatics and immunohistochemistry. Zhongguo Fei Ai Za Zhi 22(6):341–348

Wang Z et al (2020) Cancer-derived sialylated IgG promotes tumor immune escape by binding to Siglecs on effector T cells. Cell Mol Immunol 17(11):1148–1162

Welinder C et al (2013) Primary breast cancer tumours contain high amounts of IgA1 immunoglobulin: an immunohistochemical analysis of a possible carrier of the tumour-associated Tn antigen. PLoS One 8(4):e61749

Wilkin TJ (1990) Antibody markers in predicting type 1 diabetes: a review. J R Soc Med 83(10):632–636

Wu L et al (2014) The immunoglobulin heavy chain VH6-1 promoter regulates Ig transcription in non-B cells. Cancer Cell Int 14(1):1–8

Wu L et al (2020) High levels of immunoglobulin expression predict shorter overall survival in patients with acute myeloid leukemia. Eur J Haematol 105(4):449–459

Wurster U et al (1994) Passage of intravenous immunoglobulin and interaction with the CNS. J Neurol Neurosurg Psychiatry 57:21–25

Xu Y et al (2016) IgG silencing induces apoptosis and suppresses proliferation, migration and invasion in LNCaP prostate cancer cells. Cell Mol Biol Lett 21:27

Yamakawa M et al (1994) Protection of thyroid cancer cells by complement-regulatory factors. Cancer 73(11):2808–2817

Yan M et al (2016) Immunoglobulin G expression in human sperm and possible functional significance. Sci Rep 6:20166

Yang B et al (2013) Correlation of immunoglobulin G expression and histological subtype and stage in breast cancer. PLoS One 8(3):e58706

Yoshimi K et al (2002) IgG-immunostaining in the intact rabbit brain: variable but significant staining of hippocampal and cerebellar neurons with anti-IgG. Brain Res 956(1):53–66

Zhang Y, Pardridge WM (2001) Mediated efflux of IgG molecules from brain to blood across the blood–brain barrier. J Neuroimmunol 114(1–2):168–172

Zhang M et al (2006a) The role of natural IgM in myocardial ischemia-reperfusion injury. J Mol Cell Cardiol 41(1):62–67

Zhang M et al (2006b) Identification of the target self-antigens in reperfusion injury. J Exp Med 203(1):141–152

Zhang S et al (2010) Immunoglobulin gene locus events in epithelial cells of lactating mouse mammary glands. Cell Mol Life Sci 67(6):985–994

Zhang J et al (2011) The international journal of biochemistry expression and distribution of immunoglobulin G and its receptors in the human nervous system. Int J Biochem Cell Biol 43(4):556–563

Zhang L et al (2012) Expression of immunoglobulin G in esophageal squamous cell carcinomas and its association with tumor grade and Ki67. Hum Pathol 43(3):423–434

Zhang J et al (2013) Neuron-derived IgG protects neurons from complement-dependent cytotoxicity. J Histochem Cytochem 61(12):869–879

Zheng H et al (2007a) Expression and secretion of immunoglobulin alpha heavy chain with diverse VDJ recombinations by human epithelial cancer cells. Mol Immunol 44(9):2221–2227

Zheng H et al (2007b) Immunoglobulin alpha heavy chain derived from human epithelial cancer cells promotes the access of S phase and growth of cancer cells. Cell Biol Int 31(1):82–87

Zheng J et al (2009) Immunoglobulin gene transcripts have distinct VHDJH recombination characteristics in human epithelial cancer cells. J Biol Chem 284(20):13610–13619

Zhu Z et al (2017) Immunoglobulin M, a novel molecule of myocardial cells of mice. Int J Biochem Cell Biol 88:172–180

Genetic Characteristics of Non B Cell-Derived Immunoglobulin Genes

3

Miaoran Xia, Chi Zhang, Lin Xiao, and Xiaoyan Qiu

Abstract

It is widely acknowledged that immunoglobulins (Igs) are produced solely by B-lineage cells. The Ig gene is created by the rearrangement of a group of gene segments [variable (V), diversity (D), and joining (J) segments rearrangement, or V(D)J recombination], which results in the vast diversity of B cell-derived Ig responsible for recognising various antigens. Ig subsequently undergoes somatic hypermutation (SHM) and class switch recombination (CSR) after exposure to antigens, thus converting the low-affinity IgM to IgG, IgA, or IgE antibodies. IgM and IgD are primarily expressed in naïve B cells that have not been exposed to antigens, they do not undergo somatic hypermutation; hence, their variable region sequences remain the same as those in the germline. In contrast, IgG, IgA, and IgE are expressed in antigen-stimulated memory B cells or plasma cells, and thus, they often possess high-frequency mutations in their variable region sequences. Since the discovery that Ig can be produced by non-B cells, Qiu's group has investigated and compared the genetic characteristics of B cell-derived Ig and non-B cell-derived Ig. These findings demonstrated that non-B cell-derived Ig shares certain similarities with B cell-derived Ig in that the sequence of its constant region is identical to that of B cell-derived Ig, and its variable region is also strictly dependent on the rearrangement of V, D, and J gene segments. Moreover, akin to B cell-derived Ig, the V regions of IgM and IgD are rarely mutated, while IgG, IgA, and IgE produced by cancer cells are frequently mutated. However, the non-B cell-derived Ig V region sequence displays unique characteristics. (1) Unlike the vast diversity of B cell-derived Igs, non-B cell-derived Igs exhibit restricted diversity; cells from the same lineage always select the same V(D)J recombination patterns; (2) Both mRNA and proteins of RAG1/RAG2 recombinase have been detected in Ig positive cancer cell lines and normal tissues. But Ig recombination could also be found in RAG1$^{-/-}$ and RAG2$^{-/-}$ mice, suggesting that they are not necessary for the rearrangement of non-B

M. Xia (✉)
Department of Immunology, School of Basic Medical Sciences, Capital Medical University, Beijing, China
e-mail: mxia@ccmu.edu.cn

C. Zhang · X. Qiu
Department of Immunology, School of Basic Medical Sciences, Peking University, Beijing, China
e-mail: azuretimm@bjmu.edu.cn; qiuxy@bjmu.edu.cn

L. Xiao
Department of Clinical Laboratory, National Cancer Center/National Clinical Research Center for Cancer/Cancer Hospital, Chinese Academy of Medical Sciences and Peking Union Medical College, Beijing, China

X. Qiu et al. (eds.), *Non B cell-Derived Immunoglobulins*, Advances in Experimental Medicine and Biology 1445, https://doi.org/10.1007/978-981-97-0511-5_3

cell-derived Igs. These features of non-B cell-derived Igs suggest a potentially undiscovered mechanism of V(D)J recombination, ligation, and SHM in non-B cells, which necessitates further investigation with advanced technology in molecular biology.

Keywords

Non-B cell-derived immunoglobulin · V(D)J rearrangement · Ig heavy chain · Ig light chain · Monoclonality · Oligoclonality · Somatic hypermutation · Class switching

3.1 The Generation of Immunoglobulin Diversity

3.1.1 A Brief Introduction to the Discovery of Ig Diversity

Immunoglobulin (Ig) was first identified as an 'antitoxin' by Behring and Kitasato in 1890, with Ig being acknowledged as being formed in the body following the introduction of a bacterial infection or toxin (Behring 1890a, b). The 'antitoxin' can neutralise a specific type of bacteria or toxin, especially preventing re-infection by the same type of bacteria. Moreover, specific 'antitoxin' (later termed 'antibody') can be induced by almost any microbial infection. It was then discovered that a large number of antibodies exist in each individual's blood, and the antibody repertoire presents infinite diversity in response to different antigens. Until 1957, Burnet proposed the 'clonal selection hypothesis', which is widely accepted. The hypothesis is that each antibody-producing cell produces antibodies of only one specificity, and cells proliferate in response to antigens when clone expansion occurs, thus selectively increasing antibody abundance (Burnet 1957). How can a finite number of genes encode an infinite number of antibodies? Burnet also hypothesised that the Ig V region genes could be inherited as sets of gene segments rather than being fully encoded in the genome, and these gene segments are joined by random DNA rearrangement to form a unique variable region reading frame prior to Ig production, which allows the antibody repertoire in vivo to present infinite diversity.

This hypothesis was further accepted after the V(D)J recombination was confirmed by Suzumu Tonegawa, who won the Nobel Prize in Medicine and Physiology in 1987 for this study. In 1974, Tonegawa et al. revealed duplicated germline genes through a series of DNA-hybridisation experiments using various κ-chain mRNAs, which helped to explain observed diversity in the antibody repertoire (Tonegawa et al. 1974). Tonegawa et al. provided experimental evidence for the somatic rearrangement of immunoglobulin genes in 1976. In that study, DNA extracted from plasmacytoma cell line-MOPC 321 and a BALB/c mouse early embryo (as control) were digested with a restriction enzyme, and the whole Igκ mRNA and its 3′-end half of mRNA were purified from MOPC 321. Then the radio-labelled whole Igκ mRNA and its 3′-end half of mRNA were used as specific probes to detect the V- and C-region of κ chains, respectively. The result revealed distinct Ig gene expression patterns in the two types of cells. In the genome of mouse embryonic cells, two well-separated segments hybridised with either the variable (V) or constant (C) gene sequence were detected; however, the two sequences were found on the same DNA fragment in the genome of MOPC 231, demonstrating that they had recombined to form a continuous V–C gene (Hozumi et al. 1976). This experiment has been used as a classic criterion for the judgement that immunoglobulin genes are rearranged in the B-lymphocyte lineage cells but not other cells. It is worth mentioning that the probe Tonegawa used is specific for MOPC 231 cells. As each Ig-producing cell can express its unique V region sequences, further evidence is required to claim that V(D)J rearrangement cannot occur in other types of cells.

3.1.2 The DNA Segments of the Immunoglobulin Gene and Their Rearrangement

It is now widely known that the immunoglobulin gene system is comprised of three separate gene loci known as Lκ, Lλ, and H chain (L, the abbreviation for the light chain; H, the abbreviation for the heavy chain), and each of these loci contains variable and constant genes. The Lκ, Lλ, and H chain genes are on human chromosomes 2, 22, and 14, respectively. For light chains, the V region is encoded by two separate DNA segments, namely the Vκ and Jκ (or Vλ and Jλ) segments. The V gene segments encode the first 95–101 amino acids of the light chain, and the J gene segments are responsible for directing the synthesis of the rest of the 13 amino acid residues of the V region. Similarly, the VH region is encoded by three separate DNA segments for the heavy chain: the VH, D, and JH, with the D gene segment lying between the VH and JH gene segments.

The V, D, J, and C gene segments are far apart in the unrearranged DNA. When the recombination process begins of IGH, a D gene segment is first joined to a JH gene segment, and then a VH gene segment is rearranged to DJH. Finally, the formed VHDJH is joined to the C gene by RNA splicing (Fig. 3.1). However, due to the lack of D fragments, in the recombination process of IGL, only a Vκ gene segment is directly rearranged to a Jκ gene segment, or a Vλ gene segment is directly rearranged to a Jλ gene segment. Finally, the formed VκJκ or VλJλ is joined to the C gene by RNA splicing.

There are multiple copies of the V, D, and J gene segments in germline DNA, which numbers are shown in Table 3.1 (Backhaus 2018). Randomly selecting just one copy of the V, D, and J gene segments results in a great diversity of Ig molecules during recombination. Notably, the heavy chain contains a series of C regions arrayed one after the other, each corresponding to a different isotype. In classic immunology theory, a cell first expresses IgM, then expresses other isotypes, such as IgG, through isotype switching.

Apart from the V(D)J combinatorial diversity, which accounts for a substantial part of the diversity of the heavy and light chain V regions, there are other ways accounting for the enormous Ig diversity. Firstly, random insertion and deletion of nucleotides are introduced at the joint ends during V–J or V–D–J recombination. Secondly, a high frequency of somatic mutations in V regions [especially tending to cluster in the complementarity determining regions (CDRs)] upon antigen stimulation of the B cell can create further diversity, a process called affinity maturation of the antibody population.

3.2 The Discovery of Rearranged V(D)J Gene Transcripts in Non-B Cells

It was first discovered that IgG-like substances are present in breast and colon carcinoma cells by Qiu in 1996 (Qiu et al. 1996), but it remained unclear if rearranged V(D)J genes existed in the non-B cancer cells. Subsequently, Qiu et al. confirmed Ig heavy chain gene rearrangement in the genomes of breast cancer cells and HT-29 (colon cancer cell line) through PCR and Southern blotting (Qiu et al. 1997). Gene transcripts of Ig heavy chain constant regions were also detected in five cancer cell lines by nested reverse transcription (RT)-PCR (Kimoto 1998) and in hepatocellular carcinoma by cDNA microarray (Okabe et al. 2001). Subsequently, Zheng et al. discovered a novel immunoglobulin gene, SNC73, in normal mucosal tissues and colorectal cancer via subtractive hybridisation, which encodes a peptide identical to the constant region of an IgA molecule in the carboxyl-terminus (Zheng et al. 2001).

Despite these evidences for Ig transcripts (mainly Ig constant regions) in non-B cancer cells, it was still accepted that these transcripts are sterile without DNA rearrangement or class switching. In other words, these transcripts are not functional. However, increasing reports showed that the entire length of Ig gene transcripts existed in non-B cells.

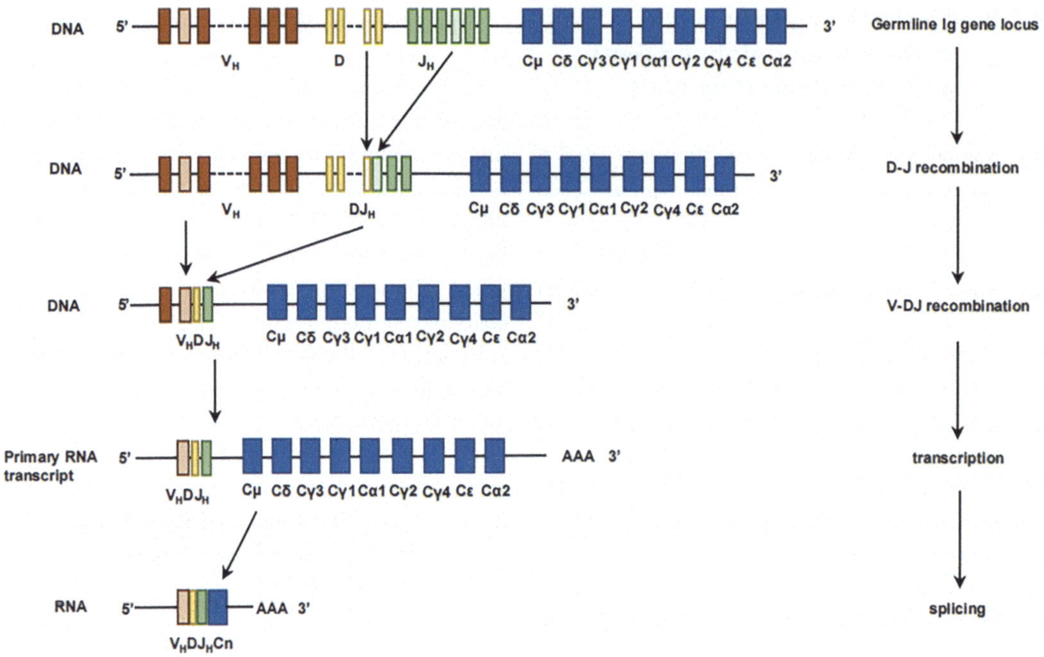

Fig. 3.1 V(D)J recombination of the Ig heavy chain (IGH) from germline gene segments. The immunoglobulin loci are organised into gene segments: the variable (V), diversity (D), joining (J), and constant (C) gene segments. The variable (V) region is generated by random recombination of the V, D, and J gene segments

Table 3.1 The numbers of functional Ig V gene segments of human heavy and light chains (Backhaus 2018)

	Gene locus		
Segments	IGH (Heavy chain)	IGK (κ light chain)	IGL (λ light chain)
Variable (V)	38–46	34–38	29–33
Diversity (D)	23	0	0
Joining (J)	6	5	4–5
Constant (C)	9	1	4–5

In fact, non-B cells can produce Ig, and its Ig production also depends on functional Ig gene rearrangement, which has been confirmed by a large number of experimental evidence. Qiu et al. reported the finding of complete IgG transcripts from tumour cells isolated from epithelial cancers and cell lines in 2003. The group detected the transcripts by in situ hybridisation (probe for IgG Fc cDNA), Northern blot analysis (probe for IgG Fc cDNA), and single-cell RT-PCR (primers for Ig gene of variable region and CH1) (Qiu et al. 2003). Consistently, Babbage et al. found potentially functional V(D)J transcripts in breast cancer cell lines (BT474, MDA MB-231, MCF-7, SKBR3, T47D, and ZR75–1) by nested RT-PCR, and they found these transcripts were stably expressed in sequential cultures (Babbage et al. 2006). It is worth noting that these cell lines were all EBV and CD19/CD20 negative and expressed activation-induced cytidine deaminase (AID) essential for mutational and switch activity (Babbage et al. 2006). In recent years, Qiu's group analysed over 700,000 Ig variable region sequences of heavy and light chains from many types of non-B cells sorted from humans or mice. They confirmed that the Ig gene in non-B cells widely exhibits functional Ig gene rearrangement characteristics through immune repertoire analysis with Sanger, NGS, or PacBio sequencing. Qiu's group has created a database (NBIGV) constituted by a non-B cell-derived Ig variable region repertoire, including 727,989 VHDJH and VκJκ recombination sequences of non-B cell-derived Igs sequenced from mouse samples (freely available at http://ubigv.org) (Zhang et al. 2021).

3.3 Non-B Cell-Derived Igs Also Display Insertion/Deletion at Joining Ends and Are Potentially Functional

As mentioned above, during V(D)J recombination, the diversity of immunoglobulins is further increased by the insertion and deletion of nucleotides between the junctions of the V, D, and J gene segments. The added nucleotides are known as P-nucleotides and N-nucleotides (N/P insertions). These nucleotides get their names because P-nucleotides constitute palindromic sequences added to the ends of the gene segments, and N-nucleotides are non-template-encoded. Similarly, nucleotides can also be deleted at gene segment junctions. As these processes are random, the reading frame of the coding sequence is often disrupted, thereby giving rise to a nonfunctional protein, known as a nonproductive rearrangement (Janeway et al. 2001).

N/P insertions were also observed in the gene sequences of non-B cell-derived Igs. 93% of the VHDJH rearrangements showed classic N/P addition in Ig expressed by the liver epithelial cells of μMT mice (lacking B cells), whereas only 5% of the VκJκ rearrangements had N/P addition (Shao et al. 2016). Zheng et al. analysed sequencing results from samples of eight different epithelial cancers and two cancer cell lines [HT-29 (a colon carcinoma cell line) and Hela S3 (a breast carcinoma cell line)] (Zheng et al. 2009). They also found typical VHDJH recombination with an N or P insertion, and the JH gene recombination occurred primarily at TG nucleotide sequences, as in B cell-derived Ig.

Ig transcripts were identified as nonfunctional if the insertion/deletion at joining ends or mutations introduced stop codons into the V region. In their study, Zheng et al. found that only 4 of the 35 VHμDμJHμ rearrangements and 7 of the 54 VHγDγJHγ rearrangements resulted in nonfunctional V region genes (Zheng et al. 2009). In the sorted liver epithelial cells from 5 μMT mice by Shao et al., 82 of the 86 VHDJH rearrangements and all 35 VκJκ rearrangements were functional (Shao et al. 2016). Generally, the non-B cell-derived IgH and Igκ were functional with rates above 80% in most reports, similar to the B cell-derived Igs.

3.4 Non-B Cell-Derived Ig Rearrangements Are Highly Biased in Individuals and Conserved Amongst Populations

It is well recognised that the V(D)J gene rearrangement generates substantial combinatorial diversity in mature B cells. Additionally, the insertion and deletion of nucleotides in the recombination process introduce further diversity at the joints between different gene segments. It is estimated that at least 10^{11} different receptors are expressed by naïve B cells. Finally, somatic hypermutation introduces point mutations into the rearranged V-region genes of activated B cells and therein generates additional diversity.

In contrast, non-B cell-derived Ig rearrangements are highly biased in individuals (Fig. 3.2) and conserved amongst populations (Table 3.2). In the report of Zheng et al. three sets of VHμDJHμ recombination were predominant: VH3-15/D3-10/JH4 found in 5 of 26 (19.2%) cases of carcinomas from the breast and colon, VH6-1/D6-13/JH4 found in 5 of 26 (19.2%) cases of carcinomas from the lung and oral cavity, and VH4-30-2/D3-22/JH4 found in 4 of 26 (15.4%) cases of carcinomas from the breast and colon. Of the VHγDJHγ sequences studied, two sets were predominant: VH5-51/D3-9/JH4 found in 19 of 38 (50%) cases of carcinomas of the breast, colon, oral and lung, and VH3-30/D6-19/JH4 found in 4 of 38 (10.5%) cases of carcinomas of the breast, colon, and lung (Zheng et al. 2009). Advanced immune repertoire sequencing, combined with multiplex-PCR and next-generation sequencing, has provided a comprehensive

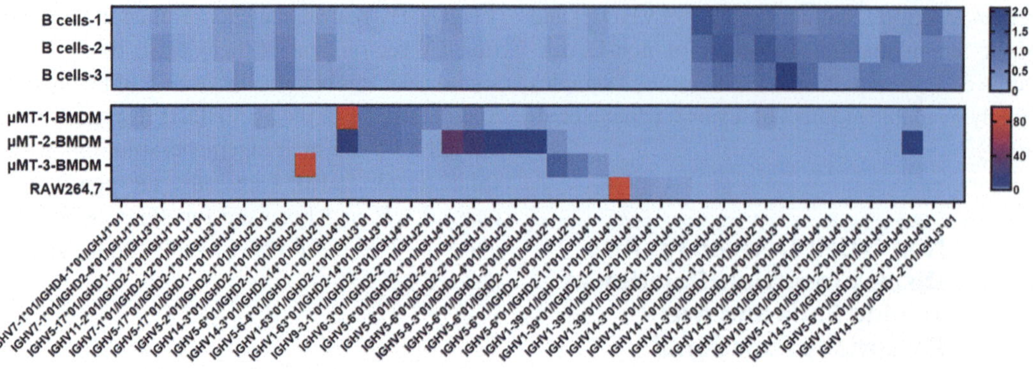

Fig. 3.2 The restricted diversity of non-B cell-derived Ig. The repertoire of B cell-derived Ig exhibited huge diversity, whereas that of non-B cell-derived-Ig is much more biased. In a study (Gong et al. 2021) that detected Ig repertoire in mouse B cells, bone marrow-derived macrophages (BMDM) from μMT mice and macrophage cell line RAW264.7 by next-generation sequencing, the results show that V(D)J recombinations in B cells are highly diverse while those in BMDMs and RAW264.7 are much more restricted

view of IgH and IgL mRNA sequences in colorectal cancer cells (Geng et al. 2019) and acute myeloid leukaemia (AML) myeloblasts (Xia et al. 2022). Although all five Ig classes (IgA, IgD, IgE, IgG, and IgM) were detected, both the colorectal cancer cells and AML myeloblasts most prominently express IgG, and their VHDJH rearrangements were biased with individual specificity rather than being generally diverse as those in B cells. Although no identical VHDJH patterns were shared by different patients in the two studies, AML-derived IGK was more conserved amongst different AML samples, the frequently shared Vκ-Jκ patterns were IGKV3-20*01/IGKJ1*01, IGKV2D-28*01/IGKJ1*01, and IGKV4-1*01/IGKJ1*01 (Xia et al. 2022).

It was hypothesised that this limited diversity might be a result of the clonal proliferation of tumour cells. However, Igs with biased V(D)J patterns were also found in normal tissues/cells. Niu et al. found that Igγ expressed by human eye tissues mainly used the VH4-61/D5-18/JH4 pattern (Niu et al. 2011). Surprisingly, Huang et al. found that the mouse kappa light chain V segments Vbv9 and Vcw9 were frequently used by mouse epithelial cells, neurons, and spermatogenic cells (Zhang et al. 2010; Huang et al. 2008, 2009). That is to say, in some cases, non-B cell-derived Ig from different lineage cells could utilise the same V(D)J patterns (Table 3.2). Interestingly, in the liver epithelial cells from μMT mice, the μ, γ, and κ transcripts showed restricted V(D)J patterns. In contrast, the α and δ transcripts displayed diverse VHDJH rearrangements similar to B cell-derived Igs (Shao et al. 2016). By next-generation sequencing, Geng et al. detected all IgH classes in normal human colorectal epithelial cells. Several VHDJH rearrangements were documented, VH1-8/D7-27/JH4 and VH1-18/D4-17/JH4, for instance, were used by normal cells from more than one sample (Geng et al. 2019). Several identical patterns of VHDJH rearrangement were observed in bone marrow-derived macrophages (BMDMs) from WT mice as well as μMT mice, while B cell-derived Ig displayed diversity as expected (Fig. 3.2). Moreover, an oligoclonal rearrangement pattern (V1-39*01/D1-1*01/J4*01) was also detected in the macrophage cell line-RAW264.7 (Fig. 3.2) (Gong et al. 2021). These findings suggest that the limited diversity of non-B cell-derived Ig is a specific characteristic of all non-B cells (not only tumour cells).

The diversity of Ig is generated by not only V(D)J rearrangement but also the random insertion/deletion during rearrangement and somatic hypermutation in germinal centres. Thus, even choosing the same V, D, and J segments, B cell-derived Igs from different clones have different nucleotide sequences. However, the sequences of

Table 3.2 Non-B cell-derived Ig rearrangements are conserved amongst populations

Igs	V(D)J	Expression profile
Igκ	Vhv9/Jκ1	Mice-mammary gland epithelial cells (Zhang et al. 2010) Mouse brain neurons (Huang et al. 2008) Mouse primary spermatocytes (Huang et al. 2009)
	IGKV1-5*03/ IGKJ3*01	Human CD33⁺ AML myeloblasts (Wang et al. 2015) Mature myeloid cells (Wang et al. 2015)
	IGKV1-NL1*01/ IGKJ5*01	Human CD33⁺ AML myeloblasts (Wang et al. 2015) AML cell lines (NB4, KG-1) (Wang et al. 2015)
Igγ	VH5-51/D3-9/JH4	Human oral cancer (Zheng et al. 2009) Human colon cancer (Zheng et al. 2009) Human lung cancer (Zheng et al. 2009) Human breast cancer (Zheng et al. 2009) Human oral basal cell ameloblastoma (Zhu et al. 2008)
	VH3-48/D4-7/JH4	AML cell lines (HL-60, HEL, THP-1, OCI-AML3) (Wang et al. 2015)
	VH3-33/D6-19/ JH5	Human oral cancer (Zheng et al. 2009) Human oral squamous cell carcinoma (Zhu et al. 2008)
Igμ	VH6-1/D6-13/JH4	HT-29 (Zheng et al. 2009) Human oral cancer (Zheng et al. 2009)

non-B cell-derived Igs always show oligoclonality. More interestingly, the variable regions of non-B cell-derived Igs share highly consistent sequences between different cells and even individuals. Identical sequences were found in the BMDMs from different WT mice (Gong et al. 2021). By cloning and sequencing the CDR3 of the IgG heavy chain, Zheng et al. reported the identical V(D)J recombination in HeLa S3 and HT-29. They also found that HT-29 had a V(D)J recombination sequence identical to HeLa MR (a breast carcinoma cell line) (Zheng et al. 2009). Subsequently, Zhu et al. reported rearranged Igγ1 gene transcripts in basal cell ameloblastoma with dominant VH5-51/D3-9/JH4 usage, whose sequence was 100% identical to the previously published sequence from a lung cancer cell line by Zheng et al. (Zhu et al. 2008). These population-shared sequences are believed to have a role in tumorigenesis and tumour progression.

3.5 Somatic Hypermutation of Functional V Genes in Non-B Cells

Somatic hypermutation (SHM), which introduces a vast number of point mutations into the V-region exon, occurs after Ig genes have been recombined and transcribed. In humans and mice, SHM occurs only after mature B cells have been activated by the specific antigen and gives rise to the affinity maturation of the antibody. As IgM and IgD are primarily expressed in naïve B cells that have not been exposed to antigens, these Ig genes do not undergo somatic hypermutation. Conversely, IgG, IgA, and IgE expressed in antigens-stimulated memory B cells or plasma cells often have a high frequency of mutations in their variable region sequences.

Ig light and heavy chains demonstrate almost no mutations in normal cells, including epithelial cells, neurons, and spermatogenic cells (Zhang et al. 2010; Huang et al. 2008, 2009). However, Ig molecules show a high frequency of mutations in tumour cells, such as squamous cell carcinoma, including lung squamous cell carcinoma, oesophageal squamous cell carcinoma, and oral squamous cell carcinoma (Zheng et al. 2009). It is known that mutations in B cell-derived VDJ or VJ are antigen-induced, characterised by mutations primarily concentrated in the CDRs rather than the framework regions (FRs), while mutations of non-B cell-derived Igs could be randomly distributed in FRs or CDRs (Zhu et al. 2008). Furthermore, identical mutation sites exist among different cell lines or samples (Babbage et al.

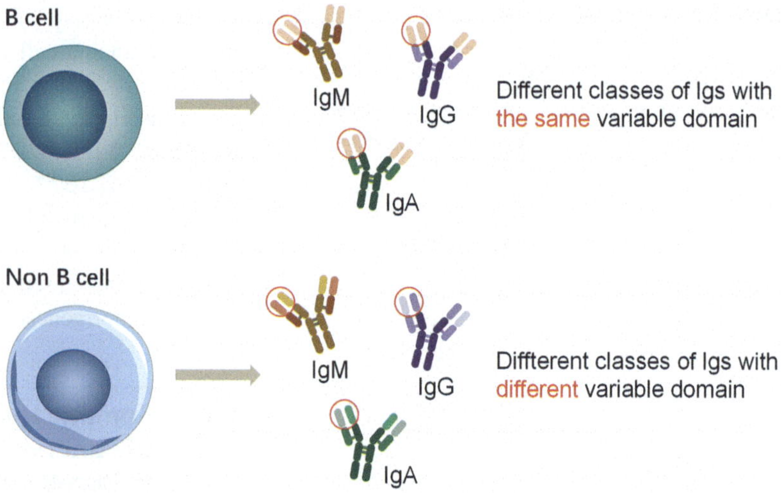

Fig. 3.3 Different classes of Igs in B cells and non-B cells. In classic theory, B cells can generate different classes of Igs with the same variable domains but different constant regions via the class-switching process. However, it has been found that the non-B cells always produce different classes of Igs using different variable domains. The red circles indicate the variable domains of Igs

2006; Zhu et al. 2008). Similarly, in almost all AML tumour cells, mutations occurred in the expressed Igγ (65/65) and κ-type light chains (102/105). In contrast, no or only a few mutations were detected in the Igμ (0/45) and κ light chains in normal monocytes and neutrophils (6/86) (Qiu et al. 2013; Huang et al. 2014; Wang et al. 2015). Using next-generation sequencing, Xia et al. found that AML cells-derived Igκ exhibited a higher mutation rate (above 2%) than the normal PBMCs-derived Igκ (below 2%). Notably, serine mutation hotspots of the Vκ3–20 light chain frequently appeared in different AML patients, suggesting that the kappa light chain may be involved in the occurrence and progression of AML (Xia et al. 2022). These findings suggest a different SHM mechanism in non-B cell-derived Ig, which needs further exploration.

3.6 The Conventional Concept of Class Switching Does Not Apply to the IgG and IgA Produced by Non-B Cells

Traditionally, the first two antigen receptors expressed by B cells are IgM and IgD, and the first antibody produced in an immune response is always IgM. Class switching occurs later in the immune response to generate IgG, IgE, and IgA antibodies, which provide antibodies with the same antigen specificity but different effector capacities. During the process, the CH gene Cμ is replaced by Cγ, Cε, or Cα, respectively, without changing the V-region. It was believed that one mature B cell could only produce one type of antibody. However, a recent study reported that a single B cell can express several classes of IgH chain with distinct V(D)J patterns simultaneously (Shi et al. 2019).

Although it has been found that non-B cells can express all five IgH classes, it is still unclear whether non-B cells use the conventional class-switching mechanism. Some clues point to a 'no' answer. AID, essential for both class switching and somatic hypermutation, was expressed in six breast cancer cell lines by nested RT-PCR (Babbage et al. 2006). However, further attempts to detect AID in many other cancer cells using RT-PCR and western blotting failed (Zheng et al. 2007). Secondly, in many studies, the selected V, D, and J segments and V(D)J recombination patterns differed among the μ, γ, and α heavy chains. Unlike B cells, where Igγ, and Igα were considered to be switched from Igμ with only changes in constant region, each class of heavy chains in non-B cells showed its own unique V(D)J pattern (Fig. 3.3). More studies need to be conducted to address this question, especially at the single-cell level.

3.7 Mechanism of Ig Gene Rearrangement in Non-B Cells is in a RAG1/RAG2-Independent Manner

In classic theory, the rearrangement process is catalysed by the recombination activating gene 1/ recombination activating gene 2 (RAG1/RAG2) recombinase following the 12/23 rule, which was discovered by David Schatz and Marjorie Oettinger from David Baltimore's laboratory (McBlane et al. 1995; Oettinger et al. 1990; Schatz et al. 1989). Intriguingly, both RAG1 and RAG2 could be detected at mRNA and protein levels in Ig-positive cancer cell lines, including lung cancer (A549), colon cancer (HT-29, SW480), cervical cancer (Hela MR, Hela S3), liver cancer (Bcl-7402), prostate cancer (PC3), gastric cancer (MGC), breast cancer (MCF7) and nasopharyngeal carcinoma (CNE-1) (Qiu et al. 2003; Zheng et al. 2007; Chen et al. 2007), and even normal tissues like human podocytes (Jing et al. 2018). However, on the other hand, Ig recombination could also be found in $RAG1^{-/-}$ and $RAG2^{-/-}$ mice (unpublished data), which indicated that the Ig gene recombination might be controlled by other unknown ways independent of the RAG genes.

3.8 Prospect

We have reviewed the genetic characteristics of non-B cell-derived Ig and contrasted them with those of B cell-derived Igs. The genetic characteristics of non-B cell-derived Ig suggest their novel functions beyond acting as diverse antibodies to recognise antigens. Their conserved sequences may contribute to unique structures and physicochemical properties, as illustrated in Chap. 5. There remains much to be investigated. (1) How are the V, D, and J gene segments distributed in non-B cell genomes? (2) What mechanism is involved in the V(D)J recombination of non-B cell-derived Igs in a RAG1/2-independent manner? (3) Why do Igs from different individuals share the same mutation sites and junction sequences? (4) What mechanism causes non-B cell-derived Ig to exhibit limited diversity? More studies are needed to address these questions, particularly at the genomic level and at single-cell resolution.

References

Babbage G et al (2006) Immunoglobulin heavy chain locus events and expression of activation-induced cytidine deaminase in epithelial breast cancer cell lines. Cancer Res 66(8):3996–4000

Backhaus O (2018) Generation of antibody diversity. In: Antibody engineering. Intech. https://doi.org/10.5772/intechopen.72818

Behring E (1890a) Ueber das Zustandekommen der Diphtherie-Immunität undder Tetanus-Immunität bei Thieren. Dtsch Med Wschr 16:1113–1114

Behring E (1890b) Untersuchungen über das Zustandekommen der Diphtherie-Immunität bei Thieren. Dtsch Med Wschr 16:1145–1147

Burnet F (1957) A modification of Jerne's theory of antibody production using the concept of clonal selection. Aust J Sci 20:67–69

Chen Z et al (2007) Immunoglobulin G expression in carcinomas and cancer cell lines. FASEB J 21(11):2931–2938

Geng Z et al (2019) Human colorectal cancer cells frequently express IgG and display unique Ig repertoire. World J Gastrointest Oncol 11(3):195–207

Gong X et al (2021) Macrophage-derived immunoglobulin M inhibits inflammatory responses via modulating endoplasmic reticulum stress. Cell 10:2812

Hozumi N et al (1976) Evidence for somatic rearrangement of immunoglobulin genes coding for variable and constant regions. Proc Natl Acad Sci USA 73(10):3628–3632

Huang J et al (2008) Expression of immunoglobulin gene with classical V–(D)–J rearrangement in mouse brain neurons. Int J Biochem Cell Biol 40(8):1604–1615

Huang J et al (2009) Expression of immunoglobulin gene with classical V–(D)–J rearrangement in mouse testis and epididymis. J Histochem Cytochem 57(4):339–349

Huang J et al (2014) Rearrangement and expression of the immunoglobulin μ-chain gene in human myeloid cells. Cell Mol Immunol 11(1):94–104

Janeway C et al (2001) The generation of diversity in immunoglobulins. In: Immunobiology: the immune system in health and disease, 5th edn. Garland Science, New York

Jing Z et al (2018) Expression of immunoglobulin G in human podocytes, and its role in cell viability and adhesion. Int J Mol Med 41(6):3296–3306

Kimoto Y (1998) Expression of heavy-chain constant region of immunoglobulin and T-cell receptor gene transcripts in human non-hematopoietic tumor cell lines. Genes Chromosomes Cancer 22(1):83–86

McBlane J et al (1995) Cleavage at a V(D)J recombination signal requires only RAG1 and RAG2 proteins and occurs in two steps. Cell 83(3):387–395

Niu N et al (2011) Expression and distribution of immunoglobulin G and its receptors in an immune privileged site: the eye. Cell Mol Life Sci 68(14): 2481–2492

Oettinger M et al (1990) RAG1 and RAG2, adjacent genes that synergistically activate V(D)J recombination. Science 248(4962):1517–1523

Okabe H et al (2001) Genome-wide analysis of gene expression in human hepatocellular carcinomas using cDNA microarray: identification of genes involved in viral carcinogenesis and tumor progression. Cancer Res 61(5):2129–2137

Qiu X et al (1996) Analysis of the immunoglobulin-like protein and immunoglobulin gene in malignant tumor cells. Chin J Immunol 5:296

Qiu X et al (1997) Structural analysis of Ig gene existed in the epithelial malignant cells. Chin J Cancer Biother 4:157–158

Qiu X et al (2003) Human epithelial cancers secrete immunoglobulin g with unidentified specificity to promote growth and survival of tumor cells. Cancer Res 63(19):6488–6495

Qiu X et al (2013) Immunoglobulin gamma heavy chain gene with somatic hypermutation is frequently expressed in acute myeloid leukemia. Leukemia 27(1):92–99

Schatz D et al (1989) The V(D)J recombination activating gene, RAG1. Cell 59(6):1035–1048

Shao W et al (2016) Identification of liver epithelial cell-derived Ig expression in mu chain-deficient mice. Sci Rep 6:23669

Shi Z et al (2019) More than one antibody of individual B cells revealed by single-cell immune profiling. Cell Discov 5:64

Tonegawa S et al (1974) Evidence for somatic generation of antibody diversity. Proc Natl Acad Sci USA 71(10):4027–4031

Wang C et al (2015) IGK with conserved IGKappaV/IGKappaJ repertoire is expressed in acute myeloid leukemia and promotes leukemic cell migration. Oncotarget 6(36):39062–39072

Xia M et al (2022) Next-generation sequencing revealed a distinct immunoglobulin repertoire with specific mutation hotspots in acute myeloid leukemia. Biology (Basel) 11(2):161

Zhang S et al (2010) Immunoglobulin gene locus events in epithelial cells of lactating mouse mammary glands. Cell Mol Life Sci 67(6):985–994

Zhang C et al (2021) NBIGV-DB: a dedicated database of non-B cell derived immunoglobulin variable region. Gene 772:145378

Zheng S et al (2001) Structure and expression of colorectal cancer related immunoglobulin novel gene SNC73. Natl Med J China 81(8):485–488

Zheng H et al (2007) Expression and secretion of immunoglobulin alpha heavy chain with diverse VDJ recombinations by human epithelial cancer cells. Mol Immunol 44(9):2221–2227

Zheng J et al (2009) Immunoglobulin gene transcripts have distinct VHDJH recombination characteristics in human epithelial cancer cells. J Biol Chem 284(20):13610–13619

Zhu X et al (2008) Immunoglobulin mRNA and protein expression in human oral epithelial tumor cells. Appl Immunohistochem Mol Morphol 16(3):232–238

The Gene Rearrangement and Transcriptional Regulation of Non B Cell-Derived Immunoglobulin

4

Teng Ma, Jie Zheng, Peng Hao, Xiaohui Zhu, and Xinmei Huang

Abstract

Traditionally, immunoglobulin (Ig) expression has been attributed solely to B cells/plasma cells with well-documented and accepted regulatory mechanisms governing Ig expression in B cells. Ig transcription is tightly controlled by a series of transcription factors. However, increasing evidence has recently demonstrated that Ig is not only produced by B cell lineages but also by various types of non-B cells (non-B-Ig). Under physiological conditions, non-B-Ig not only exhibits antibody activity but also regulates cellular biological activities (such as promoting cell proliferation, adhesion, and cytoskeleton protein activity). In pathological conditions, non-B-Ig is implicated in the development of various diseases including tumour, kidney disease, and other immune-related disorders. The mechanisms underline Ig gene rearrangement and transcriptional regulation of Ig genes in non-B cells are not fully understood. However, existing evidence suggests that these mechanisms in non-B cells differ from those in B cells. For instance, non-B-Ig gene rearrangement occurs in an RAG-independent manner; and Oct-1 and Oct-4, rather than Oct-2, are required for the transcriptional regulation of non-B derived Igs. In this chapter, we will describe and compare the mechanisms of gene rearrangement and expression regulation between B-Ig and non-B-Ig.

Keywords

Immunoglobulin · Ig · Octamer-binding proteins · Intronic enhancer · Heavy chain · Light chain · V(D)J rearrangement · Plasma cells · Promoter

T. Ma (✉)
Cancer Research Center, Beijing Chest Hospital, Capital Medical University, Beijing, China
e-mail: mateng@bjxkyy.cn

J. Zheng
Hematologic Disease Laboratory, Department of Stem Cell Transplantation, Beijing Children's Hospital, Capital Medical University, Beijing, China
e-mail: zhengjie@bch.com.cn

P. Hao
Telebrain Medical Technology Co., Ltd., Beijing, China
e-mail: haopeng@naolubrain.com

X. Zhu
Department of Obstetrics and Gynecology, Center for Reproductive Medicine, Peking University Third Hospital, Beijing, China
e-mail: zhuxiaohui@bjmu.edu.cn

X. Huang
Department of Respiratory and Critical Care Medicine, Nanjing Drum Tower Hospital, The Affiliated Hospital of Nanjing University Medical School, Nanjing, China
e-mail: huangxm2017@pku.edu.cn

X. Qiu et al. (eds.), *Non B cell-Derived Immunoglobulins*, Advances in Experimental Medicine and Biology 1445, https://doi.org/10.1007/978-981-97-0511-5_4

Ig consists of two identical heavy chains and two identical light chains; the variable (V) region exons of Ig heavy or light chain are assembled from discontinuous VH, diversity (D), and joining (JH) (for the heavy chain), or VL and JL (for the light chain) segments by the process of V_H(D) J_H, or V_LJ_L rearrangement, respectively. In B cells, the rearrangement processes involve the accessibility of V, D, and J segments to two proteins encoded by recombination activating genes, RAG1 and RAG2, as well as other transcription factors such as E2A, early B-cell factor (EBF), B-cell activator protein (BSAP), and other factors. The VHDJH or V_LJ_L rearrangement is tightly controlled in a tissue-specific and allele-specific manner, in which the RAG1/RAG2 are the specific components of the V(D)J recombinase and essential for generating the primary antigen receptor repertoire. The chromatin structure accessibility at particular gene segments and the interaction between multiple *cis*-regulatory elements within or near Ag receptor loci are involved in the precise targeting of the rearrangement machinery (Chi et al. 2020; Schebesta et al. 2002).

4.1 Mechanism of Ig Gene Rearrangement in B Cells and Non-B Cells

4.1.1 Mechanism of Ig Gene Rearrangement in B Cells

The development of B lymphocytes includes the primary stage (in the bone marrow) and antigen-dependent stage (in peripheral immune organs). The cells in primary stages, in the order they occur, include progenitor cells, pro-B cells, pre-B cells, and immature B cells. The antigen-dependent stages consist of germinal centre B cells and plasma cells (Melchers 2015; Meffre et al. 2000, 2001; Defrance et al. 2002). At different stages of B cell development, the expression of Ig genes also demonstrates stage specificity and is strictly controlled by the *cis* interaction between various regulatory elements and transcription factors.

In the bone marrow, B cell development undergoes the stages in the following sequence: progenitor cells, pro-B cells, pre-B cells, and immature B cells. At the pro-B cell stage, two recombinant activated gene products, namely RAG1 and RAG2, are first expressed and the two proteins form a complex, which can recognise the recombinant activated sequence (RSS) on both sides of the D segment and the 5′ end RSS of JH segment on IGH locus, and then cleave DNA. Subsequently, the classical non-homologous end-joining (C-NHEJ) joins RAG-initiated rearrangement signal sequence (RSS) breaks to complete V(D)J rearrangement (Fig. 4.1). In this process, Ku70/Ku80 complex binds to the RAG-initiated DSBs and recruits DNA–PKcs–Artemis complex to open the coding end-associated hairpins. At this time, terminal deoxytransferase (TDT) is induced to express, which can recognise the exposed DNA 3′ end, and add an uncertain number of nucleotides to the exposed DNA 3′ end in a template-independent manner (to increase the diversity of Ig). Then, the DH and JH are connected together by DNA ligase to form DHJH. It is worth noting that nucleotide insertion or loss occurs in this process, further increasing the diversity of Ig. Next, using the same process and mechanism as mentioned above, B cells connect any VH with DHJH to form a complete Ig heavy chain variable region reading frame called VHDJH (Fig. 4.1). When pro-B cells develop into the pre-B cell stage, IgH site rearrangement has been completed and Ig light chain rearrangement begins due to the re-expression of RAG1 and RAG2, and the pre-B cell receptor complex (pre-BCR) is expressed. At the immature B cell stage, the B cell receptor (BCR) complex is composed of IgM, Igα, and Igβ expressed on the cell surface. When both IgM and IgD are co-expressed on the cell surface, B cells enter the mature B stage and these cells have completed the antigen-independent phase, then leave the bone marrow and migrate to the surrounding lymphoid organs. The development of B cells is regulated by a complex and sequentially expressed network of transcription factors. During the pro-B cell phase, the transcription factors PU.1, Pax5, and Ikaros

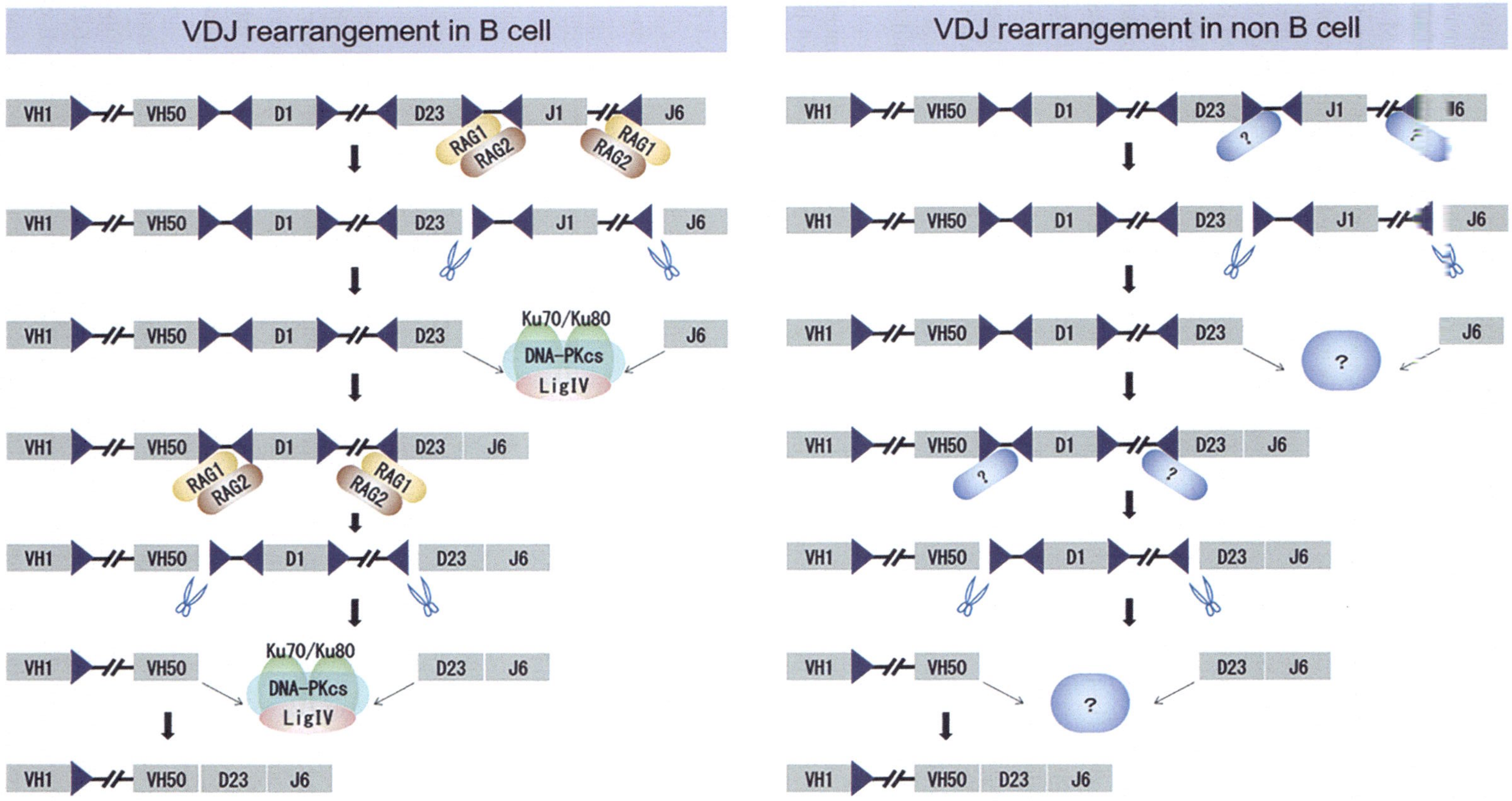

Fig. 4.1 The VDJ rearrangement in B cells and non-B cells. The left panel shows the three distinct phases of RAG-initiated V(D)J rearrangement: RAG1/RAG2 recognition of the recombination sites, induction of double-strand breaks by RAG1/2, and classical non-homologous end-joining (c-NHEJ) repair of the broken DNA by DNA-PKcs/Ku70/Ku80 and DNA Ligase IV (LigIV), etc. The right panel shows the RAG1/RAG2-independent mechanisms involved in VDJ rearrangement in non-B cells. The repair machineries for the double-strand breaks during VDJ rearrangement are yet to be identified

are necessary for Ig gene rearrangement, whilst E2A, EBF, and BSAP play a role in the pro-B/pre-B transition phase at the beginning of VHDJH rearrangement (Schebesta et al. 2002). PU.1 is an Ets family protein required for the activity of the 3′ kappa enhancer; BSAP is an isoform of the transcription factor Pax5; Ikaros is a zinc finger transcription factor required for the development of the earliest B cell progenitors, and at later stages for V(D)J rearrangement and BCR expression; E2A belongs to a large family of transcription factors that share homology in their basic DNA binding domain and their helix–loop–helix dimerisation motif (bHLH); EBF is a cell type-specific transcription factor that is expressed at all antigen-independent stages of B cell differentiation and participates in the generation of Ig-expressing B cells.

In peripheral immune organs, the V region sequence of the Ig gene will change when B cells are stimulated by a specific antigen. For instance, a specific antigen can induce the re-expression of RAG1 and RAG2, resulting in the replacement of the V gene fragment, or somatic hypermutations on the original V gene fragment with the catalysis conducted by the activation-induced deaminase (AID) to produce antibodies with higher affinity.

4.1.2 Mechanism of Ig Gene Rearrangement in Non-B Cells

Whilst the Ig rearrangement mechanism and its sequence are clear in B cells, they remain elusive in non-B cells. For example, it is not even known at which stage of non-B cell development Ig gene rearrangement occurs. As a cutting-edge exploration, Qiu et al. found that both RAG1 and RAG2 were expressed in some cancer cells (Qiu et al. 2003). However, it's worth mentioning that Shao et al. found that Ig gene rearrangement did not depend on RAG1/RAG2 in normal non-B cells, which is in contrast to the absolute dependence of this process in B cells. Ma et al. found both Ig gene rearrangement and Ig expression occurred in many non-B cells of either RAG1 or RAG2-deficient mice; moreover, the Ig expression level increased responding to the stimulation of pro-inflammatory factors. Recently, Zhang et al. analysed usage and pattern of Ig gene rearrangement in RAG1 or RAG2-deficient mice, and found that non-B cells could express different classical Ig, including IgG, IgM, and IgA, with the IgM heavy chain expression being the highest and kappa chain being the main light chain (unpublished data), moreover, as previously reported, the rearranged Ig genes showed restricted diversity. These findings suggest that in addition to RAG1 and RAG2, other unidentified recombinases might mediate Ig gene rearrangement in normal non-B cells. In addition, Deng et al. found that the Ig light chain VκJκ rearrangement pattern changed in the cultured mesangial cells stimulated by *Staphylococcus aureus* (Deng et al. 2018), but the mechanism awaits elucidation.

Some transcription factors responsible for Ig rearrangement and expression have also been identified in non-B cancer cells. E2A and Oct-1 were widely expressed in eight cancer cell lines including HT-29 and SW480 (colon cancer cell lines), A549 and H1299 (lung cancer cell lines), HeLa, HeLa S3, and HeLa MR (cervix cancer cell lines), and U2OS (a human bone osteosarcoma epithelial cell line). EBF and Oct-2, the B-cell-specific transcription factors, were expressed in a few non-B cell lines (Zhu et al. 2010).

4.2 Ig Gene Transcription Regulation in B Cells and Non-B Cells

The Ig transcriptions are regulated by various tissue-specific *cis*-acting promoters and enhancers including promoters of the V gene, the intronic enhancer (Eμ) of the Ig heavy chain gene, the 3′ Ig heavy chain enhancer, the intronic enhancer of the κ gene, the 3′ κ enhancer, and the λ light chain enhancers (Eλ).

It has been demonstrated that transcription factors required for the B-Ig transcription include octamer-binding proteins (Oct-1 and Oct-2) and coactivator OCA-B which interacts with either Oct-1 or Oct-2 to potentiate transcriptional activation of normally switched immunoglobulin heavy-chain loci (Kim et al. 1996), ETS family proteins (Ets-1 and PU.1) which bind to the Ig heavy-chain gene enhancer (Rivera et al. 1993), STATs family proteins in which STAT5 and IL-7 signalling control cell survival and the developmental ordering of immunoglobulin gene rearrangements by suppressing premature Igk rearrangement in pro-B cells whereas STAT6 is required for IL-4-induced germline Ig gene transcription and switch rearrangement (Linehan et al. 1998; Malin et al. 2010), NF-κB/REL proteins which play a key role in κ light chain gene expression (Sen 2004), C/EBP proteins which bind to the E sites in the IgH enhancer (Cooper et al. 1992), TFE/Myc proteins which bind the Ig enhancer muE3 motif (Beckmann et al. 1990), BCL-6 which represses the IL-4-dependent induction of immunoglobulin (Ig) germline epsilon transcripts (Harris et al. 1999), and BLIMP-1 which represses PAX5 and activates immunoglobulin heavy chain gene transcription, etc. (Martins et al. 2008; Lin et al. 2002; Ernst et al. 1995) (Fig. 4.2).

4.2.1 The Mechanism of Transcriptional Regulation on the Ig Promoter Region of B Cells

The mechanisms of transcriptional regulation of the Igμ gene have been extensively studied and a diverse array of proteins are involved in Igμ transcription. As shown in Fig. 4.2, rearranged Igμ genes contain a core promoter, a regulatory promoter, and an enhancer; the core promoters contain TATA boxes and function as transcription initiation factors; the regulatory promoters are located upstream of the core promoter. The 5′ enhancer in the promoter region and the 3′ enhancer constitute the regulatory elements for rearranged kappa chain gene expression. The octamer element is highly conserved among the VH and VL promoters. Oct-1, Oct-2, and their coactivators OCA-B bind the octamer element and the upstream heptamer in VH and VL promoters. It is reported that Oct-1 is mainly expressed in early developing B cells, responsible for initiating the expression of Ig gene, while Oct-2 is specifically expressed in mature B cells, responsible for transcriptional regulation of mature B cells. An enhancer is located in the J-Cμ intron which harbours elements including C/EBP, μE1, μE5, μE2, μa, μE3, μB, μE4, and octamer elements, which are binding sites for the transcription factors including EBP, TFE, E2-A products, Ets-1, PU.1, and Oct-1/Oct-2, respectively. For the V kappa enhancer elements, the NF-κB protein binds with the enhancer, promoting Ig gene transcription.

4.2.2 The Mechanism of Transcriptional Regulation on the Ig Promoter Region of Non-B Cells

At present, although the expression regulation mechanism of non-B-Ig is not fully understood, there have been some reports indicating that the expression regulation mechanism, at least on promoter region, of non-B-Ig has its unique characteristics. As early as 2010, Zhu et al. reported a mechanism regulating the transcription of the Ig gene heavy chain promoter region in lung cancer and colon cancer cell lines, and the key transcription factor involved in the regulation was Oct-1 (Zhu et al. 2010), which was also found to participate in the regulation of IgA heavy chain expression in nasopharyngeal carcinoma by Cao et al. in 2014 (Duan et al. 2014). Recently, Huang et al. discovered that epithelial tumour stem cells widely produced a type of IgG with a unique sialylation modification, and its transcription is regulated by stem cell transcription factors Sox2 and Oct-4 (Huang et al. 2023).

4.2.2.1 Oct-1 Can Drive Ig Transcription in Non-B Cells

Jie Zheng and colleagues found that the non-B cancer cell-expressed Igs show distinct features from B-Igs, in that a very restricted set of VDJ rearrangements, such as VH6–1 was frequently used in V(D)J rearrangement of non-B-Igμ, and VH4–59 was frequently used in V(D)J rearrangement of non-B-Igγ (Zheng et al. 2009). To understand the tran-

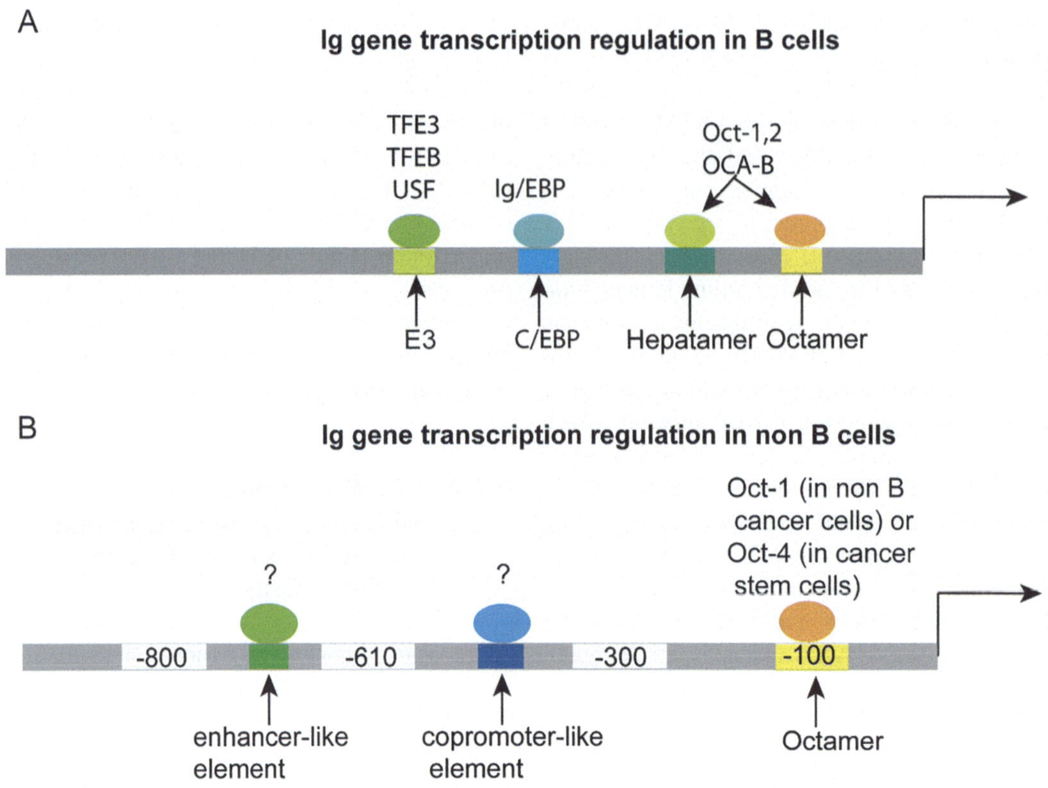

Fig. 4.2 Ig gene transcription regulation in B cells and non-B cells. (**a**) The VH1 promoter composition and known protein binding sites. The octamer element, heptamer element, C/EBP and E3 elements, and their respective binding proteins are located upstream of the transcription start site as indicated. (**b**) The VH4–59 promoter and Oct-1/Oct-4 binding site in non-B cells. The octamer element, copromoter-like element (between −300 and −610 bp) and enhancer-like element (between −610 and −800 bp) are located upstream of the transcription start site as indicated

scription mechanism of Ig in non-B cancer cells, they have investigated the transcriptional regulation mechanism on the Ig promoters of IGVH6–1 and IGVH4–59, respectively.

4.2.2.1.1 The Regulation Mechanism of Ig Transcription on VH6–1 Promoter

The human VH6 gene is one of 50 functional germline VH region genes that are subdivided into 7 families based on homology. VH6 is a single-member family and the one most proximal to the D genes. The VH6 promoter is the first VH region promoter activated during ontogeny. In 1994, Sun et al. discovered that a small fragment of the VH6–1 promoter was very active in both B cells and HeLa cells (Sun et al. 1994a, b), suggesting that the VH6–1 promoter did not possess B cell specificity and that a VH6 transcript may be required for the opening of the chromatin in this area upstream of the D gene region. However, in 2008, Zheng et al. found that IgM heavy chains preferentially selected the VH6–1 segment for expression in some epithelial cancer tissues and cancer cell lines, including HeLa cells (Hu et al. 2012; Shao et al. 2016). They captured cancer cells by laser capture microdissection (LCM) from colon, breast, oral, and lung cancer tissues, respectively, the rearranged Ig gene transcripts and repertoires were analysed, and found that all seven of the VH families were amplified, except VH2. VH6–1 accounted for 18.7% of all potentially functional cancer-derived VHμ genes (Zheng et al. 2009). The VH6–1 frequency was remarkably higher than expected compared with normal B-cell-derived VH (about 0.1%) (Brezinschek et al. 1995) and tumour-infiltrating B-cell-derived VH (below 5%) (Nzula et al. 2003).

The VH promoter harbours a conserved TATA box at approximately −30 bp and the highly conserved octamer element at approximately −70 bp relative to the transcription start site. For Vκ promoters, the octamer element is also usually located within 100 bp of the transcription initiation site. Notably, the promoter region of the VH6 gene family contains an octamer sequence with a single nucleotide substitution (AgGCAAAT) at 108 bp upstream of the ATG translation start site and 81 bp upstream of the transcription initiation site, which differs from other VH gene promoters (Shao et al. 2016).

Zhu et al. investigated the VH6–1 promoter activity in ten non-B cell lines. First, a 1.2-kb DNA fragment upstream of VH6–1 was amplified by PCR and cloned into a pGL3 luciferase reporter construct, then, the construct was transfected into non-B cancer cell lines to analyse promoter activity of VH6–1. Markedly, they found VH6–1 promoter activity exhibited in these cancer cell lines (Wu et al. 2014). Furthermore, the mutation experiment found that the octamer element in the VH6–1 of non-B cells is not essential for its promoter activity because the activity mainly remained after the atypical octamer sequence in VH6–1 (AgGCAAAT) was mutated, which is different in B cells, suggesting the existence of other elements responsible for VH6–1 promoter activity. However, no other regulatory elements were found upstream of the VH6–1 promoter.

Oct-1 is a 90–100 kDa ubiquitous transcription factor expressed in various cell types, whereas Oct-2 is composed of 60 kDa (Oct-2A) and 75 kDa (Oct-2B) proteins and is specifically expressed in B lymphoid cells (Poellinger et al. 1989; Kemler et al. 1990). Oct-1 and Oct-2 show a 150–160 amino acids homology in sequence, and this conserved domain is named POU domain, which is required for binding DNA. It is acknowledged that Oct-2 binding to octamer sites confers B cell specificity on Ig promoters in both a qualitatively and quantitatively dependent way (Poellinger et al. 1989). However, in this study, Zhu et al. found that unlike VH6–1 transcription in B cells, the transcription factor that drives VH6–1 transcription in non-B cells is not Oct-2, but Oct-1 (Wu et al. 2014).

4.2.2.1.2 VH4–59 Promoter in Regulation of Ig Transcription in Non-B Cells

VH4–59 was also preferentially used by Igγ in non-B cells (Zheng et al. 2009). Zhu et al. cloned a 5′-flanking 1.2-kb DNA fragment upstream of the Ig VH4–59 gene and found the segment showed high promoter activity in six non-B cancer cell lines, including HeLa, HeLa MR, HeLa S3, HT-29, U2OS, and Daudi (a lymphoma cell line). Moreover, they also demonstrated that the octamer sequence (ATGCAAAT) is important but not indispensable for VH promoter activity in non-B cancer cells. Notably, they first revealed two novel regulatory elements (−800 to −610 bp and −610 to −300 bp) in cancer cells, but not in the positive control cell line-Daudi cells (Zhu et al. 2010). Therefore, the VH4–59 promoter in cancer cells shows a 'Janus-faced' characteristic in comparison to its counterpart in B cells: it possesses the same canonical activities as that in B cells, and it also has different regulatory elements from that in B cells. Further investigations are warranted to gain a panoramic view.

Similar to the VH6–1 promoter in non-B cells, it was revealed that the transcription factor binding to the octamer sequence of VH4–59 promoter is Oct-1, not Oct-2, by using both the electrophoretic mobility shift assay (EMSA) and chromatin immunoprecipitation (ChIP). Consistently, Oct-1 over-expression significantly potentiated both the VH4–59 promoter activity and the IgG expression in HeLa S3 and HT-29 cells. Conversely, Oct-1 knockdown significantly decreased IgG expression in HeLa S3 cells. Therefore, Oct-1 is the crucial transcription factor for Ig gene transcription in non-B cells. This finding is in accordance with the ubiquitous expression profile of Oct-1 and its presence on promoters of histone H2B and snRNA (Zhu et al. 2010; Wu et al. 2014; Mattaj et al. 1985; LaBella et al. 1988).

In summary, both the Ig VH6–1 and the VH4–59 genes could be in the open chromatin DNA region, and the octamer sequence in the promoter region of both genes could be bound by Oct-1which drives Ig gene transcription in non-B cells. Moreover, some undefined regulatory elements may exist in the upstream 300 bp region of the VH6–1 promoter in non-B cells.

4.2.2.2 Ets-1 Can Drive Ig Iα1 Transcription in Non-B Cells

In B cell development, the Class Switch Recombination (CSR) process necessitates germline (GL) transcription of both the C region genes and the IgH genes. Igα heavy chain CSR is regulated by Iα-Cα, which is crucial for the expression of IgA. In humans, IgA1 and IgA2 are two subclasses of IgA, and their heavy chains are encoded by the Cα1 and Cα2 genes, respectively.

Ig VDJ-Cα and Iα-Cα transcripts are found to be expressed in multiple cancer cell lines (Duan et al. 2014; Zheng et al. 2007), and unlike the Ig expression pattern in B cells, rearranged VDJ region in cancer cells bears an Ig Sα region in germline configuration. Additionally, RAG1 and RAG2 are essential for Ig VDJ-Cα and Iα-Cα expression, it was found that RAG1 and RAG2 are expressed in human epithelial cancers. Furthermore, human cancer cells also secrete IgA (SIgA); moreover, Igα heavy chain expression could increase the percentage of cancer cells in the S phase.

To investigate whether the Ig Iα1 promoter could be activated in cancer cells, a 674-bp fragment containing the Ig Iα1 promoter was inserted into the pGL3-Basic vector (referred as pIα). The results revealed that pIα showed strong activity in a nasopharyngeal carcinoma cell line (CNE1) (Duan et al. 2014), and deletion and truncation mutants of pIα indicated that a crucial element for Ig Iα1 promoter activation was located 80 bp upstream of the transcription start site (TSS). Interestingly, a binding site of PU.1, a transcription factor of erythroblast transformation-specific (ETS) family, was found in the Ig Iα1 promoter. Since PU.1 was not expressed in epithelial cancer cells, another ETS family protein, Ets-1, which shares a highly homologous binding site with the PU.1, was tested for Ig Iα1 promoter activation in CNE1 and HeLa cells, two cell lines with high IgA expression levels; when the construct containing Ets-1 binding site mutation in the Ig Iα1 promoter was introduced, the luciferase activity decreased significantly, suggesting an important role of Ets-1 in Ig Iα1 promoter activation which was further corroborated by the EMSA and CHIP assays. In addition to the PU.1 binding site, there are also numerous other sites including those binding AP-2, Elk-, Sp-1, c-Jun and E47, Smad, and PEBP2/CBF in the Ig Iα1 promoter. When these sites are mutated, high promoter activity is retained, indicating their crucial regulatory roles in the process (Duan et al. 2014).

Moreover, Igκ-expressing nasopharyngeal carcinoma (NPC) cells contain the active immunoglobulin κ 3′ enhancer (3′Eκ) which could be potentiated by EBV-encoded latent membrane protein 1 (LMP1). Consistently, the Igκ light chain expression was reduced after the LMP1 was suppressed (Liu et al. 2012).

4.2.2.3 Both Sox2 and Oct-4 Together Drive Igγ Transcription in Non-B Cancer Stem Cells

Cancer stem cells (CSCs) are a subpopulation of cancer cells with a high capacity for self-renewal, pluripotency, and tumour initiation, and are regarded as the major source of cancer metastasis, recurrence, and resistance to chemotherapy and radiotherapy (Yang et al. 2020). Notably, the IgG, which bears a unique sialylated modification on the Asn162 site of the CH1 domain (SIA-IgG), was found to be highly expressed in epithelial stem-like cancer cells or recurrent castration-resistant prostate cancer cells and contributes to tumour initiation, metastasis and drug resistance, suggesting an important role of SIA-IgG in cancer stemness (Liao et al. 2015; Cui et al. 2021; Wang et al. 2020). However, our knowledge remains poor about the regulation of SIA-IgG expression in CSCs and its functional mechanisms in cancer cell stemness. To address this, Huang et al. first found that SIA-IgG could be co-expressed with Sox2 and Oct-4, two specific transcription factors, in cancer stem cells. Next, Huang et al. sought to determine the regulation of SIA-IgG expression in CSCs. A genome-wide CRISPR activation screen was conducted in a lung squamous carcinoma cell line, NCI-H520 cells by expressing a lentiviral library of 70,290 CRISPR activating gRNAs targeting 23,430 genes (Konermann et al. 2015). They expect that the activation of SIA-IgG promoting genes would lead to more SIA-IgG^{high} cells and enrichment of the corresponding gRNAs. After confirming the quality of CRISPR-Cas9 screening and observing a similar gRNA distribution pattern between the SIA-IgG^{high} and SIA-IgG^{low} cells, they applied the MAGeCK algorithm to identify promoting genes in SIA-IgG^{high} cells (Li et al. 2014). In particular, the stemness-related transcription factors Oct-4 (Gene name: POU5F1) and Sox2 were both among the top genes enriched in SIA-IgG^{high} cells, suggesting that Oct-4 and Sox2 may serve as key factors promoting SIA-IgG expression.

Subsequently, Huang et al. explored the mechanisms underlying the regulation of IgG gene expression in CSCs. Interestingly, they first revealed that both Oct-4 and Sox2 binding sites existed in the IgH promoter region, Oct-4 and Sox2 can bind to the Ig promoter region and form a complex, co-regulate SIA-IgG expression by ChIP, luciferase, and EMSA assays etc. Importantly, they revealed that both Sox2 and Oct-4 are likely bound to the IgH promoter in CSCs versus non-CSCs cells, indicating that Sox2 and Oct-4 may mediate the upregulation of IgG expression in CSCs (Huang et al. 2023).

4.3 Regulation of Ig Somatic Hypermutation and Class Switch DNA Rearrangement in B Cells and Non-B Cells

4.3.1 Regulation of Ig Somatic Hypermutation and Class Switch DNA Rearrangement in B Cells

As mentioned early, the development of B cells includes the primary stage and antigen-dependent stage. After the challenge of antigen, in order to produce high affinity and efficient antibodies, the variable region of the Ig gene will undergo somatic hypermutation (SHM), then the constant region will undergo class switch DNA rearrangement (CSR) that from the expression of μ chain to the expression of γ or α or ε chain. Therefore, variable regions of γ or α or ε chain present high-frequency mutations, while the variable region of μ chain rarely presents mutations. In the Ig production process of B cells, the enzyme AID is required for both SHM and CSR (Chandra et al. 2015). AID-target ssDNA is proposed to be generated due to supercoiling or as transcription bubbles during Ig locus transcription (Goebel et al. 2001; Romanow et al. 2000; Sadofsky 2001). Transcription regulates SHM targeting to the Ig loci (Dinesh et al. 2019). SHM occurs primarily at RGYW (R = purine base, Y = pyrimidine base, and W = A or T nucleotide) sites in the variable region. Antigen selection-induced mutations occurred more frequently in the CDRs than in the FWRs, such as a higher R/S mutation ratio.

4.3.2 Regulation of Ig Somatic Hypermutation and Class Switch DNA Rearrangement in Non-B Cells

Non-B cells were found to express 5 classes of Ig, like the B cell-derived Ig, the variable regions of non-B cell-derived γ or α or ε chain present high-frequency mutations, while its variable region of μ chain rarely present mutations. However, the mechanism of SHM in non-B-derived Ig transcription shows a distinct feature in cancer-derived V(D)J sequences from B-cell-derived V(D)J sequences, in which the mutations predominantly clustered in the FWRIII instead of in the CDRs, suggesting that most of the mutations within the cancer-derived VHγ sequences were not induced by antigen selection. In addition, according to the classical mechanism of CSR, different classes of Ig from identical cell should share the same sequence of V(D)J, but only show difference in its constant region. However, Qiu group found that different classes of Ig from identical cell never display the same sequence of V(D)J pattern, which suggest that the mechanism of SHM and CSR in non-B-derived

Ig shows a distinct feature (Zheng et al. 2009). The detailed mechanism needs to be further explored.

4.4 Future Perspectives

The study of Ig production by non-B cells is a new field, and our understanding of the rearrangement mechanism and expression regulation mechanism of Ig production by non-B cells is still superficial. Many questions need to be further answered. Specifically, under RAG deficiency conditions, there is still Ig gene rearrangement in non-B cells. What are the RAG analogues that mediate Ig gene rearrangement? The regulatory mechanism of Ig gene promoter region expression has unique regulatory characteristics in non-B cells. What is the detailed mechanism of SHM and CSR in non-B cells? Is there an epigenetic regulatory mechanism for Ig gene expression? In addition, which cytokines and signalling pathways are involved in regulating Ig gene expression? These key scientific issues need to be revealed.

References

Beckmann H et al (1990) TFE3: a helix–loop–helix protein that activates transcription through the immunoglobulin enhancer muE3 motif. Genes Dev 4(2):167–179

Brezinschek HP et al (1995) Analysis of the heavy chain repertoire of human peripheral B cells using single-cell polymerase chain reaction. J Immunol 155(1):190–202

Chandra V et al (2015) AID targeting: old mysteries and new challenges. Trends Immunol 36(9):527–535

Chi X et al (2020) V(D)J recombination, somatic hypermutation and class switch recombination of immunoglobulins: mechanism and regulation. Immunology 160(3):233–247

Cooper C et al (1992) The C/EBP family of transcriptional activators is functionally important for Ig VH promoter activity in vivo and in vitro. J Immunol 149(10):3225–3231

Cui M et al (2021) Immunoglobulin expression in cancer cells and its critical roles in tumorigenesis. Front Immunol 12:613530

Defrance T et al (2002) The life and death of a B cell. Adv Cancer Res 86:195–225

Deng H et al (2018) Expression of immunoglobulin A in human mesangial cells and its effects on cell apoptosis and adhesion. Mol Med Rep 17(4):5272–5282

Dinesh RK et al (2019) Transcription factor binding at immunoglobulin enhancers is linked to somatic hypermutation targeting. Eur J Immunol 50(3):380–395

Duan Z et al (2014) Activation of the Ig Iα1 promoter by the transcription factor Ets-1 triggers Ig Iα1–Cα1 germline transcription in epithelial cancer cells. Cell Mol Immunol 11(2):197–205

Ernst P et al (1995) Combinatorial regulation of transcription II: the immunoglobulin mu heavy chain gene. Immunity 2(5):427–438

Goebel P et al (2001) Localized gene-specific induction of accessibility to V(D)J recombination induced by E2A and early B cell factor in nonlymphoid cells. J Exp Med 194(5):645–656

Harris MB et al (1999) Transcriptional repression of Stat6-dependent interleukin-4-induced genes by BCL-6: specific regulation of iepsilon transcription and immunoglobulin E switching. Mol Cell Biol 19(10):7264–7275

Hu F et al (2012) Spontaneous production of immunoglobulin M in human epithelial cancer cells. PLoS One 7(12):e51423

Huang X et al (2023) A self-propagating c-Met-SOX2 axis drives cancer-derived IgG signaling that promotes lung cancer cell stemness. Cancer Res 83(11):1866–1882

Kemler I et al (1990) Octamer transcription factors and the cell type-specificity of immunoglobulin gene expression. FASEB J 4(5):1444–1449

Kim U et al (1996) The B-cell-specific transcription coactivator OCA-B/OBF-1/bob-1 is essential for normal production of immunoglobulin isotypes. Nature 383(6600):542–547

Konermann S et al (2015) Genome-scale transcriptional activation by an engineered CRISPR-Cas9 complex. Nature 517(7536):583–588

LaBella F et al (1988) Cell-cycle regulation of a human histone H2b gene is mediated by the H2b subtype-specific consensus element. Genes Dev 2(1):32–39

Li W et al (2014) MAGeCK enables robust identification of essential genes from genome-scale CRISPR/Cas9 knockout screens. Genome Biol 15(12):554

Liao Q et al (2015) Aberrant high expression of immunoglobulin G in epithelial stem/progenitor-like cells contributes to tumor initiation and metastasis. Oncotarget 6(37):40081–40094

Lin K-I et al (2002) Blimp-1-dependent repression of Pax-5 is required for differentiation of B cells to immunoglobulin M-secreting plasma cells. Mol Cell Biol 22(13):4771–4780

Linehan LA et al (1998) STAT6 is required for IL-4-induced germline Ig gene transcription and switch recombination. J Immunol 161(1):302–310

Liu H et al (2012) EBV-encoded LMP1 upregulates Igκ 3′ enhancer activity and Igκ expression in nasopharyngeal cancer cells by activating the Ets-1 through ERKs signaling. PLoS One 7(3):e32624

Malin S et al (2010) Role of STAT5 in controlling cell survival and immunoglobulin gene recombination during pro-B cell development. Nat Immunol 11(2):171–179

Martins G et al (2008) Regulation and functions of Blimp-1 in T and B lymphocytes. Annu Rev Immunol 26:133–169

Mattaj IW et al (1985) An enhancer-like sequence within the Xenopus U2 gene promoter facilitates the formation of stable transcription complexes. Nature 316(6024):163–167

Meffre E et al (2000) Antibody regulation of B cell development. Nat Immunol 1(5):379–385

Meffre E et al (2001) Immunoglobulin heavy chain expression shapes the B cell receptor repertoire in human B cell development. J Clin Invest 108(6):879–886

Melchers F (2015) Checkpoints that control B cell development. J Clin Invest 125(6):2203–2210

Nzula S et al (2003) Antigen-driven clonal proliferation, somatic hypermutation, and selection of B lymphocytes infiltrating human ductal breast carcinomas. Cancer Res 63(12):3275–3280

Poellinger L et al (1989) Octamer transcription factors 1 and 2 each bind to two different functional elements in the immunoglobulin heavy-chain promoter. Mol Cell Biol 9(2):747–756

Qiu X et al (2003) Human epithelial cancers secrete immunoglobulin G with unidentified specificity to promote growth and survival of tumor cells. Cancer Res 63(19):6488–6495

Rivera RR et al (1993) Ets proteins: new factors that regulate immunoglobulin heavy-chain gene expression. Mol Cell Biol 13(11):7163–7169

Romanow WJ et al (2000) E2A and EBF act in synergy with the V(D)J recombinase to generate a diverse immunoglobulin repertoire in nonlymphoid cells. Mol Cell 5(2):343–353

Sadofsky MJ (2001) The RAG proteins in V(D)J recombination: more than just a nuclease. Nucleic Acids Res 29(7):1399–1409

Schebesta M et al (2002) Transcriptional control of B-cell development. Curr Opin Immunol 14(2):216–223

Sen R (2004) NF-kappaB and the immunoglobulin kappa gene enhancer. J Exp Med 200(9):1099–1102

Shao W et al (2016) Epithelial cells are a sourcc of natural IgM that contribute to innate immune responses. Int J Biochem Cell Biol 73:19–29

Sun Z et al (1994a) Analysis of the imperfect octamer-containing human immunoglobulin VH6 gene promoter. Nucleic Acids Res 22(5):850–860

Sun Z et al (1994b) Bidirectional transcription from the human immunoglobulin VH6 gene promoter. Nucleic Acids Res 22(5):861–868

Wang Z et al (2020) Cancer-derived sialylated IgG promotes tumor immune escape by binding to Siglecs on effector T cells. Cell Mol Immunol 17(11):1148–1162

Wu L et al (2014) The immunoglobulin heavy chain VH6-1 promoter regulates Ig transcription in non-B cells. Cancer Cell Int 14(1):114

Yang L et al (2020) Targeting cancer stem cell pathways for cancer therapy. Signal Transduct Target Ther 5(1):8

Zheng H et al (2007) Expression and secretion of immunoglobulin alpha heavy chain with diverse VDJ recombinations by human epithelial cancer cells. Mol Immunol 44(9):2221–2227

Zheng J et al (2009) Immunoglobulin gene transcripts have distinct VHDJH recombination characteristics in human epithelial cancer cells. J Biol Chem 284(20):13610–13619

Zhu X et al (2010) Distinct regulatory mechanism of immunoglobulin gene transcription in epithelial cancer cells. Cell Mol Immunol 7(4):279–286

The Structure Characteristics and Function of Non B Cell-Derived Immunoglobulin

5

Jing Huang, Jingxuan Zhang, Li Zhang, Zihan Wang, Tianrui Fan, and Sha Yin

Abstract

According to classical immunology theory, immunoglobulin (Ig) is exclusively produced by differentiated B lymphocytes, which exhibit a typical tetrapeptide chain structure and are predominantly present on the surface of B cells and in bodily fluids. B-Ig is one of the critical effector molecules for humoral immune responses specifically recognising antigens and eliminating them. However, mounting evidence has demonstrated that Ig is widely expressed in non B lineage cells, especially malignant ones (referred to as non B-Ig). Interestingly, non B-Ig mainly resides in the cytoplasm and secretion, but to some extent on the cell surface. Furthermore non B-Ig not only displays a tetrapeptide chain structure but also shows free heavy chains and free light chains (FLCs). Additionally, Ig derived from non B cancer cell typically displays unique glycosylation modifications. Functionally, non B-Ig demonstrated diversity and versatility, showing antibody activity and cellular biological activity, such as promoting cell proliferation and survival, and it is implicated in cancer progression and some immune-related diseases, such as renal diseases.

Keywords

Ig · Immunoglobulin · Non B cell-derived Ig · Glycosylation · Tumour progression · Immunoregulation · Free light chain · Cytoskeleton · Immune evasion

J. Huang (✉)
Department of Immunology, School of Basic Medical Sciences, Peking University, Beijing, China
e-mail: huangjing82@bjmu.edu.cn

J. Zhang
Platform Biology, Suzhou Abogen Biosciences, Suzhou, Jiangsu, China
e-mail: zjx78@bjmu.edu.cn

L. Zhang
Division of HIV/AIDS and Sex-Transmitted Virus Vaccines, Institute for Biological Product Control, National Institutes for Food and Drug Control (NIFDC) and WHO Collaborating Center for Standardization and Evaluation of Biologicals, Beijing, China
e-mail: zhangli@nifdc.org.cn

Z. Wang
Department of Nephrology, Peking University Third Hospital, Beijing, China
e-mail: wangzihan81@bjmu.edu.cn

T. Fan
Shanghai Discovery and Development Center, Abiosciences, Shanghai, China
e-mail: fantianrui@bjmu.edu.cn

S. Yin
Xi'an Children's Hospital, Affiliated Children's Hospital of Xi'an Jiaotong University, Xi'an, Shaanxi, China
e-mail: shayin@bjmu.edu.cn

X. Qiu et al. (eds.), *Non B cell-Derived Immunoglobulins*, Advances in Experimental Medicine and Biology 1445, https://doi.org/10.1007/978-981-97-0511-5_5

5.1 Structure and Characteristics of Non B-Ig

The classical structure of Igs is a four-chain structure consisting of two identical light chains weighing about 25 kDa and two identical heavy chains weighing from 50 to 75 kDa. The N-end (about 110 aa) of both heavy and light chains form amino-terminal variable (V) regions, which display unique sequences for each Ig that participate in antigen recognition. The remaining regions are called constant regions which are identical among an individual or the same species. Because it presents a circular domain every 110 aa on Ig constant regions by forming a disulphide bond, the number of circular domains in different Ig constant regions varies. The Ig light chain only includes one circular domain, Igγ, Igα, and Igδ have three circular domains, and Igμ and Igε include four circular domains. Currently, the function of the constant region of the Ig light chain is unclear. Still, the constant region of the Ig heavy chain is well-known to activate complement, play complement-mediated cytotoxicity, or bind with IgG Fc receptors on the surface of macrophages and NK cells to play antibody-dependent cell-mediated cytotoxicity. Additionally, all classes of Igs are characterised by a distinctive set of glycoforms that reflect the wide variation in the number, type, and location of their oligosaccharide.

5.1.1 Non B-Ig Can Express All Five Classes of Ig

In 1972, Rodney Porter and Gerald Edelman were honoured with the Nobel Prize because they revealed the secrets of the basic structure of the Ig molecules (Raju 1999). Ig was divided into five classes, named IgG, IgM, IgA, IgD, and IgE, which can be distinguished by their heavy chains, which are termed γ, α, ε, μ, and δ, respectively. All these classes of Ig have a basic four-chain structure consisting of two identical heavy chains and two identical light chains. The differences are most pronounced in the constant regions of the antibody classes, leading to the triggering of different functions after binding to antigen.

Ig heavy chain proteins and their corresponding genes have also been identified in cancer cells. The Ig heavy chains expressed in cancer cells were found to include γ, α, ε, μ, and δ chains. In 1997, Ig heavy chain variable region gene transcripts were detected in T47-D, HeLa, and HT-29 cell lines using Southern blot and nested reverse transcription (RT)-PCR (Qiu and Yang 1997). Subsequently, in 1998, Kimoto also reported that the expression of gene transcripts of IgM, IgD, IgG3, IgG1, IgE, and IgA heavy chain constant region and the T-cell receptor-α existed in various carcinoma cell lines, such as SW1116, HEp2, MCF-7, and HC48 (pancreatic adenocarcinoma) (Kimoto 1998).

Like the classical Igs, non B-Ig is a roughly Y-shaped molecule based on a four-chain structure consisting of two identical light chains and two identical heavy chains linked by disulphide bridges (Fig. 5.1). Moreover, the non B-Ig also contains five classes of Ig heavy chain and two types of Ig light chain. However, cells of different lineages produce a biased class of Ig. For example, normal lung epithelial cells predominantly produce IgA and IgM (Gao et al. 2021), normal renal tubular epithelial cells mainly express IgG, renal mesangial cells secrete IgA (Deng et al. 2021), and renal podocytes tend to produce IgG4 (Jing et al. 2018). Interestingly, non B cells widely express Ig free light chain (FLC), especially Igκ FLC (Gao et al. 2021). Surprisingly, the Igκ FLC usually displays a unique VκJκ rearrangement. For instance, the free Igκ produced by mouse lung epithelial cells, breast epithelial cells, and neurons mainly displayed a Vκ9–120/Jκ rearrangement pattern (Gao et al. 2021; Huang et al. 2008; Zhang et al. 2010). At the same time, human non B cells also express the Igκ light chain with a unique Vκ4–1/Jκ rearrangement (Wang et al. 2022). These FLCs are suggested to be a novel mediator of inflammation and even cancer (Napodano et al. 2019; Ma et al. 2017). Correspondingly, free heavy chains may also exist in non B cells under stress conditions.

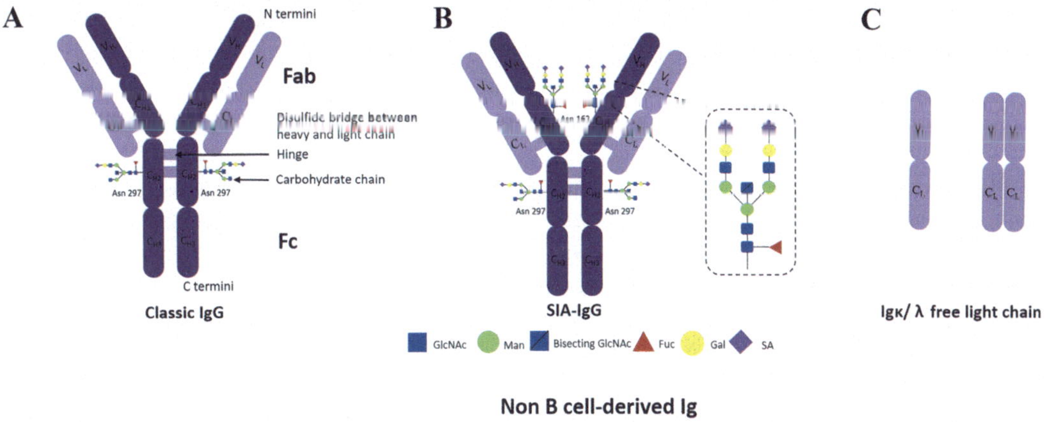

Fig. 5.1 The structure of non B-Ig. (**a**) The structure of classic IgG is a four-chain structure consisting of two identical light chains and heavy chains. The disulfide bridge between heavy and light chain, hinge and carbohydrate chain are shown. (**b**) The structure of sialylated IgG derived from non B cells. SIA-IgG has a non-consensus S/TXN glycosylation motif of TVSWN162SGAL in CH1 domain. (**c**) Igκ/λ free light chain (FLC) was widely seen in non B cells, like Vκ4–1/Jκ3-FLC in amyloid light chain amyloidosis

Qiu and other groups have compared the Ig expression levels and Ig classes between normal and cancer cells. The consensus is that cancer cells can produce higher Ig levels than normal cells (Wang et al. 2022; Qiu et al. 2003; Chen and Gu 2007). Moreover, the combination of Igγ and Igκ made the most abundant cancer Ig (Chen et al. 2009). For example, normal colonic epithelial cells express IgA and IgM at a higher frequency, but colon cancer cells in the same individual express IgG at a higher frequency. Similarly, normal hematopoietic stem cells mainly express IgM, but the malignant tumour of myeloid origin, acute myeloid leukaemia (AML) cells, mainly express IgG (Zhang et al. 2021; Huang et al. 2014). In addition, normal non B cells can express two types of light chains, Igκ chain and λ chain; however, almost all cancer cells highly express Igκ (Wang et al. 2015). Moreover, in human tissues, such as glomerular epithelium, there are four subclasses of Igγ heavy chains labelled γ1, γ2, γ3, and γ4, which give rise to the IgG1, IgG2, IgG3, and IgG4 subclasses. In mice, such as hepatic epithelium, four subclasses are denoted IgG1, IgG2a, IgG2b, and IgG3, and Igκ chains are more prevalent than λ chains in both humans and mice (Shao et al. 2016a).

5.1.2 Variable Region and Constant Domains of Non B-Ig

As highlighted above, all Ig heavy and light chains consist of variable and constant regions (Fig. 5.1). The constant region sequences of the same class of heavy chain or the same type of light chain are constant amongst the same individual or species. However, Ig variable region sequences vary even within the same individual. It was demonstrated that the constant region sequences of both Ig heavy chain and light chain derived from non B cells are entirely identical to B-Ig.

Igs can undergo somatic recombination to assemble the final Ig variable region (Bassing et al. 2002). Ig heavy chain variable region genes are assembled from the connection of three gene segments, including V (variable), D (diversity), and J (joining). Ig light chain variable region genes are constructed by joining V and J gene segments, but no D segment. In the early stage of B cell development, the rearrangement of the Ig heavy chain variable region occurs first. The sequence is that the B cell first selects any one JH segment and any DH fragment to make DH–JH joining. Then the B cell selects any VH gene segment to connect DH–JH and forms a complete Ig heavy chain VH–DH–JH rearrangement (variable region reading frame) on the genome. Then the VH–DH–JH rearrangement is connected with the gene of the constant region, transcribed and translated into the Ig heavy chain. Following Ig heavy chain expression, B cells further select any VL gene fragment of the light chain variable region gene to connect with any JL light chain to form the complete VL–JL rearrangement on the genome. Then VL–JL rearrangement is connected with the light chain constant region, transcribed and translated into the Ig light chain. As the genome V, D, and J gene fragments are randomly selected during Ig gene rearrangement, each B cell's Ig variable region sequence is unique, resulting in a significant diversity of the Ig library for each individual. In addition, since the rearrangement of the Igκ light chain variable region always precedes Igλ light chain, the Igκ light chain variable region is always dominant in the Ig library.

Like the classical B-Ig, the non B-Ig variable region also exhibits VH–DH–JH rearrangement for the heavy chain and VL–JL rearrangement for the light chain. Moreover, these rearrangements displayed an N or P insert, and the JH gene recombination occurred primarily at TG nucleotide sequences. Meanwhile, the Igκ light chain was dominant. It is worth noting that, like the B-Ig, the variable region of non B-Ig, including Igγ, Igα, and Igδ, also exhibits somatic hypermutation, and the Igε and Igμ retain the germline sequence (Huang et al. 2014; Zheng et al. 2009). However, unlike the B-Ig that displays great diversity, the non B-Ig variable region demonstrated restricted VH–DH–JH or VL–JL rearrangement patterns in humans and mice by Sanger sequencing and next-generation sequencing. For instance, many lineages of non B cells can produce Vκ4–1/Jκ3 rearrangement (Wang et al. 2022). Ig derived from identical lineages of non B cells usually exhibits a conservative VH–DH–JH or VL–JL rearrangement. For example, the most frequently encountered Ig germline rearrangement derived from non B cancer cells was VH5–51/D3–9/JH4 with a frequency of 23.7%, which far exceeded the frequency in B cells, primarily occurs most frequently with 50% in epithelial cancer cells (Zheng et al. 2009). Similar results were observed in IGK with conserved IGKV1–5 repertoire expressed in acute myeloid leukaemia (Wang et al. 2015). The V(D)J recombination in various non B cells, and all of these cells, appeared to have their unique rearrangement pattern (Zheng et al. 2009; Niu et al. 2011; Liu et al. 2012; Geng et al. 2019). The findings suggested that these non B-Ig with unique V(D)J rearrangement patterns have unique functions.

In addition, according to the classical theory, the variable region V(D)J rearrangement pattern of different types of Ig produced by a single B cell is the same, while the constant region sequence is different, which is named class switch recombination. However, Zheng et al. discovered that the variable regions of Igγ, Igα, and Igμ produced from the same non B cells displayed a limited V(D)J rearrangement pattern, but the same V(D)J rearrangement pattern among the different classes of Ig variable region was never observed (Zheng et al. 2009). It is suggested that the production of IgA or IgG might not conform with the classical theory, and the phenomenon mechanism requires further investigation.

5.1.3 Physicochemical Property of Non B-Ig

Classical Ig is the product of B lymphocytes and plasma cells, which are the effector molecules and antigen receptors for B lymphocytes and are responsible for humoral immunity. The secreted Ig serves as an antibody in the serum and secretory fluids such as saliva, bronchial and nasal secretions, etc. It is also present on the surface of B lymphocytes as the B cell receptor (BCR), which is critical for antigen recognition. However, non B-cell-derived IgG is primarily soluble, which mainly exists in the cytoplasm or is secreted into the extracellular matrix. Still, it is rarely detected in blood or other body fluids due to its hydrophobicity. Shao et al. have detected all classes of Ig levels not only in wild-type mice but also in B cell-deficient mice. And they found that, although a very low Ig level was detected in blood, a similar or even higher Ig level was seen in other tissues of B cell-deficient mice, including liver, lung, and kidney compared with wild-type mice (Shao et al. 2016a). The finding suggested that non B-Ig mainly exists in local tissue rather than body fluid. Additionally, the non B Ig exhibits strong hydrophobic characteristics and extracellular deposition as a novel extracellular matrix protein, which is a potential ligand of integrins. Moreover, non B cells, especially epithelial cancer cells, widely express an Ig free light chain, which exists in a dimer or polymer and is deposited outside the cells. Especially in the case of inflammation, the free Ig light chain can be secreted by non B cells, promoting inflammation and even inflammatory-cancer transformation (Ma et al. 2017).

It is known that protein aggregates in immunomodulatory formulations can trigger an unwanted immunogenic response. Amyloidosis is a disease of protein misfolding where secreted proteins aggregate into insoluble fibrils that are then deposited into tissues and cause functional and structural organ damage (Glenner et al. 1971; Eulitz et al. 1990). In the Ig light chain amyloidosis (known as AL amyloidosis), the N-terminal fragment of FLCs composed of the variable domain and a part of the constant region are reported to form organised fibrils. Non B-Igs are often insoluble aggregates. Especially non B-Ig FLC, which shares the same properties as the pathogenic Igκ, such as hydrophobicity, exists as the free form in the extracellular or intracellular space (Wang et al. 2022). Although these reports implicated the importance of variable domain properties, the understanding of the mechanisms is still insufficient.

5.1.4 Glycosylation of Non B-Ig

B-Igs are glycoproteins, and the glycosylation modification and position of different Ig are inconsistent. For example, the classical IgG has only one N-glycosylation at position 297 of each heavy chain, but no O-glycosylation. IgA1 has two N-glycosylation in each heavy chain, and IgA2 has five N-glycosylation. IgD, IgM, and IgE all have glycosylation sites. These three Ig heavy chains have O-glycosylation modifications, with IgM having the most significant glycosylation modification. Conversely, the light chain derived from B cells has no glycosylation sites in its constant region. The glycans of Igs are multifunctional and play vital roles in structure and pathological progression. For instance, the glycans maintain solubility and conformation, facilitate subcellular transport, secretion, and clearance, and even assist in binding to Fc receptors.

As for different Ig produced by non B cells, its glycosylation characteristics remain unclear. However, it has been reported that IgG or free Igκ produced by non B cells shows a unique N-glycosylation modification. The N-glycosylation modification endows IgG with critical biological functions. In 2009, Valliere-Douglass et al. reported that glycans on a non-consensus asparagine residue in the Asn162 of IgG1 and IgG2 CH1 were detected through HPLC-MS and peptide map analysis. This highly unusual modification was present at 0.5–2.0% levels on IgG1 and IgG2 (Valliere-Douglass et al. 2009). However, it is not a naturally expressed IgG because this special glycosylation modification is found on the heavy chain of humanised antibodies expressed in engineered cells (CHO cells), which has not received sufficient attention. Recently, Tang et al. first revealed that IgG from epithelial cancer cells and a few normal epithelial cells, such as squamous epithelial cells, also contained the non-consensus S/TXN glycosylation motif in the CH1 domain (Tang et al. 2018).

The discovery of this site was due to an antibody, RP215, which can specifically recognise epithelial cancer cell-derived IgG, but not the B cell-derived IgG (B-IgG) and non-epithelial cancer cell-derived IgG. RP215 was initially produced by Lee et al. in 1987 using cell extracts of the human OC-3-VGH ovarian cancer cell line as an antigen. It was found to specifically recognise almost all of the cancer cells but not normal cells (Lee et al. 1992). RP215 was reported to specifically recognise a carbohydrate-associated epitope localised in the variable region of the Ig heavy chains expressed by cancer cells (Lee and Ge 2009). Recently, RP215 has been confirmed to only recognise IgG, but not other proteins. Moreover, it was further revealed that RP215 primarily recognises the epithelial cancer cell-expressed sialylated IgG (SIA-IgG). These epithelial cancers include breast cancer, non-small cell lung cancer, oesophageal cancer, colorectal cancer, gastric cancer, pancreatic cancer, renal cancer, oral cancer, etc. More importantly, the epitope recognised by RP215 was the non-consensus S/TXN glycosylation motif of $TVSWN^{162}$ SGAL rather than the consensus NXT/S motif glycosylation in QFN^{297} STFR within the Igγ CH2 domain (Tang et al. 2018). The Asn162 glycosylation recognised by RP215 was identified as α-2,6-linked sialylation.

In addition to IgG heavy chains, light chains could also acquire sequence motifs available for N-glycosylation (Asn–X–Ser/Thr) through somatic recombination. In AL amyloidosis, the serum light chains were modified by N-glycosylation for one-third of κ-amyloidosis and 10% of λ-amyloidosis patients (Kumar et al. 2019). Recently, it was found that a free Igκ chain with a unique Vκ4–1/Jκ3 rearrangement pattern (Vκ4–1/Jκ3-FLC) was widely seen in non B cells, particularly in non B cancer cells (Wang et al. 2022). The Vκ4–1/Jκ3-FLC was highly homologous with FLC displayed in amyloid light chain amyloidosis, and the size of the protein is about 29 kDa, not the customary 25 kDa. After treatment with PNGase F, Vκ4–1/Jκ3-FLC significantly reduced to 25 kDa, suggesting that the free light chain in cancer patients might have N-linked glycosylations.

Based on the above findings, we speculate that in addition to the unique structure of variable region sequence, non B-Ig also has unique characteristics in glycosylation modification, including other classes of Igs. The uniqueness of these glycosylation modifications on non B-Ig needs to be further revealed.

5.2 The Biological Function of Non B-Ig

It is known that the Ig produced by B cells is located in the cell membrane and secretory fluid respectively. The Ig on the cell membrane is mainly responsible for identifying antigens and initiating B cell activation and proliferation. Secretory Ig is secreted by plasma cells differentiated from B cells, which are primarily accountable for specifically binding antigen substances, and exerts a direct neutralisation effect, activates complement and triggers the complement-dependent cytotoxicity (CDC), or binds to Fc receptors on the surface of NK cells and macrophages with its Fc fragment to further trigger the antibody-dependent cell-mediated cytotoxicity (ADCC).

Compared with the classical B-Ig, non B-Ig originates from different lineages of cells, including neurons, epithelial cells, sperm, etc., and displays functions in cell localisation, physicochemical characteristics, and variable region sequence characteristics. Non B-Ig is mainly located in the cytoplasm and secreted into the extracellular matrix, which is often hydrophobic and deposited in the extracellular matrix. Therefore, the function of non B-Ig exhibits diversity and versatility, including antibody activity and promoting cell proliferation and survival, even oncogenesis (Fig. 5.2).

5.2.1 Non B-Ig Serve as Natural Antibody

Natural antibodies are produced spontaneously by B lymphocytes without antigenic exposure from infection or vaccination. Natural antibodies are predominantly IgM. Additionally, IgG and IgA natural antibodies have also been identified. Natural antibodies provide protection against diseases (Elluru et al. 2015) and perform housekeeping to eliminate cell debris and damaging molecules (Kaveri et al. 2012). Natural antibodies are also used as drugs to treat autoimmune diseases and delivered as pools from the serum of healthy donors, the so-called IVIG (Gilardin et al. 2015).

There is a large amount of Ig in skin and mucosa, including IgM, IgA, and IgG, which have natural antibody activity. Previously, these natural antibodies were thought only derived from plasma cells in lamina propria or dermis. These natural antibodies were transported and passed through epithelial cells into mucosa or

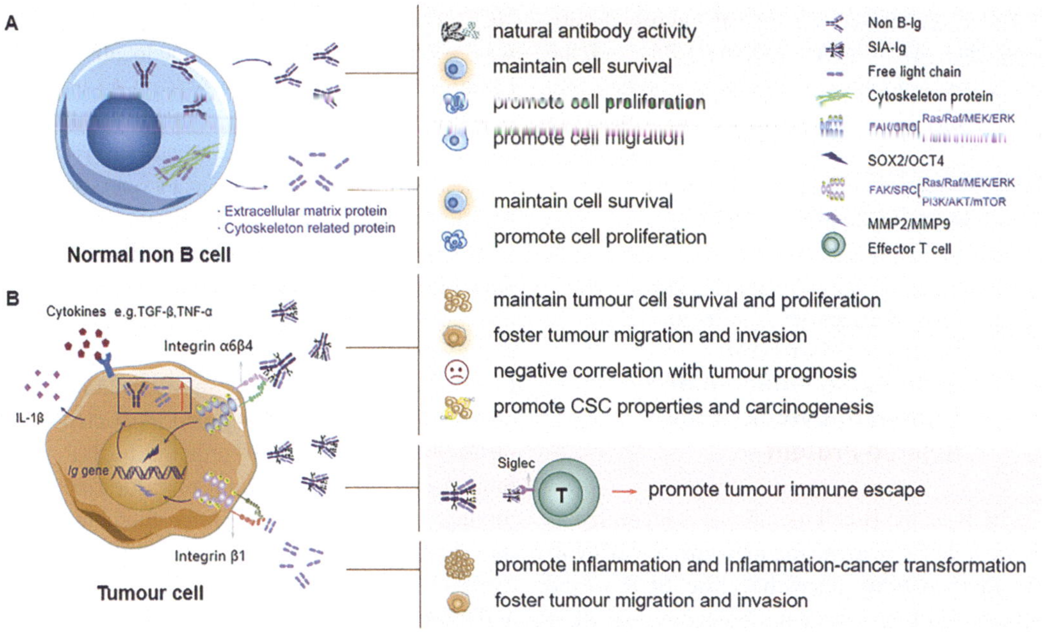

Fig. 5.2 The biological function of non B-Ig. (**a**) The biological function of non B-Ig in normal non B cell, including complete non B-Ig with tetrapeptide chain and Ig free light chain (FLC). The complete non B-Ig with tetrapeptide chain shows natural antibody activity in epithelial cells, maintains cell survival, and promotes cell proliferation and migration. FLC maintains cell survival, and promotes cell proliferation as extracellular matrix protein or cytoskeleton-related protein. (**b**) The biological function of non B-Ig with tetrapeptide chain and FLC in tumour cell. Non B-Ig with tetrapeptide chain is involved in tumour progression, including maintaining the survival and proliferation of tumour cells, promoting tumour migration and invasion, a negative correlation with tumour prognosis, and promoting CSC properties and carcinogenesis. Besides these, non B-Ig binds to sialic acid-binding immunoglobulin-type lectins (Siglecs) on effector $CD4^+$ and $CD8^+$ T cells to promote tumour immune escape. FLC promotes proliferation, migration, and invasion of tumour cells through interacting with integrin β1 to activate the integrin-FAK pathway, thereby promoting inflammation and inflammation-cancer transformation

skin surface by polymeric IgA receptor and play vital roles in protecting the ~400 m^2 of human mucosa from invasion by pathogens.

Given that the epithelium is the primary cell layer that initially encounters the microorganisms, normal epithelial cells can produce Ig. Does the Ig produced by epithelial cells have natural antibody activity and play an immense role in host defence and immune surveillance? Jiang et al. have found that IgG and IgA are expressed in normal human epidermal cells, including basal cells and the human keratinocyte line HaCaT. These non B-IgG and IgA can bind to various bacteria, such as *Escherichia coli*, *Candida albicans,* and *Staphylococcus aureus*, demonstrating potential natural antibody activity and might participate in skin innate immunity. More interestingly, the epidermal cell-derived IgA and IgG showed their own unique conserved VHDJH rearrangement pattern. The IgA mainly recognises *Staphylococcus epidermidis*, and IgG mainly recognises *Staphylococcus aureus*, suggesting that different Ig classes perform their responsibilities in local immune defence (Jiang et al. 2015).

In addition, Shao et al. revealed that mouse liver epithelial cell-expressed IgM showed natural antibody properties, such as binding to a variety of autoantigens with low affinities, including single-stranded DNA (ssDNA), double-stranded DNA (dsDNA), lipopolysaccharide (LPS), and insulin (Shao et al. 2016b). Besides, the CpG, an agonist of TLR9, which mimics bacterial infection, substantially increased IgM secretion in human epithelial cancer cells. More importantly, the epithelial cancer cell-derived

IgM also recognised a series of microbial antigens, such as ssDNA, dsDNA, LPS, and the HEp-2 cell antigen, as natural antibodies (Hu et al. 2012). These results suggest that non B-Ig from different cells, which are often exposed to antigens, have natural antibody activities to participate in the innate immune response.

5.2.2 Non B-Ig Exerts Cellular Biological Function as an Extracellular Matrix Protein or Cytoskeleton-Related Protein

The BCR on the B-cell membrane is identified as responsible for recognising antigens. B-cells will lose their viability once they lose BCR in the early stage of development, indicating that BCR displays essential cellular biological activity, also responsible for maintaining B-cell survival and immune recognition. It has been demonstrated that non B-Ig also contributes to cell proliferation, adhesion, migration, and survival as a novel extracellular matrix protein or cytoskeleton-related protein.

Since non B-Ig was initially found in cancer cells and proliferative cells by Qiu et al. they hypothesised that the biological activity of non B-Ig was related to cell proliferation and survival. They first identified that non B-Ig expressed in cancer cells promotes cell proliferation and survival using antisense oligonucleotides of Ig heavy chain and anti-human IgG antibody (Qiu et al. 2003). Then it was demonstrated that non B-Ig expressed in normal non B cells promotes cell adhesion and survival. For example, Jing et al. found that IgG was expressed in podocytes, and that podocyte-derived Ig can maintain podocyte viability and adhesive capabilities (Jing et al. 2018). Moreover, IgA was also produced by primary human glomerular mesangial cells, and inhibition of Igα1 expression by small interfering RNA (siRNA) could inhibit cell adhesion and promote cell apoptosis (Deng et al. 2018). In addition, Yin et al. also found that the expression of Ig light chain-Igκ existed in both primary hepatocytes and normal hepatocyte line NCTC1469. Hepatocyte-derived Igκ expression was elevated and displayed unique localisation in hepatocytes of concanavalin A (Con A)-induced hepatitis model. Moreover, hepatocyte-derived Igκ played an essential role in cell survival and exerted a protective effect against Con A-induced liver injury in mice (Yin et al. 2020).

It has been found that non B cells also produced Igκ free light chain, which usually showed a unique VκJκ rearrangement pattern with identical Vκ4–1/Jκ3 (Vκ4–1/Jκ3-FLC). Wang et al. found that Vκ4–1/Jκ3-FLC showed filamentous distribution in the cytoplasm and interacted with intracellular cytoskeleton protein, further promoting cell proliferation and migration by promoting the assembly of microfilaments and intermediate filaments. The Igκ FLC binds to the cytoskeletal cross-linker plectin in non B cells and promotes the assembly of microfilaments and intermediate filaments by inhibiting the ubiquitination of plectin (unpublished data). In addition, Zhu et al. have identified that the Igκ FLC was also expressed in mouse and human cardiomyocytes, predominantly located in the intercalated disc of cardiomyocytes. Importantly, cardiomyocyte-specific deletion of Igκ in adult mice (cKO) significantly resulted in hypotension, rapid decrease of cardiac contractility, and conduction defects. Histologically, Igκ deficiency in mouse cardiomyocytes led to structural disruption of the intercalated disc (ICD), loss of intracellular localisation of the adhesion connection-related E-cadherin and Connexin 43, and the plectin on ICD (unpublished data). The mechanism investigation indicated that Igκ could bind with plectin and enhance the protein stability of plectin by inhibiting the ubiquitination of plectin.

In addition, Vκ4–1/Jκ3-FLC also acts as an extracellular matrix protein, which is highly homogeneous with the Igκ in AL amyloidosis and LCDD (light chain deposition disease). Moreover, the Vκ4–1/Jκ3-FLC shares the same properties with the pathogenic Igκ described above, including hydrophobicity and free form in extracellular or intracellular space. Wang et al. have reported that colorectal cancer cell-derived FLC with Vκ4–1/Jκ3 combination pattern could interact with integrin β1 to activate the integrin-FAK pathway, further significantly promoting proliferation, migration, and invasion of colorectal cancer cells in vitro and in vivo (Wang et al. 2022). These findings suggested that the free Vκ4–1/Jκ3-FLC is a novel extracellular matrix protein.

5.2.3 Non B-Ig Is Implicated in Cancer Progression

Non B-Ig originates not solely from normal tissue cells, but also from various malignant tumour cells. Unlike traditional B-Ig, which predominantly fulfils the humoral immune defence role against incursions by external microorganisms, tumour cell-derived Ig encourages the development and progression of tumours from numerous aspects, correlating with poor prognoses in tumour patients. It might offer a new biomarker for tumour diagnosis and prognosis evaluation.

5.2.3.1 Non B-Ig Sustains the Survival and Proliferation of Tumour Cells

In 2003, Qiu and others reported for the first time that epithelial tumour cells can produce IgG. Subsequently, it was discovered that silencing IgG or blocking IgG activity with a specific antibody inhibited the growth of tumour cells and induced cell apoptosis, suggesting that tumour cell-derived IgG (tumour-IgG) exhibits growth factor-like activity (Qiu et al. 2003). Thereafter, it was reported that IgG expression in colon cancer positively correlates with proliferating cell nuclear antigen (PCNA), cyclin D1, and NF-κB (Niu et al. 2012). Additionally, the expression of IgG in oesophageal squamous cell carcinoma was positively correlated with Ki67 (Zhang et al. 2012), indicating that tumour-IgG expression was positively correlated with the expression of various tumour proliferation factors. Furthermore, it has been found that using specific siRNA against IgG in tumour cells significantly inhibited tumour cell proliferation and promoted cell apoptosis in colon cancer (Niu et al. 2012), prostate cancer (Liu et al. 2012), lung cancer (Tang et al. 2018), pancreatic ductal adenocarcinoma (Cui et al. 2020a), and other epithelial tumour cell lines. At the same time, overexpression of tumour-IgG maintains the survival and proliferation of bladder cancer cells (Liang et al. 2013).

Tumour-IgG principally promotes tumour proliferation by activating proliferation-related signalling pathways. Wen et al. found that the Fab segment of tumour IgG binds to MDA-MB-231 and MCF-7 breast cancer cells, subsequently upregulating the phosphorylation level of ERK through MEK and Src signalling pathways, thus promoting tumour cell growth (Wen et al. 2005). Wang and colleagues also discovered that IgG interacted with receptors for activated protein kinase C-1 (Rack1) and peroxiredoxin (PRDX) and further induced tumour cells to produce ROS and activated ERK signalling pathway to promote cell proliferation (Wang et al. 2013). Furthermore, tumour-IgG was found to interact with the integrin α6β4 complex and activate the FAK signalling pathways. Notably, the IgG-integrin-FAK signalling relies on the N-glycan epitope (Tang et al. 2018). The specifics are discussed in Chap. 13. Similar results were observed in different types of cancers, such as colon cancer, renal cancer, and bladder cancer, with tumour-IgG promoting cell survival and growth (Qiu et al. 2003; Sheng et al. 2016a).

Non B-IgA also appears in seromucous secretions, such as cancer cell culture supernatant and secretions of the lung, saliva, and colostrum. It defends the exposed external surfaces of the body against attack by microorganisms. In 2001, Zheng and others discovered a novel immunoglobulin gene, SNC73, using cDNAs of normal mucosal tissues and colorectal cancer. It encoded a peptide identical to the constant region of an IgA and is expressed in both normal epithelial cells and cancer cells, including colorectal cancer, gastric cancer, breast cancer, lung cancer, and liver cancer (Zheng et al. 2001, 2007a; Hu et al. 2003). Furthermore, blocking cancer-derived Igα using mouse anti-human Igα antibody suppressed the growth, viability, and capacity for proliferation of cancer cells (HeLa and CNE1). Moreover, it reduced the access percentage of the S phase from the early mitosis of synchronised cancer cells, supporting the concept that cancer-derived IgA encourages the growth of cancer cells (Zheng et al. 2007b).

5.2.3.2 Non B-Ig Fosters Tumour Migration and Invasion

Qiu and others found that IgG expression in papillary thyroid carcinoma was related to local lymph node metastasis via tissue microarray (Qiu et al. 2012). The positive signal of IgG in metastatic foci was notably stronger than that in primary foci, both in breast cancer and lung cancer, as revealed by immunohistochemical analysis of hundreds of clinical tissue samples (Liu et al. 2015; Liao et al. 2015). Moreover, using siRNA to silence IgG or blocking IgG activity using specific antibodies inhibited the migration ability of tumour cells in colon cancer, lung cancer, pancreatic ductal adenocarcinoma, breast cancer, bladder cancer, and other tumour cell lines (Tang et al. 2018; Niu et al. 2012; Liao et al. 2015; Jiang et al. 2014; Sheng et al. 2016b). Tumour cells exhibiting high expression of IgG demonstrated a stronger capacity for liver metastasis in vivo. More importantly, tumour-IgG reduced the expression of E-cadherin, a molecule associated with epithelial to mesenchymal transition (EMT), and fostered the expression of matrix metalloproteinase-2 (MMP-2) and MMP-9, which are related to cell invasion. In addition, cancer-IgG, predominantly sialylated IgG, promotes tumour migration and invasion by binding to $\alpha6\beta4$ integrin to activate the FAK pathway (Tang et al. 2018). Therefore, tumour-IgG fosters tumour cell migration and invasion via EMT and the focal adhesion pathway.

5.2.3.3 Differentiation of Cancer Cells and Prognosis of Cancer Are Correlated with Non B-IgG

There is an expanding body of evidence indicating that tumour-IgG bears a substantial negative correlation with cell differentiation in an array of cancers, such as oesophageal squamous cell carcinoma, colon cancer, pancreatic duct adenocarcinoma, lung cancer adenocarcinoma, kidney clear cell carcinoma, bladder cancer, and parathyroid carcinoma (Niu et al. 2012; Zhang et al. 2012; Cui et al. 2020a, b; Liu et al. 2015). Yang et al. reported that IgG transcript was decreased in human choriocarcinoma cell line JAR when induced with methotrexate, while no discernible alteration of IgG expression was observed in normal human extravillous trophoblast cell line TEV-1 (Yang et al. 2014). The expression of IgG in breast cancer cells was positively correlated with poor prognosis indicators, such as human epidermal growth factor receptor-2 (HER-2) and multidrug resistance gene p170 protein. However, it was positively correlated with good prognosis indexes, including oestrogen receptor (ER), progesterone receptor (PR), oestrogen regulatory peptide PS-2, and anti-cancer gene nm23 (Liao et al. 2015). Moreover, the expression of sialylated IgG derived from tumour cells is significantly correlated with the poor prognosis of lung squamous cell carcinoma and pancreatic ductal adenocarcinoma (Tang et al. 2018; Cui et al. 2020a). In addition, a high level of IgG expression is associated with monocytic differentiation, multilineage dysplasia, TET2 and KRAS mutations, and poor overall survival (Wu et al. 2020). In conclusion, the high expression of IgG is related to the low differentiation of the tumour. This might be used as a risk factor for the prognosis evaluation of various cancers.

5.2.3.4 Sialylated IgG Is Highly Expressed in Cancer Stem Cells and Promotes Carcinogenesis and Metastasis

Cancer stem cells (CSCs) are a self-renewing population localised within cancer niches and play crucial roles in tumour initiation, recurrence and metastasis. CSCs take charge of a cancer hierarchy, harbouring stem cell-like properties involving self-renewal and aberrant differentiation potential, closely associated with cancer initiation, metastasis and drug resistance. It has been found that CSCs of many epithelial cancers highly express IgG, which displayed a unique N-glycosylation modification on Asn162. Moreover, the end of its glycan showed high sialylation modification (known as sialylated IgG, SIA-IgG). Liao et al. found that SIA-IgG was highly expressed in a small population of basal/myoepithelial-like cancer cells of breast, colon, lung, and prostate in situ. In contrast, no immunostaining was found in lymphomas or mesenchymal tumours. Especially in breast cancer, the cells with high expression of sialylated tumour-IgG displayed $CD44^{+}CD24^{-/low}$ phenotype of CSCs in breast cancer (Liao et al. 2015). Tang and Huang et al. further found that SIA-IgG was highly expressed in CSCs of non-small cell lung cancer, and SIA-IgG bound to integrin $\alpha 6\beta 4$ and c-Met, then triggered cancer initiation and metastasis by activating FAK and c-Met pathway (Tang et al. 2018; Huang et al. 2023).

Since these invasive basal-like cells at the leading edge of cancer show a higher frequency of tumour-IgG staining than CD44v6 staining, sialylated tumour-IgG was predictive of cancer metastasis correlated with poor prognosis. Moreover, increased expression of sialylated tumour-IgG is related closely with other CSC properties of breast cancer, such as promoting tumour cell proliferation, division, migration, invasion capacity, as well as the mammospheres formation and resistance to chemotherapy drugs. It's worth noting that knockdown of IgG in breast cancer leads to reduced adhesion, invasion, and self-renewal and increased apoptosis (Liao et al. 2015).

5.2.3.5 Sialylated IgG Promotes Tumour Immune Escape

IgG in the tumour microenvironment (TME) is thought to be a product of B cells and exerts antitumour activity. However, SIA-IgG has been reported recently to be expressed in many cancer cells and it could be secreted into the TME. In addition to this, the SIA-IgG directly inhibited effector T cell proliferation, dependent on its sialic acid modification that exists in the CH1 domain, and then reduced $CD4^{+}$ and $CD8^{+}$ T cell infiltration into tumour tissues to promote tumour growth significantly. More importantly, SIA-IgG exerts its immunosuppressive effect by binding to sialic acid-binding immunoglobulin-type lectins (Siglecs) on effector $CD4^{+}$ and $CD8^{+}$ T cells (Wang et al. 2020). Since Siglecs are expressed at very low levels on normal T cells, while an upregulation of Siglecs, including Siglec-9, on tumour-infiltrating T cells from non-small cell lung cancer, colorectal, and ovarian cancer patients, has been observed (Stanczak et al. 2018). More importantly, the majority of intratumoural, but not peripheral blood, cytotoxic $CD8^{+}$T cells expressed Siglec-9 in melanoma, and these Siglec-9^{+} $CD8^{+}$T cells serve as a subset of effector memory cells with high functional capacity and signatures of clonal expansion, but was functionally inhibited in the presence of Siglec-9 ligands on the majority of tumour cells (Quentin et al. 2019). Also, Siglec-$7^{+}CD8^{+}$T cells, which were found in the blood and bone marrow of AML patients, and Siglec-7 repolarisation on $CD8^{+}$ T cells into the immune synapse with target cells expressing the cognate ligand, as well as sialic-acid dependent suppression of cytotoxicity and cytokine production selectively of Siglec-$7^{+}CD8^{+}$ T cells (Quentin et al. 2022). Fan et al. reported that Siglec-7 was prominently upregulated on tumour-infiltrating $CD8^{+}$ T cells and coexpressed with PD-1 and LAG-3, which indicates the exhaustion phenotype. SIA-IgG served as a ligand of Siglec-7, activating downstream signals and suppressing TCR pathways (Fan et al. 2023). All the above results support that sialylated tumour-IgG might be a novel ligand for Siglecs, which serve as a potential checkpoint molecule to participate in tumour immune evasion.

5.3 Concluding Remarks

From the discovery of non B-Ig 30 years ago to our current understanding of non B cell-derived Ig, it has become evident that the structure, physicochemical properties, and functions of non B-Ig are different from classical B-Ig. The content described in this chapter mainly focuses on understanding the non B-IgG and free Ig light chain. However, it has not fully elucidated the characteristics of other classes of Ig derived from non B cells. In the future, profound understanding will be required through a series of studies about non B-IgG and other classes of Ig to expand people's inherent understanding of Ig structure, physicochemical properties, and functions.

References

Bassing CH, Swat W, Alt FW (2002) The mechanism and regulation of chromosomal V(D)J recombination. Cell 109(Suppl):S45–S55

Chen Z, Gu J (2007) Immunoglobulin G expression in carcinomas and cancer cell lines. FASEB J 21(11):2931–2938

Chen Z et al (2009) Immunoglobulin expression in non-lymphoid lineage and neoplastic cells. Am J Pathol 174(4):1139–1148

Cui M et al (2020a) High expression of cancer-derived glycosylated immunoglobulin G predicts poor prognosis in pancreatic ductal adenocarcinoma. J Cancer 11(8):2213–2221

Cui M et al (2020b) Cancer-derived immunoglobulin G: a novel marker for differential diagnosis and relapse prediction in parathyroid carcinoma. Clin Endocrinol 92(5):461–467

Deng H et al (2018) Expression of immunoglobulin a in human mesangial cells and its effects on cell apoptosis and adhesion. Mol Med Rep 17(4):5272–5282

Deng Z et al (2021) Expression of immunoglobulin G in human proximal tubular epithelial cells. Mol Med Rep 23(5):327

Elluru S et al (2015) The protective role of immunoglobulins in fungal infections and inflammation. Semin Immunopathol 37(2):187–197

Eulitz M et al (1990) Immunoglobulin heavy-chain-associated amyloidosis. Proc Natl Acad Sci USA 87(17):6542–6546

Fan T et al (2023) Sialylated IgG in epithelial cancers inhibits antitumor function of T cells via Siglec-7. Cancer Sci 114:370–383

Gao E et al (2021) Lung epithelial cells can produce antibodies participating in adaptive humoral immune responses. bioRxiv 2021:443498

Geng Z et al (2019) Human colorectal cancer cells frequently express IgG and display unique Ig repertoire. World J Gastrointest Oncol 11(3):195–207

Gilardin L et al (2015) Intravenous immunoglobulin as clinical immune-modulating therapy. CMAJ 187(4):257–264

Glenner G et al (1971) Amyloid fibril proteins: proof of homology with immunoglobulin light chains by sequence analyses. Science 172(3988):1150–1151

Hu J et al (2003) Expression of a novel immunoglobulin gene SNC73 in human cancer and non-cancerous tissues. World J Gastroenterol 9(5):1054–1057

Hu F et al (2012) Spontaneous production of immunoglobulin M in human epithelial cancer cells. PLoS One 7(12):e51423

Huang J et al (2008) Expression of immunoglobulin gene with classical V–(D)–J rearrangement in mouse brain neurons. Int J Biochem Cell Biol 40(8): 1604–1615

Huang J et al (2014) Rearrangement and expression of the immunoglobulin μ-chain gene in human myeloid cells. Cell Mol Immunol 11(1):94–104

Huang X et al (2023) A self-propagating c-met-SOX2 axis drives cancer-derived IgG signaling that promotes lung cancer cell stemness. Cancer Res 83(11):1866–1882

Jiang C et al (2014) Immunoglobulin G expression in lung cancer and its effects on metastasis. PLoS One 9(5):e97359

Jiang D et al (2015) IgG and IgA with potential microbial-binding activity are expressed by normal human skin epidermal cells. Int J Mol Sci 16(2):2574–2590

Jing Z et al (2018) Expression of immunoglobulin G in human podocytes, and its role in cell viability and adhesion. Int J Mol Med 41(6):3296–3306

Kaveri S et al (2012) Natural IgM in immune equilibrium and harnessing their therapeutic potential. J Immunol 188(3):939–945

Kimoto Y (1998) Expression of heavy-chain constant region of immunoglobulin and T-cell receptor gene transcripts in human non-hematopoietic tumor cell lines. Genes Chromosomes Cancer 22(1):83–86

Kumar S et al (2019) Assay to rapidly screen for immunoglobulin light chain glycosylation: a potential path to earlier AL diagnosis for a subset of patients. Leukemia 33(1):254–257

Lee G, Ge B (2009) Cancer cell expressions of immunoglobulin heavy chains with unique carbohydrate-associated biomarker. Cancer Biomark 5(4):177–188

Lee C et al (1992) Studies of a tumor-associated antigen, COX-1, recognized by a monoclonal antibody. Cancer Immunol Immunother 35(1):19–26

Liang P et al (2013) Overexpression of immunoglobulin G prompts cell proliferation and inhibits cell apoptosis in human urothelial carcinoma. Tumour Biol 34(3):1783–1791

Liao Q et al (2015) Aberrant high expression of immunoglobulin G in epithelial stem/progenitor-like cells contributes to tumor initiation and metastasis. Oncotarget 6(37):40081–40094

Liu Y et al (2012) IgG gene expression and its possible significance in prostate cancers. Prostate 72(6):690–701

Liu Y et al (2015) Binding of the monoclonal antibody RP215 to immunoglobulin G in metastatic lung adenocarcinomas is correlated with poor prognosis. Histopathology 67(5):645–653

Ma J et al (2017) Free immunoglobulin light chain (FLC) promotes murine colitis and colitis-associated colon carcinogenesis by activating the inflammasome. Sci Rep 7(1):5165

Napodano C et al (2019) Free light chains and autoimmunity. Autoimmun Rev 18(5):484–492

Niu N et al (2011) Expression and distribution of immunoglobulin G and its receptors in the human nervous system. Int J Biochem Cell Biol 43(4):556–563

Niu N et al (2012) IgG expression in human colorectal cancer and its relationship to cancer cell behaviors. PLoS One 7(11):e47362

Qiu X, Yang G (1997) Ig gene analysis of epithelial malignant tumor cells. Chin J Cancer Biother 2(4):2

Qiu X et al (2003) Human epithelial cancers secrete immunoglobulin G with unidentified specificity to

promote growth and survival of tumor cells. Cancer Res 63(19):6488–6495
Qiu Y et al (2012) Immunoglobulin G expression and its colocalization with complement proteins in papillary thyroid cancer. Mod Pathol 25(1):36–45
Quentin H et al (2019) Siglec-9 regulates an effector memory CD8+ T-cell subset that congregates in the melanoma tumor microenvironment. Cancer Immunol Res 7(5):707–718
Quentin H et al (2022) Siglec-7 represents a glyco-immune checkpoint for non-exhausted effector memory CD8+ T cells with high functional and metabolic capacities. Front Immunol 13:996746
Raju T (1999) The Nobel chronicles. 1972: Gerald M Edelman (b 1929) and Rodney R Porter (1917–85). Lancet 354(9183):1040
Shao W et al (2016a) Identification of liver epithelial cell-derived Ig expression in μ chain-deficient mice. Sci Rep 6:23669
Shao W et al (2016b) Epithelial cells are a source of natural IgM that contribute to innate immune responses. Int J Biochem Cell Biol 73:19–29
Sheng Z et al (2016a) IgG is involved in the migration and invasion of clear cell renal cell carcinoma. J Clin Pathol 69(6):497–504
Sheng Z et al (2016b) Involvement of cancer-derived IgG in the proliferation, migration and invasion of bladder cancer cells. Oncol Lett 12(6):5113–5121
Stanczak M et al (2018) Self-associated molecular patterns mediate cancer immune evasion by engaging Siglecs on T cells. J Clin Invest 128(11):4912–4923
Tang J et al (2018) Lung squamous cell carcinoma cells express non-canonically glycosylated IgG that activates integrin-FAK signaling. Cancer Lett 430:148–159
Valliere-Douglass J et al (2009) Asparagine-linked oligosaccharides present on a non-consensus amino acid sequence in the CH1 domain of human antibodies. J Biol Chem 284(47):32493–32506
Wang J et al (2013) Cancer-derived immunoglobulin G promotes tumor cell growth and proliferation through inducing production of reactive oxygen species. Cell Death Dis 4:e945
Wang C et al (2015) IGK with conserved IGKappaV/IGKappaJ repertoire is expressed in acute myeloid leukemia and promotes leukemic cell migration. Oncotarget 6(36):39062–39072
Wang Z et al (2020) Cancer-derived sialylated IgG promotes tumor immune escape by binding to Siglecs on effector T cells. Cell Mol Immunol 17(11):1148–1162
Wang Q et al (2022) A widely expressed free immunoglobulin κ chain with a unique Vκ4–1/Jκ3 pattern promotes colon cancer invasion and metastasis by activating the integrin β1/FAK pathway. Cancer Lett 540:215720
Wen Y et al (2005) Antigen binding of human IgG Fabs mediate ERK-associated proliferation of human breast cancer cells. DNA Cell Biol 24(2):73–84
Wu L et al (2020) High levels of immunoglobulin expression predict shorter overall survival in patients with acute myeloid leukemia. Eur J Haematol 105(4):449–459
Yang M et al (2014) IgG expression in trophoblasts derived from placenta and gestational trophoblastic disease and its role in regulating invasion. Immunol Res 60(1):91–104
Yin S et al (2020) Hepatocyte-derived Igkappa exerts a protective effect against ConA-induced acute liver injury. Int J Mol Sci 21(24):9379
Zhang S et al (2010) Immunoglobulin gene locus events in epithelial cells of lactating mouse mammary glands. Cell Mol Life Sci 67(6):985–994
Zhang L et al (2012) Expression of immunoglobulin G in esophageal squamous cell carcinomas and its association with tumor grade and Ki67. Hum Pathol 43(3):423–434
Zhang C et al (2021) NBIGV-DB: a dedicated database of non-B cell derived immunoglobulin variable region. Gene 772:145378
Zheng S et al (2001) Structure and expression of colorectal cancer related immunoglobulin novel gene SNC73. Zhonghua Yi Xue Za Zhi 81(8):485–488
Zheng H et al (2007a) Expression and secretion of immunoglobulin alpha heavy chain with diverse VDJ recombinations by human epithelial cancer cells. Mol Immunol 44(9):2221–2227
Zheng H et al (2007b) Immunoglobulin alpha heavy chain derived from human epithelial cancer cells promotes the access of S phase and growth of cancer cells. Cell Biol Int 31(1):82–87
Zheng J et al (2009) Immunoglobulin gene transcripts have distinct VHDJH recombination characteristics in human epithelial cancer cells. J Biol Chem 284(20):13610–13619

Comparison of Non B-Ig and B-Ig

6

Xiaojun Xu, Peter J. Delves, Jing Huang, Wenwei Shao, and Xiaoyan Qiu

Abstract

Immunoglobulin (Ig) has been widely acknowledged to be produced solely by B-lineage cells. However, growing evidence has demonstrated the expression of Ig in an array of cancer cells, as well as normal cells including epithelial cells, epidermal cells, mesangial cells, monocytes, and neutrophils. Ig has even been found to be expressed in non-B cells at immune-privileged sites such as neurons and spermatogenic cells. Despite these non-B cell-derived Igs (non-B-Igs) sharing the same symmetric structures with conventional Igs (B-Igs), further studies have revealed unique characteristics of non-B-Ig, such as restricted variable region and aberrant glycosylation. Moreover, non-B-Ig exhibits properties of promoting malignant behaviours of cancer cells, therefore it could be utilised in the clinic as a potential therapeutic biomarker or target. The elucidation of the generation and regulation of non-B-Ig will certainly broaden our understanding of immunology.

Keywords

Ig · Immunoglobulin (Ig) · Non-B cell-derived Ig (non-B-Ig) · V(D)J recombination · Post-translational modifications · Somatic hypermutation · Class switching · Antigen binding

6.1 The Discovery of Immunoglobulins

The roots of the term 'antibody' can be traced back to the end of the nineteenth century (Lindenmann 1984). In 1890 Shibasaburo Kitasato and Emil von Behring convincingly demonstrated that the serum of guinea pigs immunised with heat-treated diphtheria toxin was capable of preventing the harmful effects of diphtheria toxin. The German term 'antitoxisch' (anti-toxin) was proposed for this activity (Von Behring 1890). A year later, Ehrlich (1891) was the first to use the description 'antikörper' (antibody). Subsequently, in 1900, Ehrlich put forward his side-chain theory of antibody production in which he proposed that cells would make a

X. Xu (✉) · J. Huang · X. Qiu
Department of Immunology, School of Basic Medical Sciences, Peking University, Beijing, China
e-mail: xiaojunxu@bjmu.edu.cn; huangjing82@bjmu.edu.cn; qiuxy@bjmu.edu.cn

P. J. Delves
Division of Infection and Immunity, Department of Immunology, UCL (University College London), London, UK
e-mail: p.delves@ucl.ac.uk

W. Shao
Academy of Medical Engineering and Translational Medicine, Tianjin University, Tianjin, China
e-mail: wenwei.shao@tju.edu.cn

X. Qiu et al. (eds.), *Non B cell-Derived Immunoglobulins*, Advances in Experimental Medicine and Biology 1445, https://doi.org/10.1007/978-981-97-0511-5_6

large variety of receptors on their cell surface capable of binding foreign antigens (Ehrlich 1900). Although this theory was not entirely accurate (it became clear much later that circulating antibodies of different specificities are produced by different individual B cells), the fundamentals of antibody production and specificity were therefore established by the start of the twentieth century. Subsequent biochemical characterisation of the molecules responsible for protection against toxins and other antigens revealed that they were globulins and thus were designated as immune globulins (immunoglobulins) (Tiselius and Kabat 1938).

Following on from these earlier observations, Astrid Fagraeus demonstrated that antibodies are produced by plasma cells (Fagraeus 1947). However, at that time, it was not known that plasma cells are derived from lymphocytes. A few years later it was observed that lymphocytes could be divided into Bursa of Fabricius-derived 'B cells' (Glick et al. 1956) and thymus-derived 'T-cells' (Miller 1961). Subsequently, James ('Jim') Gowans identified lymphocytes as the source of both humoral and cellular immunity (Gowans et al. 1962). Removal of the Bursa of Fabricius from chickens completely abolishes their ability to produce both plasma cells and detectable levels of circulating antibodies (Cooper et al. 1965). However, mammals lack a Bursa of Fabricius, so further studies were required to identify bone marrow as the source of B cells in mammals (Miller and Mitchell 1968). Bone marrow was proved to be the source of mammalian B cells later. There are two major populations of B cells, designated B1 and B2 cells (Kantor 1991), both of which possess a transmembrane form of immunoglobulin which functions as the B cell receptor (BCR) for antigen. B cells with BCRs of appropriate specificity are selected by antigen in a process referred to as clonal selection (Burnet 1976). Upon activation through the BCR, and with additional intracellular signals generated from costimulatory molecules such as CD19 (Callard et al. 1992), B cells differentiate into plasma cells which secrete a soluble version of the BCR, i.e. antibody.

It was not until the end of the twentieth century that it was reported that immunoglobulins can also be produced by cells other than B cells (i.e. non-B cells). In 1996, Qiu et al. observed the existence of 'immunoglobulin (Ig)-like' protein in non-B cancer cells from human tissue samples (Qiu and Yang 1996). Qiu, together with several other groups, subsequently established both the existence of non-B-Ig in epithelial cells, epidermal cells, mesangial cells, neurons, spermatogenic cells, podocytes, monocytes, and neutrophils (Huang et al. 2014), and in immature myeloblasts from acute myeloid leukaemia patients (Qiu et al. 2013; Wang et al. 2015), and the production of Ig by a variety of tumour cells including those from cancers of the breast (Kang et al. 2012; Ma et al. 2013; Welinder et al. 2013; Yang et al. 2013), colon (Hu et al. 2012; Niu et al. 2012), cervix (Niu et al. 2012; Chen and Gu 2007), lung (Chen and Gu 2007; Xiao et al. 2005; Liu et al. 2015), larynx (Wang et al. 2013), nasopharynx (Hu et al. 2011), pancreas (Li et al. 2011; Wan et al. 2015; Cui et al. 2020a), liver (Hu et al. 2012; Chen and Gu 2007), prostate (Liu et al. 2012; Pan et al. 2013; Qin et al. 2020), mouth (Zhu et al. 2008; Wang and Gan 2021), thyroid (Qiu et al. 2012), parathyroid (Cui et al. 2020b), oesophagus (Zhang et al. 2012), kidney (Sheng et al. 2016), and bladder (Liang et al. 2013). Moreover, Ig production has also been demonstrated in immune-privileged sites (Dunn et al. 1995; Hedger and Hettiarachchi 1994; Yoshimi et al. 2002; Naegele et al. 1991; Kaur et al. 2017; Niu et al. 2011). However, although some non-B-Ig was found localised on the cell membrane and formed a BCR-like structure with Igα and Igβ to mediate the activation of non-B cells (Hu et al. 2012), different from the subcellular localisation of B-Ig, most non-B-Ig was not localised on the cell membrane, but widely existed within the cells, such as in the cytoplasm, nucleus, and extracellular matrix (Shao et al. 2016; Tang et al. 2018). The discovery of non-B-Igs and their wide spectrum of distribution may extend our understanding of the functions of Ig and reveal previously unknown roles in pathology.

6.2 Immunoglobulin Biochemistry

Biochemical studies on antibody molecules by Porter (1950, 1958) using the proteolytic enzyme papain demonstrated that they could be split into two fragments. Protein crystals could be produced from some of these fragments and consequently, this component of the antibody molecule became known as the 'Fragment Crystallisable (Fc)'. To crystallise a protein, the preparation must be relatively homogeneous; a mixture of different proteins cannot readily be crystallised. This suggested that the Fc fragments were constant in nature. However, unlike intact antibody molecules, these fragments were unable to bind antigen. In addition to the Fc fragments, other cleavage products were produced which were capable of binding antigen but were not amenable to crystallisation due to their variable nature; these became known as 'Fragment Antigen-Binding (Fab)'. Around the same time as Porter's studies, using reducing agents such as urea, Gerald Edelman revealed that antibodies are multi-chain molecules with disulphide bonds linking together the different chains (Edelman and Poulik 1961; Edelman and Gally 1964). Thus, it was established biochemically that the fundamental structure of an immunoglobulin was a tetramer of four polypeptide chains with each chain containing an N-terminal antigen-binding variable (V) region and a C-terminal non-antigen-binding constant (C) region. Two chains were identical to each other and relatively large, hence referred to as heavy (H) chains. The other two chains were also identical to each other but were about half of H chains in size and therefore designated as light (L) chains. The two H chains are linked together by disulphide bonds via cysteine residues, and each H chain is also linked to one of the L chains via one disulphide bond, respectively. This produces the classic 4-chain structure of the immunoglobulin molecule with inter-chain disulphide bonds.

Protein crystals including both the Fc and the Fab were also found from non-B-Ig treated with papain. Non-B-Ig also presents a 4-peptide-chain structure, including two identical H chains and two identical L chains, both H chain and L chain contain an N-terminal antigen-binding variable (V) region and a C-terminal non-antigen-binding constant (C) region (Wang et al. 2020). Similar to conventional B-Ig, the light chains of non-B-Ig are also connected with the heavy chains, or the two heavy chains are connected by disulphide bonds (Qiu et al. 2003). Interestingly, it is also noteworthy that non-B cells often express free Ig light chains (FLC), especially Igκ. FLC is hydrophobic, can exist in the form of monomer, dimer or polymer, and combines with integrin beta1 on the cell surface to promote cell adhesion and migration, particularly to promote the growth and metastasis of tumour, without exerting antibody activity (Wang et al. 2022).

6.3 The Immunoglobulin Heavy and Light Chains

There are five different major classes of antibody based upon the constant region sequences of the heavy chains; IgG, IgA, IgM, IgD, and IgE, and their H chains are termed by the lower case Greek letters γ, α, μ, δ, and ε. IgG has four (IgG1, IgG2, IgG3, and IgG4 in humans; IgG1, IgG2a, IgG2b, and IgG3 in mice), and IgA has two (IgA1 and IgA2), subclasses; IgM, IgD, and IgE do not have subclasses. The five different main classes have overlapping but distinct functions.

The two heavy (H) chains are approximately 55 kDa (450 amino acids for IgG, IgA, and IgD) or 70 kDa (550 amino acids for IgM and IgE) each. The number of disulphide bonds linking the two H chains together varies between the different main classes and subclasses of Ig. There is a single inter-heavy-chain disulphide bond in IgM, IgD, and IgE, two inter-heavy-chain disulphide bonds in IgG1, IgG4, IgA1, and IgA2, four in IgG2, and 11 in IgG3. The IgG and IgA heavy chains contain a region of relative flexibility referred to as the hinge region, allowing these Ig molecules to adopt a range of configurations including the Y-shaped structure often depicted in textbooks.

The two light chains are approximately 25 kDa (211–217 amino acids) each. Their constant region can be either of two different types; kappa (κ) or lambda (λ). As already outlined, in a single antibody molecule the two light chains have an identical polypeptide sequence and therefore both light chains will have either a κ or λ constant region, never both simultaneously. The variable region will also be identical between the two light chains in a single antibody molecule. In human antibodies κ light chains are used twice as frequently as λ type ones.

After recent 20 years' study, it has been well established, by Qiu's and other groups, that non-B cells can also express five classes of Ig heavy chains and two classes of Ig light chains, showing a cell type-specific pattern. For example, normal colonic epithelial cells can express IgA, IgG, IgM, and IgD (Geng et al. 2019); normal pulmonary epithelial cells mainly express IgA and IgM (Gao et al. 2021) and normal renal tubular epithelial cells mainly express IgG (Deng et al. 2021); normal cardiac myocytes express IgG, IgA, IgM, and IgE. Interestingly, malignant tumour cells of any cell lineage origin highly express IgG rather than other types of Ig (Cui et al. 2021). Similar to B-Ig light chain, it mainly expresses Igα and little Igλ, but the Igλ will increase significantly in chronic inflammation.

6.4 Post-translational Modifications

All antibodies are glycoproteins, comprising carbohydrate as well as amino acids, and glycosylation can influence antibody function (Jennewein and Alter 2017). In all four subclasses of IgG, carbohydrate is enzymatically attached by an N-linkage to asparagine (Asn) 297 in the CH2 domain of the heavy chain. Varying numbers of N-linked glycans are also present in all the other classes of antibody, including both subclasses of IgA. In addition, O-linked glycans are present in IgD and IgA1 (van de Bovenkamp et al. 2016). Furthermore, there may be glycosylation of the variable regions of the heavy and/or light chains, e.g. in 15–25% of IgG molecules. A small number of germline V genes possess a consensus sequence for the addition of N-linked carbohydrates but, much more commonly, a glycosylation consensus sequence is generated in other germline genes during somatic hypermutation (van de Bovenkamp et al. 2018).

The carbohydrate sequence on the two heavy chains within a single antibody molecule can differ, as can the sequence between the two light chains if their variable regions are glycosylated. Therefore, the mantra of 'two identical heavy chains, two identical light chains' only holds true for the polypeptide sequence. Other post-translational modifications such as glycation (non-enzymatic glycosylation) and disulphide bond modification may also be present in antibody molecules (Hmiel et al. 2015).

In recent years, an important discovery is that IgG produced by epithelial tumour cells exhibits a unique glycosylation modification. A new N-glycosylation modification appears at site Asn162 of CH1, and its glycan end is highly sialylated. This modification does not exist in IgG produced by B cells, normal epithelial cells and tumour cells other than those of epithelial origin. Moreover, IgG produced by epithelial tumour cells depends on its sialylation modification. On one hand, it promotes the biological behaviour of tumour stem cells, such as tumour initiation, drug resistance, and metastasis (Tang et al. 2018; Liao et al. 2015). On the other hand, it can promote tumour immune escape by inhibiting T cell proliferation (Wang et al. 2020).

IgA nephropathy (IgAN) is the most common primary glomerulonephritis in the world, which is characterised by IgA–IgG glomerular immune deposition. These immune deposits are believed to originate from immune complexes formed in the circulation. Currently, IgA containing an abnormal glycoform (galactose-deficient IgA1, Gd IgA1) is considered to be the initiating factor for the formation of circulating immune complexes, as it is recognised by IgG autoantibodies and then forms pathogenic IgA1–IgG immune complexes depositing in the glomerular mesangial region. However, Deng et al. found that normal renal mesangial cells can produce high-level galactose deficient IgA1 which is different from the IgA produced by normal B cells (Deng et al.

2018). Glycosylation characteristics of non-B-Ig still await further analysis.

6.5 Polymerised Forms of Immunoglobulin

Although the basic four-chain structure of the immunoglobulin molecule is biochemically a tetramer, this basic antibody unit is inclined to be referred to as a monomer by immunologists. These monomers polymerise, which is facilitated by special structures such as the tailpiece, an 18 amino acid extension in the C-terminus of the heavy chain of both IgM and IgA (Pasalic et al. 2017). In polymerised IgM and IgA molecules, all antibody units have the same polypeptide sequence because they are produced by the same plasma cell. IgM molecules that are released into the circulation are usually in the form of pentamers with 5 units of IgM being associated with a protein called J chain. The J chain is replaced by a sixth IgM molecule to form a hexamer in smaller amounts under the circumstance when the J chain is insufficient, and is also released into the circulation (Pasalic et al. 2017). At mucosal surfaces, IgA is released from plasma cells as a J chain-associated dimer (secretory IgA) (Brandtzaeg 2013), although IgA released into the blood circulation is not polymerised and therefore is present as a monomer, like IgG, IgD, and IgE.

The polymerised IgM and IgA could be transported across mucosal epithelial cells, which involves a type of Fc receptor called the poly-IgR (Turula and Wobus 2018). The receptor becomes cleaved as a mechanism of releasing the antibody onto the mucosal surface, with a fragment of the receptor still attached to the antibody. This fragment is referred to as the secretory component.

It has been demonstrated that non-B-Ig also displays the basic four-chain structure. Both IgM and IgA, especially high level of IgA in the bronchoalveolar lavage fluid (BALF) of B cell-deficient mice, have been found to be secreted into skin and mucosa and exert natural antibody activity. However, at present, it has not been identified whether polymers of non-B-Ig, such as IgM pentamer, IgA dimer, or any other polymers, exist or not.

6.6 The Antigen-Binding Site

Within the variable domain of both the heavy and the light chains, there are three areas of sequence that have the greatest variability, which are referred to as the hypervariable regions (HVR1, HVR2, HVR3) and largely correspond to the location of the variable region amino acids that are complementary to and directly interact with an antigen. These hypervariable sequences are situated within the loops that connect the strands of the Ig domains and constitute the complementarity-determining regions (CDR1, CDR2, and CDR3) which together form the antigen-binding site of an antibody molecule (Wu and Kabat 1970).

Most antibodies utilise all six CDRs (three from VH, three from VL) in their interaction with antigen, although sometimes not all six CDRs are employed. The amino acids in the CDRs interact with molecular structures on the antigen, which are referred to as the epitope, although only 20–33% of the amino acids in the CDRs actually contact the epitope (Padlan 1994). CDR3 is the longest and most diverse one of the three CDRs and therefore usually the most important one with regard to antigen binding (Shi et al. 2014). The epitope that is recognised by the antigen-binding site has an area of approximately 20 Å × 30 Å (Davies et al. 1990) and, for protein antigens, typically involves approximately 18 or 19 amino acids (Stave and Lindpaintner 2013). It has been estimated that over 90% of protein epitopes recognised by antibodies are discontinuous in that the contact residues only come close together once the protein folds into its tertiary structure (Walter 1986). These discontinuous epitopes are sometimes referred to as conformational epitopes, although this terminology is not optimal since all chemicals have a conformation. The remaining <10% of protein epitopes recognised by antibodies are continuous, in other words composed of amino acids that are adjacent to each other in the primary sequence of the polypeptide. In addition to protein antigens, carbohydrates, lipids, nucleic acids, or indeed virtually any other chemical structure can be recognised by antibodies of appropriate specificity.

Antibody binding to antigen relies entirely on non-covalent interactions—a combination of hydrophobic, van der Waals, electrostatic, and hydrogen bonds. Thus, antibody–antigen interactions are reversible and obey the Law of Mass Action. The strength of binding is referred to as the affinity and antibodies typically have affinities, expressed as equilibrium dissociation constants (KD), ranging from 10^{-6} to 10^{-12}. An antibody with a KD of 10^{-9} is considered to be of high affinity, and therefore an antibody with a KD of 10^{-12} would be of very high affinity. The binding of one antigen-binding site (one Fab arm) to its epitope on the antigen is referred to as the intrinsic affinity. Evolution has, however, provided antibodies with at least two physically linked Fab arms and therein two antigen-binding sites. Pathogens (whether they are viruses, bacteria, fungi, protozoa, or multicellular parasites) virtually always possess physically linked multiple copies of each antigen on their surfaces. This, particularly given the flexibility of IgG and IgA molecules due to the hinge region, permits the antibody to simultaneously employ both its antigen-binding sites and therefore effectively to own two valencies (i.e. is bivalent) when encountering pathogens, and thereby greatly enhances the overall strength of antibody–antigen binding and this total functional affinity is called avidity. This enhancement becomes even more dramatic in polymerised antibodies, e.g. dimeric secretory IgA possesses four antigen-binding arms (i.e. it is tetravalent) and pentameric circulating IgM with the potential to use all ten of its antigen-binding sites (decavalent). Indeed, in cases where conformational considerations permit IgM to utilise all ten antigen-binding sites, the binding strength to antigen is ten million times higher than that of any single antigen binding site of the same antibody molecule.

When the amino acid sequence of non-B-Ig was analysed, it was found that, consistent with B-Ig, the non-B-Ig variable region also contains three hypervariable regions and four frame regions, among which the three CDRs (CDR1, CDR2, and CDR3) are the key domains binding to antigens (or interacting proteins). A free light chain of Vκ4–1/Jκ3 (Vκ4–1/Jκ3-FLC) is widely expressed in various non-B cells, especially in secretory form at a high level in cancer cells. The secretory Vκ4–1/Jκ3-FLC can bind to integrin β1, and promote the activation of FAK pathway, cell migration, and tumour cell metastasis. Wang et al. further analysed the binding domain between the Vκ4–1/Jκ3-FLC and integrin β1. After mutating the three CDR1, CDR2, and CDR3, respectively, they found that CDR1 and CDR2, but not the CDR3 on which B-Ig mainly depends in binding antigen, are the key domains for the Vκ4–1/Jκ3-FLC binding to integrin β1. Using surface plasmon resonance (SPR) experiment, Wang et al. also found that the binding affinity between the Vκ4–1/Jκ3-FLC and integrin β1 was at a medium-high level, with a KD value of 10^{-8} (Wang et al. 2022).

6.7 Genetic Recombination Generates Antibody Diversity

Cloning and sequencing of immunoglobulin (Ig) genes identified variable (V), diverse (D), joining (J), and constant (C) gene fragments recombined in B cells to encode antibody molecules (Tonegawa et al. 1977). V gene sequences, some of which are pseudogenes, can be divided into different families according to sequence homology. The Ig gene and the gene encoding the T cell receptor (TCR) are unique because a plethora of recombinations from about 200 gene sequences could be generated, which can encode millions of different polypeptide sequences. The recombination is mediated by many enzymes, including those encoded by recombinant activator genes 1 and 2 (RAG1 and RAG2 protein). The substrates of RAG1 and RAG2 enzymes are nucleotide recombinant signal sequences (RSS) located on both sides of V, D, and J genes. These recombination events are necessary for the subsequent transcription of both heavy and light peptide chains of full-length Ig.

According to the current clonal selection theory, due to the random selection of VDJ recombination, each B cell-derived-Ig has its unique variable region sequence, resulting in an Ig library of infinite varieties for each individual. Similarly, the non-B-Ig variable region is also encoded by the Ig gene after V(D)J rearrangement. However, the non-B-Ig variable region does not show infinite diversity. In spite of cell lineage source, the non-B-Ig heavy chain or light chain variable region shows limited V(D)J rearrangement patterns which are cell lineage-specific, even identical V(D)J rearrangement pattern can be shared among different individuals (Wang et al. 2015; Zheng et al. 2009; Zhang et al. 2021).

The RAG1 and RAG2 proteins, which were previously considered to be lymphoid-specific, have also been detected in cancer cells (Chen and Gu 2007; Qiu et al. 2003). Further study suggested that the VHDJH transcripts of cancer-derived Ig share some features with those of B-Ig, such as V(D)J recombination, N-region insertion, and a high mutation rate. However, it was unexpected to find that functional V(D)J rearrangement patterns display in either RAG1 or RAG2 knockout mice, which suggested that V(D)J rearrangement in non-B cells does not depend on RAG1 and RAG2. Up to date, the mechanism of Ig gene rearrangement in non-B cells remains unclear.

6.8 Class Switching

B cells initially produce the IgM and the IgD which function as the BCR on naive B cells. In order to switch to IgG, IgA or IgE production after B cells are stimulated by antigens, the CD40 molecule on the cell surface of B cells needs to be engaged by the CD40 ligand (CD40L, CD154) molecule on the cell surface of helper T cells (Th) (Elgueta et al. 2009). Different populations of helper T cells (e.g. Th1, Th2, Th9, Th17, Th22, or Tfh [T follicular helper]) release cytokines which determine the class of antibody the B cell should switch to (Stavnezer and Schrader 2014). This class switching requires the enzyme activation-induced cytidine deaminase (AID) which, acting together with other molecules, mediates the recombination of the previously rearranged VDJ gene segment from one C gene to another (Leeman-Neill et al. 2018). This class switch recombination is facilitated by switch (S) nucleotide motifs that lie on the 5′ end of each CH gene sequence (except for Cδ, thereby providing an explanation for why IgM and IgD are co-expressed as the BCR on the cell surface of naive B cells).

It has been found that non-B cells can express all the 5 classes of Igs, i.e. IgM, IgG, IgA, IgD, and IgE (Shao et al. 2016; Tonegawa et al. 1977; Huang et al. 2008, 2009). However, unlike B cells, which initially produce the IgM and the IgD in response to antigen stimulation and switch to IgG, IgA, or IgE production, non-B cells initially only produce IgG, IgA, or both of IgA and IgM, but rarely produce both IgM and IgD simultaneously. In addition, different Ig heavy chains with entirely different variable region sequences were often found expressed by the same non-B cell, which is not consistent with the class switching theory in B cells (Zheng et al. 2009). It is noteworthy that Shi et al. have reported that the variable regions of the same B cell are not the same even if the same B cell expresses different class heavy chains, suggesting that the classical class switching theory needs to be modified (Shi et al. 2019).

6.9 Somatic Hypermutation

It is well accepted that antibodies are highly antigen-specific, although this statement is generally correct, it warrants further elaboration. In particular, IgM antibodies utilising VH genes that have not undergone somatic hypermutation usually exhibit a certain degree of specificity. Many natural antibodies produced by B1 cells appear to be encoded by germline VH gene sequences without specific antigen stimulation, binding to numerous different common pathogens. In this case, the binding of bivalent IgM and IgD of BCR to antigen may demonstrate a relatively low affinity during clonal selection, but its strength is sufficient to activate B cells. This induces cell division and subsequent somatic hypermutation, leading to the affinity maturation of the response (Mishra and Mariuzza 2018). With further clonal selection, this leads to the reaction focusing on the stimulating antigens that produce IgG, IgA, or IgE antibodies conversion with higher affinity. Therefore, the V region genes of IgG, IgA, or IgE produced by B cells often demonstrate somatic high-frequency mutations, while the V region genes of IgM and IgD often do not show mutations. During somatic hypermutation, nucleotide substitution occurs in CDR, giving rise to the nonsense mutation, the silent mutation, the change of amino acid has no or adverse effect on the affinity for antigen binding, or more importantly, some mutations rendering BCRs with enhanced antigenic affinity. B cells bearing high-affinity BCRs will preferentially compete for antigens during clonal selection, dominating the reaction.

Having analysed thousands of variable region sequences of non-B-Ig from mice and humans, Qiu's group discovered that, similar to B-Ig, the variable region of non-B-IgM is rarely mutated, but it is extraordinary to notice that non-B-IgG, non-B-IgA, or non-B-IgE demonstrate high-frequency mutations even under steady-state conditions. In particular, non-B-IgG, non-B-IgA, or non-B-IgE from B cell-deficient mice also exhibited high-frequency mutations. However, unlike the somatic hypermutation of B-Ig, the hot spots of the mutation are not entirely confined to the CDRs region. When non-B cells were stimulated with antigens, non-B-IgG variable regions also showed low-frequency mutations. The mechanism of high-frequency mutation in non-B-Ig variable region requires further exploration.

AID had also been identified in breast cancer cell lines for mutational and switch activity, indicating the existence of either a de novo rearrangement and modification of VH genes in carcinoma cells or an assimilation of lymphocyte-derived chromatin. In breast cancer cell lines, constitutive AID activation potentially contributes to the aberrant mutational activity; somatic mutations of VH genes without intraclonal variation, suggesting that the mutational activity ceased; both pre- and post-isotype switched VH transcripts as well as Iγ-Cγ sterile transcripts were detectable (Babbage et al. 2006).

6.10 Switching from BCR to Secreted Immunoglobulin

Following antigen binding, B cells become activated and differentiate into the antibody-secreting plasma cell. The change from the transmembrane immunoglobulin (BCR) to the secreted (soluble) version is brought about by alternative splicing which results in the loss of a hydrophobic transmembrane sequence (Rogers et al. 1981). Without this transmembrane sequence, the antibody molecule is released into the surrounding interstitial fluid and distributes throughout the body once it reaches the surface of the plasma cell.

In contrast to B-Ig, non-B-Ig is typically localised in the cytoplasm or nucleus, and exists primary in a membrane-binding form on the cell membrane (such as binding to integrin). When cells are exposed to hypoxia or stimulated by cell metabolites, a large number of non-B-Ig will be released to the outside of cells, or secreted in soluble form, or deposited in hydrophobic form. For example, Deng et al. discovered that a large quantity of IgA was secreted into the extracellular matrix after renal mesangial cells were stimulated by bacterial entertoxin (Deng et al. 2018). Hu et al. demonstrated that IgM was secreted into the culture supernatant in large quantities and exerted natural antibody activity after HeLa cells were stimulated by TLR9 agonist (CPG), (Hu et al. 2012). However, the mechanism of non-B-Ig switching from intracellular localisation to secretion still needs to be investigated.

6.11 Ig Gene Expression Regulation

The Ig expression is controlled at the DNA, RNA, and protein level, initially producing transmembrane antibody on the surface of B cells and subsequent soluble antibody released from plasma cells. Gene transcription after the recombination process involves multiple regulatory elements present within the Ig gene loci, and those elements of most importance are discussed here.

The transcription factor Pax5, expressed in B cells, CNS, and testes, plays a key role in directing lymphoid progenitors to differentiate into B cells, and therefore maintains expression throughout B cell development and maturation until the phase of differentiation into plasma cells (Gibson et al. 2006). In addition to directing the expression of B cell-specific genes such as CD19, CD79a (Igα), and Blk, Pax5 transactivates RAG-mediated V to DJ recombination in the Ig H chain locus (Zhang et al. 2006) is required for the class switch recombination that produces the IgG, IgA, and IgE classes of antibody (Calderón et al. 2021).

OCT-2, another transcription factor that is also present in both the CNS and B cells, binds to octamer motifs within the Ig gene promoter sequences where it interacts with the B cell-specific coactivator BOB.1 to activate Ig gene transcription. BOB.1 appears to play an important role in mediating the formation of germinal centres following antigen-induced B cell activation. Consequently, homozygous deletion of BOB.1 in mice prevents the formation of germinal centres and thereby severely limits B cell differentiation.

Protein antigens are particularly associated with the production of the IgG1 and IgG3 classes of antibody, but continuous stimulation with these antigens can lead to class switch to IgG4. Class switch recombination is partly controlled by signals transmitted from cytokine receptors binding their partner cytokines. For example, IL-4 directs switching to the C gene, subsequently producing IgE, whereas IL-10 is involved in class switching to the Cγ1 and Cγ3 genes, giving rise to IgG1 and IgG3. Further examples include a role for TGF-β in class switch recombination to Cα with subsequent production of the IgA class of antibody, and a role for gamma interferon in the class switching (in the mouse) that is associated with IgG2a and IgG3 production (Malisan et al. 1996). Thymic stromal lymphopoietin, a cytokine which is released by damaged epithelial cells in patients with allergy, can activate populations of dendritic cells that direct the differentiation of Th2 cells and enhance the production of IgE by secreting IL-4, IL-9, and IL-13 subsequently (Wu et al. 2017). Recent studies have also demonstrated that infection with pathogens may be involved in regulating the expression of cancer-derived Ig. Furthermore, cytokines such as TGF-β and TNF-α have been found to regulate Ig expression in cancer cells in a dose-dependent manner (Duan et al. 2014, 2016).

As early as 2010, Zhu et al. reported a mechanism regulating the transcription of the Ig gene heavy chain promoter region in lung cancer and colon cancer cell lines, and the key transcription factor involved in the regulation was Oct-1 (Zhu et al. 2010), which was also found to participate in the regulation of IgA heavy chain expression in nasopharyngeal carcinoma by Cao et al. in 2014. Recently, Huang et al. discovered that epithelial tumour stem cells widely produced a type of IgG with a unique sialylation modification, and its transcription is regulated by stem cell transcription factors Sox2 and Oct-4 (Huang et al. 2023).

6.12 Function of Immunoglobulins

Antibodies are bifunctional molecules, with the variable regions of the heavy and light chains mediating the recognition of a specific antigen, and the constant region of the heavy chain linking the antibody to other components of the immune system.

Antibodies can be found in two forms: either associated with cell surfaces (transmembrane) or secreted (soluble). The transmembrane form of an antibody acts as the B cell receptor (BCR) on B cells. Naïve B cells have a BCR composed of IgM and IgD molecules with identical antigen specificity which is achieved through the use of different heavy chain constant regions, whilst the heavy chain variable regions and light chains remain identical. On the other hand, secreted (soluble) antibodies can attach to cells via their heavy chain constant region, which can be recognised by Fc receptors present on the cell surface of a variety of cell types (Hogarth 2015). Alternatively, secreted antibodies can specifically recognise a cell surface antigen through their antigen-binding site.

Secreted antibodies play a crucial role in the immune response by neutralising bacterial toxins, inhibiting the entry of viruses, and preventing pathogenic bacteria colonisation of body surfaces. However, they often serve to focus other components of the immune response, such as complement and Fc-receptor-bearing cells, onto the pathogen. When an antibody binds to an antigen, it can activate the complement system via the classical pathway, facilitate phagocytosis, promote eosinophil activation, stimulate the activity of natural killer (NK) cells and other cells for antibody-dependent cellular cytotoxicity (ADCC), and bind as immune complexes to red blood cells (RBCs). Moreover, antibody binding can activate platelets, endothelial cells, and epithelial cells. In particular, IgE antibodies can sensitise tissue mast cells and circulating basophils for the release of inflammatory mediators in response to an antigen trigger, leading to an allergy (Bulfone-Paus et al. 2017).

While B cell-derived Ig is a crucial component of humoral immunity, non-B cell-derived Ig serves a range of functions that display diversity and versatility. These functions include natural antibody activity, maintaining cell survival, and involvement in oncogenesis and tumour progression. Several non-B-Ig molecules expressed in normal human epidermal cells, such as IgG and IgA, show potential natural antibody activity and might participate in skin innate immunity (Jiang et al. 2015). Additionally, natural IgM produced by epithelial cells and macrophages may be involved in the innate immune response to pathogen infection (Hu et al. 2012; Gong et al. 2021). Podocyte-derived IgG maintains podocyte viability and adhesive capabilities, whereas IgA derived from primary human glomerular mesangial cells promotes cell adhesion and inhibits cell apoptosis (Jing et al. 2018). Moreover, hepatocyte-derived Igκ is critical for hepatocyte survival and exerts a protective effect against ConA-induced liver injury in mice (Yin et al. 2020).

It is important to note that non-B cell-derived Ig plays a multifaceted role in tumour development, acting as a growth factor that directly promotes malignant behaviour such as cell proliferation, migration, invasion, and resistance to apoptosis (Qiu et al. 2013; Liu et al. 2015; Cui et al. 2020a; Zheng et al. 2007). Cancer cells with high IgG expression were also found to display cancer stem cell-like properties, including expression of CD44v6, high sphere-forming capability, and resistance to chemotherapy (Liao et al. 2015). Non-B cell-derived IgG can also interact with many membrane proteins involved in cell–cell adhesion, focal adhesion, and hemidesmosomes. In addition, it was found that IgG with a unique sialylated modification is specifically expressed in epithelial cancer cells, which can significantly inhibit T cell proliferation (Wang et al. 2020). This sialylated IgG can also dampen the cytotoxic activity of NK cells by inhibiting the ADCC function of NK cells (Scallon et al. 2007).

6.13 Perspectives

The discovery of non-B-Ig has greatly expanded our understanding of immunology and opened up vast possibilities for potential clinical applications. A comparison of B Ig and non-B-Ig is listed in Table 6.1. Although non-B-Igs share the same symmetrical four peptide structures with conventional Igs, their distinctive characteristics, such as restricted variable region sequences, aberrant glycosylation, and protumorigenic function, make them a possible target for cancer therapy.

The exploration of the mechanism of genetic regulation of non-B-Ig will undoubtedly broaden our understanding of immunology and lead to new therapeutic approaches. Further research in this field could have important implications for developing targeted cancer therapies and improving patient outcomes.

Table 6.1 Comparison of B-Ig and non-B-Ig

	B-Ig	Non-B-Ig
The discovery	Termed as 'antitoxisch' in 1890	
	Designated as immunoglobulin by biochemical characterisation in 1938	
	Produced by plasma cells (1947)	'Ig-like' protein reported in non-B cancer cell (1996)
	Bursa of Fabricius derived lymphocytes as the source of humoral immunity (1965)	Found in a variety of tumour cells (twenty-first century)
	Bone marrow as the source of B cells in mammals (1968)	Detected in immune-privileged sites and normal cells: Epithelial cells, epidermal cells, mesangial cells, neurons, spermatogenic cells, podocytes, monocytes, neutrophils, and immature myeloblasts
	B1 and B2 (1991)	
Biochemistry	4-chain structure with inter-chain disulfide bonds: 2 H chain and 2 L chain, containing V and C region for each	Four-chain structure with inter-chain disulfide bonds: 2 H chain and 2 L chain, containing V and C region for each
		Free Ig light chains (FLC), especially Igκ
Heavy and light chains	Five different classes: IgG, IgA, IgM, IgD, and IgE	Five different classes: IgG, IgA, IgM, IgD, and IgE
	Heavy chain is approximately 450 amino acids for IgG, IgA and IgD, or 550 amino acids for IgM and IgE	IgG highly expressed in malignant tumour cells of any cell lineage-origin
	1 disulfide bond in IgM, IgD, and IgE, 2 disulfide bonds in IgG1, IgG4, IgA1, and IgA2, 4 disulfide bonds in IgG2, and 11 disulfide bonds in IgG3. Hinge region in IgG and IgA	IgA and IgM expressed in pulmonary epithelial cells; IgG expressed in renal tubular epithelial cells; IgA, IgG, IgM, and IgE expressed in cardiac myocytes; IgA, IgG, IgM, and IgD expressed in colonic epithelial cells
	κ light chains are used twice as frequently as λ light chains	It mainly expresses Igκ and little Igλ; Igλ increased significantly in chronic inflammation
	25 kDa for κ or λ light chain	
Post-translational modifications	N-linked glycans at Asn297 in the CH2 domain of the heavy chain of IgG and IgA	N-glycosylation and sialylation modification at site Asn162 of CH1 of IgG produced by epithelial tumour cells
	O-linked glycans in IgD and IgA1 heavy chain	Galactose-deficient IgA1 in normal renal mesangial cells
	Glycosylation in variable regions (15–25% of IgG molecules)	
Polymerised forms	It forms pentamer for IgM and dimer for IgA through J chain, while IgD, IgG, and IgE form monomers	It has not been identified whether polymers formed

(continued)

Table 6.1 (continued)

	B-Ig	Non-B-Ig
The antigen-binding site	CDR1, CDR2, and CDR3 together form the antigen-binding site	CDR1, CDR2, and CDR3 are key domains to binding
	CDR3 as the longest and most diverse CDR, is the most important with regard to antigen binding	CDR1 and CDR2, but not the CDR3, are the key domains for the binding of the Vκ4–1/Jκ3-FLC to integrin β1
	Affinities ranging from 10^{-6} to 10^{-12} KD	Affinity between the Vκ4–1/Jκ3-FLC and integrin β1 was 10^{-8} KD
	Non-covalent interactions: a combination of hydrophobic, van der Waals, electrostatic, and hydrogen bonding forces	
	90% of protein epitopes recognised by antibodies are discontinuous	
Genetic recombination and immunoglobulin diversity	RAG1 and RAG2 enzymes mediate V(D)J recombination	V(D)J rearrangement required
	Random selection of VDJ recombination	Limited V(D)J rearrangement patterns
		V(D)J rearrangement doesn't depend on either RAG1 or RAG2
Class switching	Both IgM and IgD are produced by naïve B cell	It is rare to produce IgM and IgD simultaneously
	Guided by helper T cells, B cells began to express IgG, IgA, or IgE after antigen stimulation	Different Ig heavy chains can be expressed by the same non-B cell with entirely different variable region sequences
	The same B cell can express different class heavy chains with diverse variable regions	
	AID mediate the class switching	
Somatic hypermutation	IgM antibodies utilise VH genes without somatic hypermutation	The variable region of non-B-IgM is rarely mutated
	The V region genes of IgG, IgA, or IgE produced by B cells often show somatic high-frequency mutations	Non-B-IgG, non-B-IgA or non-B-IgE have high-frequency mutations
	Somatic hypermutation occurs in CDR	Somatic hypermutation spots are not completely limited to the CDRs
		Non-B-IgG showed low-frequency mutations after antigen challenge
		AID identified in malignant epithelial cells raises a potential for inducing mutational activity
Switching from BCR to secreted immunoglobulin	Alternative splicing results in changing to secreted Ig from BCR after antigen binding	Non-B-Ig is usually localised in the cytoplasm or nucleus, and even on the cell membrane
		Non-B-Ig can be released to the outside of cells stimulated by cell metabolites or hypoxia
Function of immunoglobulins	BCR as a transmembrane form can bind to antigen specifically	Non-B-Ig shows natural antibody activity and participates in innate immunity
	Acting as neutralizing antibodies	The podocyte-derived IgG maintains podocyte viability and adhesive capabilities

(continued)

Table 6.1 (continued)

	B-Ig	Non-B-Ig
	Secreted antibodies can activate complement	IgA derived from primary human glomerular mesangial cells promotes cell adhesion and inhibits cell apoptosis
	Facilitate phagocytosis, promote eosinophil activation, ADCC	Hepatocyte-derived Igκ plays an important role in hepatocyte survival
	Bind as immune complexes to RBC, or activate platelets, endothelial cells, and epithelial cells	As a growth factor, it promotes tumour proliferation, migration, invasion, and resistance to apoptosis
		Interact with many membrane proteins involved in cell–cell adhesion, focal adhesion, and hemidesmosomes
		Epithelial cancer cell-derived IgG inhibits T cell proliferation and ADCC of NK

References

Babbage G et al (2006) Immunoglobulin heavy chain locus events and expression of activation-induced cytidine deaminase in epithelial breast cancer cell lines. Cancer Res 66(8):3996–4000

Brandtzaeg P (2013) Secretory IgA: designed for antimicrobial defense. Front Immunol 4:222

Bulfone-Paus S et al (2017) Positive and negative signals in mast cell activation. Trends Immunol 38(9):657–667

Burnet FM (1976) A modification of Jerne's theory of antibody production using the concept of clonal selection. CA Cancer J Clin 26(2):119–121

Calderón L et al (2021) Pax5 regulates B cell immunity by promoting PI3K signaling via PTEN down-regulation. Sci Immunol 6(61):eabg5003

Callard RE et al (1992) CD19 regulation of human B cell responses. B cell proliferation and antibody secretion are inhibited or enhanced by ligation of the CD19 surface glycoprotein depending on the stimulating signal used. J Immunol 148(10):2983–2987

Chen Z, Gu J (2007) Immunoglobulin G expression in carcinomas and cancer cell lines. FASEB J 21(11):2931–2938

Cooper MD et al (1965) Delineation of the thymic and bursal lymphoid systems in the chicken. Nature 205:143–146

Cui M et al (2020a) High expression of cancer-derived glycosylated immunoglobulin G predicts poor prognosis in pancreatic ductal adenocarcinoma. J Cancer 11(8):2213–2221

Cui M et al (2020b) Cancer-derived immunoglobulin G: a novel marker for differential diagnosis and relapse prediction in parathyroid carcinoma. Clin Endocrinol 92(5):461–467

Cui M et al (2021) Immunoglobulin expression in cancer cells and its critical roles in tumorigenesis. Front Immunol 12:613530

Davies DR, Padlan EA, Sheriff S (1990) Antibody–antigen complexes. Annu Rev Biochem 59:439–473

Deng H et al (2018) Expression of immunoglobulin A in human mesangial cells and its effects on cell apoptosis and adhesion. Mol Med Rep 17(4):5272–5282

Deng Z et al (2021) Expression of immunoglobulin G in human proximal tubular epithelial cells. Mol Med Rep 23(5):327

Duan Z et al (2014) Activation of the Ig Ialpha1 promoter by the transcription factor Ets-1 triggers Ig Ialpha1-Calpha1 germline transcription in epithelial cancer cells. Cell Mol Immunol 11(2):197–205

Duan Z et al (2016) AID expression increased by TNF-alpha is associated with class switch recombination of Igalpha gene in cancers. Cell Mol Immunol 13(4):484–491

Dunn JA et al (1995) Transient immunoglobulin-like molecules are present in the subplate zone and cerebral cortex during postnatal development. Cereb Cortex 5(6):494–505

Edelman GM, Gally JA (1964) A model for the 7S antibody molecule. Proc Natl Acad Sci USA 51:846–853

Edelman GM, Poulik MD (1961) Studies on structural units of the gamma-globulins. J Exp Med 113:861–884

Ehrlich P (1891) Experimentelle Untersuchungenuber Immunitat. II. Ueber Abrin. Dtsch Med Wochenschr 17:1218–1219

Ehrlich P (1900) On immunity with special reference to cell life. Proc R Soc Lond 66:424

Elgueta R et al (2009) Molecular mechanism and function of CD40/CD40L engagement in the immune system. Immunol Rev 229(1):152–172

Fagraeus A (1947) Plasma cellular reaction and its relation to the formation of antibodies in vitro. Nature 159(4041):499

Gao E et al (2021) Lung epithelial cells can produce antibodies participating in adaptive humoral immune responses. bioRxiv 2021:443498

Geng ZH et al (2019) Human colorectal cancer cells frequently express IgG and display unique Ig repertoire. World J Gastrointest Oncol 11(3):195–207

Gibson SE et al (2006) Expression of the B cell-associated transcription factors PAX5, OCT-2, and BOB.1 in acute myeloid leukemia: associations with B-cell antigen expression and myelomonocytic maturation. Am J Clin Pathol 126(6):916–924

Glick B, Chang TS, Jaap RG (1956) The bursa of Fabricius and antibody production. Poult Sci 35:224–225

Gong X et al (2021) Macrophage-derived immunoglobulin M inhibits inflammatory responses via modulating endoplasmic reticulum stress. Cell 10(11):2812

Gowans JL et al (1962) Initiation of immune responses by small lymphocytes. Nature 196:651–655

Hedger MP, Hettiarachchi S (1994) Measurement of immunoglobulin G levels in adult rat testicular interstitial fluid and serum. J Androl 15(6):583–590

Hmiel LK et al (2015) Post-translational structural modifications of immunoglobulin G and their effect on biological activity. Anal Bioanal Chem 407(1):79–94

Hogarth PM (2015) Fc receptors: introduction. Immunol Rev 268(1):1–5

Hu D et al (2011) Heterogeneity of aberrant immunoglobulin expression in cancer cells. Cell Mol Immunol 8(6):479–485

Hu F et al (2012) Spontaneous production of immunoglobulin M in human epithelial cancer cells. PLoS One 7(12):e51423

Huang J et al (2008) Expression of immunoglobulin gene with classical V–(D)–J rearrangement in mouse brain neurons. Int J Biochem Cell Biol 40(8):1604–1615

Huang J et al (2009) Expression of immunoglobulin gene with classical V–(D)–J rearrangement in mouse testis and epididymis. J Histochem Cytochem 57(4):339–349

Huang J et al (2014) Rearrangement and expression of the immunoglobulin mu-chain gene in human myeloid cells. Cell Mol Immunol 11(1):94–104

Huang X et al (2023) A self-propagating c-Met-SOX2 Axis drives cancer-derived IgG signaling that promotes lung cancer cell stemness. Cancer Res 83(11):1866–1882

Jennewein MF, Alter G (2017) The immunoregulatory roles of antibody glycosylation. Trends Immunol 38(5):358–372

Jiang D et al (2015) IgG and IgA with potential microbial-binding activity are expressed by normal human skin epidermal cells. Int J Mol Sci 16(2):2574–2590

Jing Z et al (2018) Expression of immunoglobulin G in human podocytes, and its role in cell viability and adhesion. Int J Mol Med 41(6):3296–3306

Kang S et al (2012) In situ identification and localization of IGHA2 in the breast tumor microenvironment by mass spectrometry. J Proteome Res 11(9):4567–4574

Kantor A (1991) A new nomenclature for B cells. Immunol Today 12(11):388

Kaur G et al (2017) An overview of a Sertoli cell transplantation model to study testis morphogenesis and the role of the Sertoli cells in immune privilege. Environ Epigenet 3(3):dvx012

Leeman-Neill RJ, Lim J, Basu U (2018) The common key to class-switch recombination and somatic hypermutation: discovery of AID and its role in antibody gene diversification. J Immunol 201(9):2527–2529

Li X et al (2011) The presence of IGHG1 in human pancreatic carcinomas is associated with immune evasion mechanisms. Pancreas 40(5):753–761

Liang PY et al (2013) Overexpression of immunoglobulin G prompts cell proliferation and inhibits cell apoptosis in human urothelial carcinoma. Tumour Biol 34(3):1783–1791

Liao Q et al (2015) Aberrant high expression of immunoglobulin G in epithelial stem/progenitor-like cells contributes to tumor initiation and metastasis. Oncotarget 6(37):40081–40094

Lindenmann J (1984) Origin of the terms 'antibody' and 'antigen'. Scand J Immunol 19(4):281–285

Liu Y et al (2012) IgG gene expression and its possible significance in prostate cancers. Prostate 72(6):690–701

Liu Y et al (2015) Binding of the monoclonal antibody RP215 to immunoglobulin G in metastatic lung adenocarcinomas is correlated with poor prognosis. Histopathology 67(5):645–653

Ma C et al (2013) Immunoglobulin G expression and its potential role in primary and metastatic breast cancers. Curr Mol Med 13(3):429–437

Malisan F et al (1996) Interleukin-10 induces immunoglobulin G isotype switch recombination in human CD40-activated naive B lymphocytes. J Exp Med 183(3):937–947

Miller JF (1961) Immunological function of the thymus. Lancet 2(7205):748–749

Miller JF, Mitchell GF (1968) Cell to cell interaction in the immune response. I. Hemolysin-forming cells in neonatally thymectomized mice reconstituted with thymus or thoracic duct lymphocytes. J Exp Med 128(4):801–820

Mishra AK, Mariuzza RA (2018) Insights into the structural basis of antibody affinity maturation from next-generation sequencing. Front Immunol 9:117

Naegele JR et al (1991) Expression of a unique 56-kDa polypeptide by neurons in the subplate zone of the developing cerebral cortex. Proc Natl Acad Sci USA 88(2):330–334

Niu N et al (2011) Expression and distribution of immunoglobulin G and its receptors in an immune privileged site: the eye. Cell Mol Life Sci 68(14):2481–2492

Niu N et al (2012) IgG expression in human colorectal cancer and its relationship to cancer cell behaviors. PLoS One 7(11):e47362

Padlan EA (1994) Anatomy of the antibody molecule. Mol Immunol 31(3):169–217

Pan B et al (2013) Suppression of IGHG1 gene expression by siRNA leads to growth inhibition and apoptosis induction in human prostate cancer cell. Mol Biol Rep 40(1):27–33

Pasalic D et al (2017) A peptide extension dictates IgM assembly. Proc Natl Acad Sci USA 114(41):E8575–E8584

Porter RR (1950) The formation of a specific inhibitor by hydrolysis of rabbit antiovalbumin. Biochem J 46(4):479–484

Porter RR (1958) Separation and isolation of fractions of rabbit gamma globulin containing the antibody and antigenic combining sites. Nature 182(4636):670–671

Qin C et al (2020) Cancer-driven IgG promotes the development of prostate cancer though the SOX2-CIgG pathway. Prostate 80(13):1134–1144

Qiu X, Yang G (1996) Existance of Ig-like protein in malignant tumor cells. J Norman Bethune Univ Sci 22:575–577

Qiu X et al (2003) Human epithelial cancers secrete immunoglobulin G with unidentified specificity to promote growth and survival of tumor cells. Cancer Res 63(19):6488–6495

Qiu Y et al (2012) Immunoglobulin G expression and its colocalization with complement proteins in papillary thyroid cancer. Mod Pathol 25(1):36–45

Qiu X et al (2013) Immunoglobulin gamma heavy chain gene with somatic hypermutation is frequently expressed in acute myeloid leukemia. Leukemia 27(1):92–99

Rogers J et al (1981) Gene segments encoding transmembrane carboxyl termini of immunoglobulin gamma chains. Cell 26:19–27

Scallon BJ et al (2007) Higher levels of sialylated Fc glycans in immunoglobulin G molecules can adversely impact functionality. Mol Immunol 44(7):1524–1534

Shao W et al (2016) Identification of liver epithelial cell-derived Ig expression in mu chain-deficient mice. Sci Rep 6:23669

Sheng Z et al (2016) IgG is involved in the migration and invasion of clear cell renal cell carcinoma. J Clin Pathol 69(6):497–504

Shi B et al (2014) Comparative analysis of human and mouse immunoglobulin variable heavy regions from IMGT/LIGM-DB with IMGT/HighV-QUEST. Theor Biol Med Model 11:30

Shi Z et al (2019) More than one antibody of individual B cells revealed by single-cell immune profiling. Cell Discov 5:64

Stave JW, Lindpaintner K (2013) Antibody and antigen contact residues define epitope and paratope size and structure. J Immunol 191(3):1428–1435

Stavnezer J, Schrader CE (2014) IgH chain class switch recombination: mechanism and regulation. J Immunol 193(11):5370–5378

Tang J et al (2018) Lung squamous cell carcinoma cells express non-canonically glycosylated IgG that activates integrin-FAK signaling. Cancer Lett 430:148–159

Tiselius A, Kabat EA (1938) Electrophoresis of immune serum. Science 87(2262):416–417

Tonegawa S et al (1977) Cloning of an immunoglobulin variable region gene from mouse embryo. Proc Natl Acad Sci USA 74(8):3518–3522

Turula H, Wobus CE (2018) The role of the polymeric immunoglobulin receptor and secretory immunoglobulins during mucosal infection and immunity. Viruses 10(5):237

van de Bovenkamp FS et al (2016) The emerging importance of IgG fab glycosylation in immunity. J Immunol 196(4):1435–1441

van de Bovenkamp FS et al (2018) Adaptive antibody diversification through N-linked glycosylation of the immunoglobulin variable region. Proc Natl Acad Sci USA 115(8):1901–1906

Von Behring EKS (1890) Ueber das Zustandekommen der Diphtherie-Immunitat and der Tetanus-Immunität bei Thieren. Dtsch Med Wochenschr 16:1113–1114

Walter G (1986) Production and use of antibodies against synthetic peptides. J Immunol Methods 88(2):149–161

Wan X et al (2015) Pancreatic expression of immunoglobulin G in human pancreatic cancer and associated diabetes. Pancreas 44(8):1304–1313

Wang LM, Gan YH (2021) Cancer-derived IgG involved in cisplatin resistance through PTP-BAS/Src/PDK1/AKT signaling pathway. Oral Dis 27(3):464–474

Wang H et al (2013) Prognostic significance of immunoglobulin M overexpression in laryngeal squamous cell carcinoma. Acta Otolaryngol 133(10):1080–1087

Wang C et al (2015) IGK with conserved IGKappaV/IGKappaJ repertoire is expressed in acute myeloid leukemia and promotes leukemic cell migration. Oncotarget 6(36):39062–39072

Wang Z et al (2020) Cancer-derived sialylated IgG promotes tumor immune escape by binding to Siglecs on effector T cells. Cell Mol Immunol 17(11):1148–1162

Wang Q et al (2022) A widely expressed free immunoglobulin kappa chain with a unique Vkappa4-1/Jkappa3 pattern promotes colon cancer invasion and metastasis by activating the integrin beta1/FAK pathway. Cancer Lett 540:215720

Welinder C et al (2013) Primary breast cancer tumours contain high amounts of IgA1 immunoglobulin: an immunohistochemical analysis of a possible carrier of the tumour-associated Tn antigen. PLoS One 8(4):e61749

Wu TT, Kabat EA (1970) An analysis of the sequences of the variable regions of Bence Jones proteins and myeloma light chains and their implications for antibody complementarity. J Exp Med 132(2):211–250

Wu YL et al (2017) Intrinsic transcriptional heterogeneity in B cells controls early class switching to IgE. J Exp Med 214(1):183–196

Xiao T et al (2005) An approach to studying lung cancer-related proteins in human blood. Mol Cell Proteomics 4(10):1480–1486

Yang B et al (2013) Correlation of immunoglobulin G expression and histological subtype and stage in breast cancer. PLoS One 8(3):e58706

Yin S et al (2020) Hepatocyte-derived Igkappa exerts a protective effect against ConA-induced acute liver injury. Int J Mol Sci 21(24):9379

Yoshimi K et al (2002) IgG-immunostaining in the intact rabbit brain: variable but significant staining of hippocampal and cerebellar neurons with anti-IgG. Brain Res 956(1):53–66

Zhang ZX et al (2006) Transcription factor Pax5 (BSAP) transactivates the RAG-mediated V(H)-to-DJ(H) rearrangement of immunoglobulin genes. Nat Immunol 7(6):616–624

Zhang L et al (2012) Expression of immunoglobulin G in esophageal squamous cell carcinomas and its association with tumor grade and Ki67. Hum Pathol 43(3):423–434

Zhang C et al (2021) NBIGV-DB: a dedicated database of non-B cell derived immunoglobulin variable region. Gene 772:145378

Zheng H et al (2007) Immunoglobulin alpha heavy chain derived from human epithelial cancer cells promotes the access of S phase and growth of cancer cells. Cell Biol Int 31(1):82–87

Zheng J et al (2009) Immunoglobulin gene transcripts have distinct VHDJH recombination characteristics in human epithelial cancer cells. J Biol Chem 284(20):13610–13619

Zhu X et al (2008) Immunoglobulin mRNA and protein expression in human oral epithelial tumor cells. Appl Immunohistochem Mol Morphol 16(3):232–238

Zhu X et al (2010) Distinct regulatory mechanism of immunoglobulin gene transcription in epithelial cancer cells. Cell Mol Immunol 7(4):279–286

Part II

Physiological and Pathological Significance of Non B-Ig in Different Tissues

7 Functions and Clinical Relevance of Liver-Derived Immunoglobulins

Fanlei Hu, Wenwei Shao, and Xiaoyan Qiu

Abstract

Liver is the largest internal organ of the body with vital functions. In addition to its endocrine and exocrine activities, liver also plays a pivotal role in the immune system, including haematopoietic functions. Liver parenchymal cells, which are epithelial cells, have been found to possess innate immune functions by expressing pattern-recognition receptors (PRRs), producing complement components, and secreting cytokines. Intriguingly, in recent years, it has been discovered that liver epithelial cells also produce immunoglobulins (Igs), which have long been thought to be produced exclusively by B cells. Notably, even liver epithelial cells from B lymphocyte-deficient mice, including SCID mice and μMT mice, could also produce Igs. Compelling evidence has revealed both the physiological and pathological functions of liver-derived Igs. For instance, liver epithelial cells-derived IgM can serve as a source of natural and specific antibodies that contribute to innate immune responses, while liver-produced IgG can act as a growth factor to promote cell proliferation and survival in normal hepatocytes and hepatocarcinoma. Similar to that in B cells, the toll-like receptor 9 (TLR9)-MyD88 signaling pathway is also actively involved in promoting liver epithelial cells to secrete IgM. Liver-derived Igs could potentially serve as biomarkers, prognostic indicators, and therapeutic targets in the clinical setting, particularly for liver cancers and liver injury. Nevertheless, despite significant advances, much remains unknown about the mechanisms governing Ig transcription in liver cells, as well as the detailed functions of liver-derived Igs and their involvement in diseases and adaptive immunity. Further studies are still needed to reveal these underlying, undefined issues related to the role of liver-derived Igs in both immunity and diseases.

Keywords

Liver · Immunoglobulins · Innate immunity · Adaptive immunity · TLR9

F. Hu (✉)
Department of Rheumatology and Immunology, Peking University People's Hospital and Beijing Key Laboratory for Rheumatism Mechanism and Immune Diagnosis, Beijing, China
e-mail: fanleihu@bjmu.edu.cn

W. Shao
Academy of Medical Engineering and Translational Medicine, Tianjin University, Tianjin, China
e-mail: wenwei.shao@tju.edu.cn

X. Qiu
Department of Immunology, School of Basic Medical Sciences, Peking University, Beijing, China
e-mail: qiuxy@bjmu.edu.cn

X. Qiu et al. (eds.), *Non B cell-Derived Immunoglobulins*, Advances in Experimental Medicine and Biology 1445, https://doi.org/10.1007/978-981-97-0511-5_7

7.1 The Non-immunological and Immunological Functions of Liver

Liver, comprising about 2% of an adult's body weight, is the largest internal organ in the body. It functions as both an endocrine and exocrine organ, playing a pivotal role in mediating communication between the gut and blood (Vernon et al. 2020). The liver's multifaceted functions and its involvement in metabolism have been widely recognised. For instance, it synthesises essential proteins such as albumin and blood clotting factors, manufactures triglycerides and cholesterol, produces glycogen, synthesises bile, stores energy, and eliminates toxins. In addition, the liver can produce carbohydrates and convert glucose into glycogen which can be stored in both the liver and muscle cells (Cordero-Espinoza et al. 2018). The liver also converts ammonia into urea and participates in lactic acid metabolism. Furthermore, due to its unique metabolism and relationship to the gastrointestinal tract, the liver serves as a critical organ for the detoxification of xenobiotics and drugs (Kalra et al. 2023).

Besides being one of the body's vital organs, the liver is also a gland responsible for the secretion of hundreds of chemicals and vitamins necessary for the body's survival (Robinson et al. 2016). Hepatocytes are the primary effector cells responsible for these liver functions including actively synthesising proteins, cholesterol, bile salts, fibrinogen, phospholipids, and glycoproteins (Fig. 7.1).

The immune functions of the liver have been gradually recognised. The liver's anatomical location and blood supply are advantageous for its immune functions due to its critical location downstream of the gut, enabling it to remove gut-derived pathogens from the bloodstream (Kubes et al. 2018). Moreover, the portal vein, which carries a high concentration of pathogen-derived molecules, supplies the majority of blood to the liver, and the slow velocity of blood flow in the liver contributes to the surveillance of pathogens (Oda et al. 2003).

Generally, studies have shown that the lymphoid repertoire in the liver immune system is composed of approximately 5% B cells, 23% T cells, 20–30% natural killer cells (NK), and 50% natural killer T cells (NKT) (Doherty et al. 2000). Moreover, liver sinusoidal endothelial cells (LSEC) (Onoe et al. 2005; Connolly et al. 2010), Kupffer cells (Erhardt et al. 2007), and

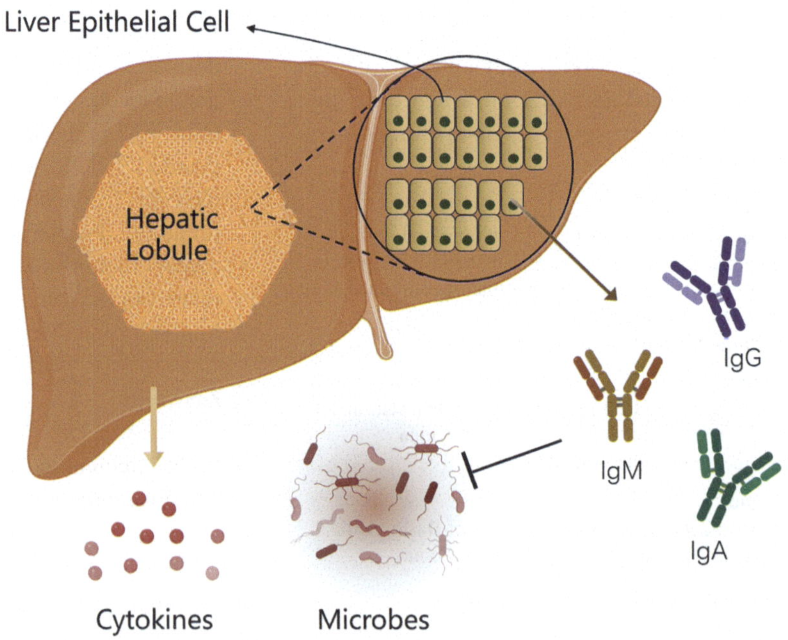

Fig. 7.1 The structure and function of liver tissue, and the expression of Igs in normal liver epithelial cells

hepatic dendritic cells serve as classical liver antigen-presenting cells (APCs) that present antigens to T cells. Besides, under pathophysiological circumstances, biliary epithelial cells and hepatocytes (Qian et al. 2001) may be considered as non-conventional APCs. Various liver diseases are associated with aberrant immunity, such as autoimmune liver diseases, primary biliary cirrhosis, autoimmune hepatitis, and alcoholic hepatitis.

Emerging evidence suggests that the liver is an essential part of the immune system, primarily serving innate immune functions (Fig. 7.1). Hepatocytes play a crucial role in innate immunity by expressing pattern-recognition receptors (PRRs) and producing complement components (Gao 2016). These immune proteins produced by hepatocytes are activated by two groups of cytokines and their downstream signals: IL-6 and IL-22, which activate STAT3, and IL-1β and TNF-α, which activate NF-κB (Zhou et al. 2016). Besides hepatocytes, the non-parenchymal cells, including endothelial cells, stellate cells, Kupffer cells, and lymphocytes, make up 30% of the liver cells. Both myeloid and lymphoid cells in the liver, such as macrophages, dendritic cells, myeloid-derived suppressor cells, NK, NKT, and γδT Cells, constitute a rich source for the innate immune system (Tian et al. 2016). Besides the resident immune cells in the liver, immune cells like monocytes and neutrophils could also be recruited from the bloodstream by adhesion molecules in the liver, such as CD44, hyaluronan, Mac-1, ICAM-1, VLA4, VAP-1, and CCR2 (Wong et al. 1997; Fox-Robichaud et al. 2000; McDonald et al. 2008).

However, the role of liver in adaptive immunity has been less extensively investigated. Several studies focused on T cells and their functions in liver cancers, hepatic infections, and autoimmune liver diseases (Shuai et al. 2016). It has been illustrated that the liver immune hyporesponsiveness is associated with adaptive immunity, involving liver sinusoidal endothelial cells, hepatocytes, dendritic cells, and Kupffer cells (Kubes et al. 2018). Under steady-state conditions, these liver resident immune cells functionally inhibit adaptive immunity and maintain immune hyporesponsiveness due to their low expression of MHC, the absence of costimulatory molecules, and the constitutive expression of anti-inflammatory cytokines (Qian et al. 2001; Knolle et al. 1998, 1999; Sana et al. 2014). In recent years, liver parenchymal cells have also been found to produce immunoglobulins (Igs) with distinct functions (Fig. 7.1). These findings have inspired further investigation into the involvement of the liver in adaptive immunity and other vital processes.

In this chapter, we will briefly summarise the expression, characteristics, mechanisms, functions, and clinical relevance of liver cell-derived Igs and discuss the perspectives of related research and clinical applications.

7.2 Expression of Liver Epithelial Cell-Derived Igs

In recent years, it has been discovered that liver parenchymal cells have the capacity to express Igs, including IgG, IgM, IgD, and IgA. Moreover, these liver-derived Igs have been shown to exhibit physiological and pathological functions distinct from those of B cell-derived Ig (B-Ig).

7.2.1 Expression of Igs in Liver Cancer Cells

The presence of IgG heavy chain mRNA was first recorded in human hepatocellular carcinoma in 2003 by Qiu et al. (2003). Furthermore, after excluding contamination from infiltrating B cells in the liver, IgG was also identified by Western blotting and immunocytochemical staining in liver cancer cells, suggesting the existence of both secreted and cytoplasmic forms of IgG and IgM (Qiu et al. 2003; Hu et al. 2003, 2012). Moreover, Chen et al. also detected the expression of Igγ and Igκ in hepatocytes of cirrhotic liver tissues, and the germline transcripts for both Iγ-Cγ and Iα-Cα were identified in liver cancer cells (Chen et al. 2007), implying the occurrence of functional Ig gene rearrangement in these cancer cells. In 2007, Geng et al. demonstrated the expression of SNC73, the heavy chain of IgA1, in liver cancer cell lines using immunochemistry and RT-PCR (Geng et al. 2007).

To further investigate the IgG expression in hepatocellular carcinoma (HCC), Yi et al. utilised an HCC tissue array and cancer cell lines to confirm the IgG secretion by malignant hepatocytes. They discovered that the human liver cell line L-02, the rat hepatocyte cell line BRL-3A, and the human hepatoma cell line HepG-2 were all capable of producing IgG, thereby ruling out the potential of B cell contamination. Furthermore, MCCH97, a HCC cell line with high metastatic capacity, displayed a superior ability to secrete IgG (Yi et al. 2015).

7.2.2 Expression of Igs in Normal Liver Epithelial Cells

Normal liver epithelial cells were also identified as capable of producing Igs. As early as 2002, Zheng et al. detected the transcripts of Ig α-1 in normal liver tissues (Hu et al. 2002, 2003). In 2014, Yu et al. reported that normal adult human and rat hepatocytes could express IgG at both the mRNA and protein levels.

To systematically study the expression of Ig in normal liver epithelial cells, Shao et al. utilised the Balb/c mice and B cell-deficient mice, including severe combined immunodeficient (SCID, T cell and B cell-deficient) mice and μMT (B cell-deficient) mice as models. The researchers discovered that sorted liver epithelial cells expressed rearranged transcripts of Ig heavy chains, including μ, δ, γ, and α, as well as light chains, including both κ and λ (Shao et al. 2016a). Interestingly, the serum levels of IgM and IgG in μMT mice which lack mature B cells due to the disruption of membrane IgM and BCR reduced to 1/100 of the concentration in WT mice, but the levels of IgM, IgG1, IgG2b, and Igκ in the liver lysate were the same as those of WT mice. Intriguingly, the levels of IgA, IgD, Igλ, and IgG2a in the liver lysate of μMT mice were much higher than those in WT mice (Shao et al. 2016a). This finding suggests that B cells are necessary to maintain Ig levels in peripheral blood but are less important for maintaining Ig levels in tissues and organs. As expected, except for IgE, all classes of Ig are found in liver epithelial cells. It is worth noting that the findings introduced above are also present in SCID mice which is both T cell and B cell deficiency. However, the detailed mechanisms need to be further investigated.

It is well known that the Ig produced by B cells presents infinite diversity in the same individual. However, as previously reported, Ig produced by non-B cells typically display restricted diversity and patterns of conservation within the same individual or the same lineage, as illustrated by the Ig produced by mouse liver epithelial cells. Moreover, the variable regions of IgM, IgG, and IgA produced by liver epithelial cells from a given individual do not exhibit identical V(D)J rearrangement pattern, suggesting that the mechanism of IgG and IgA production by these cells does not follow the conventional class switching hypothesis (Shao et al. 2016a).

RAG1 and RAG2, enzymes indispensable for V(D)J recombination, are expressed at both the mRNA and protein levels in liver cancer cell lines (Chen et al. 2007). Furthermore, the transcriptional factor Pax5, which activates Rag expression and controls the accessibility of Ig variable genes, has also been found to be expressed in liver cancers (Geng et al. 2007). Moreover, liver cancer cell lines have been revealed to express endogenous early B cell factor (EBF) (Cao et al. 1997; Zheng et al. 2001), which is considered as a driver of both B lineage determination and commitment by influencing genes that code for survival signal proteins essential for developmental checkpoints, such as pre-BCR and BCR components. However, further study is needed to elucidate whether the liver-derived Ig is also expressed through the classical RAG-dependent pathway or via other non-canonical mechanisms.

In conclusion, these pieces of evidence demonstrate that the liver is capable of expressing different classes of Ig, including IgM, IgA, IgG, and IgD, in both physiological and pathological conditions.

7.3 The Function of Liver Epithelial Cell-Derived Igs

For a long time, the functions of liver-derived Igs remained unknown. Only recently have the physiological and pathological roles of the Igs expressed by both normal liver cells and liver cancer cells been gradually recognised. The following section summarises the functions that have been identified so far:

7.3.1 Liver Epithelial Cell-Derived IgM Serves as a Source of Natural Antibody That Contributes to Innate Immune Responses

IgM possesses numerous unique properties, such as pattern recognition capabilities and multiple complement binding sites. Additionally, IgM is present in high concentrations in the blood (typically around 1.5 mg/mL) and is the first antibody elicited in immune response to immunisation or infection. These characteristics make IgM particularly well-suited for its role in anti-microbial immunity. IgM can be classified as natural IgM and immune IgM, which are produced with or without antigen stimulation, respectively. Natural IgM is predominantly generated by innate-like B1 cells in the absence of pathogen encounters, whereas immune IgM is produced by both innate-like B1 and adaptive B2 cells following antigen exposure. Natural IgM constitutes the majority of total circulating IgM and most natural IgM is germline-encoded and polyreactive, binding with low affinity to various antigens, such as microbial pathogens, and contributing to early immunity prior to the onset of the adaptive humoral response. IgM has always been considered to be produced only by B cells.

In recent years, it has been identified that liver epithelial cells could also produce IgM. Shao and Hu et al. found that liver epithelial cells spontaneously produce IgM, and these IgM are polyreactive and can bind to single-stranded DNA (ssDNA), double-stranded DNA (dsDNA), lipopolysaccharide (LPS), and insulin with low affinity. These liver epithelial cell-derived natural IgM could also bind to different types of bacteria such as *E. coli*, *Salmonella typhi*, *Proteus vulgaris*, *Staphylococcus aureus*, *Shigella*, and *Paratyphoid bacilli* (Shao et al. 2016b), which fully conforms to the classical concept of natural antibody.

Toll-like receptors (TLRs) are transmembrane proteins that recognise various pathogen-related molecular patterns, such as lipopolysaccharide, sodium urate crystal, viral double-stranded RNA, etc., and initiate the body's inflammatory immune response. There are at least ten TLRs present in mammals responsible for cell-extrinsic and -intrinsic activities. TLR9 detects the "CpG" motif, unmethylated cytosine–phosphate–guanine (CpG) dinucleotides. CpG is common in bacterial and viral DNA but uncommon in the vertebrate genome (Akira et al. 2003; Hemmi et al. 2000).

Myeloid differentiation primary response gene 88 (MyD88) is the canonical adaptor for inflammatory signalling pathways downstream of members of the TLR. It was reported that CpG oligodeoxynucleotides (ODN) could directly induce IgM secretion by B cells through the TLR9-MyD88 signalling pathway (Krieg et al. 1995). Moreover, this critical pathway could promote Th1-biased Ig class switching in B cells (He et al. 2004). In general, once activated, TLR9 and its adaptors, MyD88, activate intracellular signalling, resulting in the production of cytokines and Igs, mainly IgM (Landers et al. 2005).

It has been demonstrated that the TLR9-MyD88 signalling pathway mediates natural IgM production in liver epithelial cells. Hu et al. found that liver epithelial cells could express most TLRs and MyD88 (Hu et al. 2012). Moreover, they discovered that the TLR9-MyD88 signalling pathway is not only an essential regulator of B cell-mediated IgM secretion but also a crucial modulator of Ig expression in the liver. Like IgM expression in B cells, CpG stimulation, which mimics bacterial infection or direct bacterial immunisation could significantly promote liver epithelial cells to produce natural IgM in mice. This suggests that epithelial cells could secrete IgM and defend against bacterial invasion. More importantly, the preferential usage of the variable region changed, and the somatic mutation occurred after CpG stimulation, eventually leading to elevated natural IgM activity. All of these indicate that liver epithelial cell-derived IgM contributes to innate immune function. Besides, when MyD88 was knocked out, the upregulated natural IgM in liver epithelial cells following CpG stimulation was abolished, highlighting the importance of the TLR9-MyD88 signalling pathway in the production of liver epithelial cell-derived IgM (Shao et al. 2016b). Similar results were observed in liver epithelial cancer cells, and chloroquine could abolish this TLR9-MyD88-mediated IgM secretion (Hu et al. 2012). These findings suggest a potential role of the TLRs pathway in liver-derived Ig production (Fig. 7.2). However, whether other pathways, such as the cGAS-STING pathway, are involved in liver Ig production remains to be elucidated.

In summary, all these suggest that liver epithelial cell-derived IgM contributes to innate immune responses. Yet, whether they could directly target and kill the microbes needs to be further studied.

7.3.2 Liver Epithelial Cell-Derived Igs Acts as a Growth Factor in Promoting Cell Proliferation and Survival in Normal Hepatocytes and Hepatocarcinoma

Following hepatectomy surgery, a discernible augmentation in the expression of IgG within the hepatic tissue was observed, in comparison to the pre-surgical state. Subsequently, the IgG levels reverted to their original state upon the cessation of hepatocyte regeneration (Lei et al. 2014). It was found that patients with HCC who tested positive for IgG had higher levels of core-fucosylated α fetoprotein (AFP-L3), a higher incidence of portal vein tumour thrombus, and larger tumours (Hu et al. 2002). Compared with healthy controls, liver cancer patients showed significantly increased IgG levels. Similar results were also attained in a series of HCC cell lines, including HepG2, Hep3B, and Huh7 (Hu et al. 2002). Moreover, IgG expression knock-down using siRNA significantly inhibited cell prolifer-

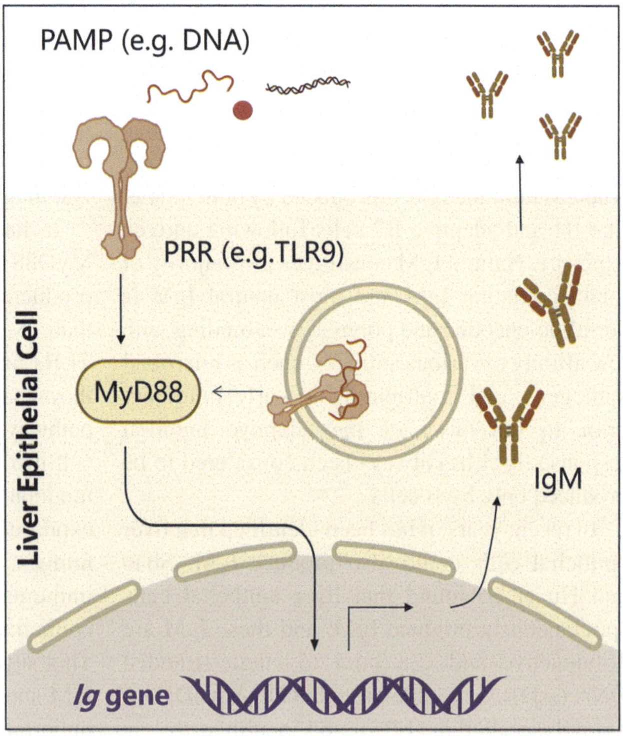

Fig. 7.2 The TLRs pathway participates in liver-derived Ig production

ation and promoted apoptosis in L02 and HepG2 liver cancer cell lines. In addition, the migration and metastasis of these cells were also inhibited after IgG silencing. These findings suggest that liver-derived IgG is actively involved in liver cell growth and regeneration.

To further investigate the role of hepatocyte-derived Ig, Yin et al. conditionally knocked out the Igκ light chain in liver epithelial cells. The study revealed that the absence of Igκ accentuated liver injury upon ConA challenge, as demonstrated by a higher Ishak score, increased levels of aspartate aminotransferase (AST) and alanine transaminase (ALT), as well as elevated levels of inflammatory cytokine MCP-1. Furthermore, knockout of *Igκ* led to accelerated activation of the mitochondrial death pathway and caspase-3 cleavage (Yin et al. 2020). These demonstrate a critical role of hepatocyte-derived Igκ in hepatocyte survival and maintenance in mice.

7.4 Clinical Significance of Liver-Derived Igs

7.4.1 Liver-Derived Igs May Serve as Potential Biomarkers or Prognostic Indicators for Liver Cancers

Since Ig has been found in numerous liver diseases, it may become a potential marker for different liver diseases. It has been reported that human liver carcinomas could express high levels of IgG and IgM, and these Igs may serve as the potential biomarker or prognostic indicator for the disease. Yi et al. reported a new marker for hepatitis B virus-related HCC, namely elevated core-fucosylated IgG (IgG-L3) (Yi et al. 2015). Compared with healthy individuals, patients with liver cirrhosis and HBV carriers, HCC patients demonstrated higher levels of IgG-L3. Through cross-sectional observations and longitudinal evaluation, they reported that serum IgG-L3 revealed better diagnostic performances than AFP, the conventional screening marker for HCC. These findings indicated the associations between HCC-derived IgG and the progression of HCC and suggested the potential of measuring IgG levels in assisting HCC diagnosis (Yi et al. 2015). Moreover, these liver-derived Igs may serve as a therapeutic target for treating liver cancers since silencing IgG expression in liver cancer cells significantly inhibited their proliferation and metastasis.

7.4.2 Liver-Derived Igs Can be the Target for Therapy of Liver Cancer and Autoimmune Hepatitis (AIH)

Growing evidence has demonstrated that Ig can be highly expressed by a wide range of cancer cells. Particularly, epithelial cancer stem cells could express a unique IgG with novel glycosylation on the CH1 Asn162 site of the heavy chain. Since the functions of IgG rely on its unique glycosylation modification, this could further promote the biological behaviour of cancer stem cells, such as tumour occurrence, recurrence, metastasis, and drug resistance (Tang et al. 2018), and facilitate tumour immune escape (Wang et al. 2020). Notably, sialylated IgG antibodies, demonstrative of heightened expression within cancerous cells, manifest selective enrichment in patients afflicted by hepatocellular carcinoma who undergo immune checkpoint blockade (ICB) therapy (Wu et al. 2023). This enrichment potentially unveils a novel mechanism contributing to resistance against ICB treatment, characterised by negative feedback. This observation suggests a compelling prospect wherein liver-derived sialylated IgG could serve as an avenue for enhancing the efficacy of ICB therapy. These findings suggest that the Ig produced by liver cancer cells may be used as a novel target for tumour immunotherapy.

In addition, it is conceivable that liver-derived Igs may assume an integral function in the context of autoimmune hepatitis (AIH), a protracted inflammatory hepatic ailment distinguished by escalated gammaglobulin levels, the existence of autoantibodies, and hepatic tissue inflammation.

These autoantibodies have been ascribed to be derived from B cell lineage. Nevertheless, given the hepatic parenchymal cells' capability to generate immunoglobulins, a supposition arises that the immunoglobulins synthesised by these hepatic parenchymal cells might also contribute to the intricate processes governing AIH.

7.5 Perspectives of Fundamental Research and Clinical Applications of Liver-Derived Igs

Accumulated experimental data have confirmed the concept that non-B cells can produce Ig. In recent years, liver epithelial cells have been found to be able to express Ig. At present, it has been proved that Ig produced by liver epithelial cells can exert the function of antibodies under physiological conditions and maintain the proliferation and survival of liver epithelial cells, demonstrating both immune and cellular biological functions. Under pathological conditions, Ig expressed by liver epithelial cells promotes the occurrence of liver cancer and may also participate in immune-related inflammatory liver diseases such as autoimmune hepatitis. It is well worth exploring the physiological and pathological significance of Ig expressed by liver epithelial cells in greater depth in the future.

References

Akira S et al (2003) Recognition of pathogen-associated molecular patterns by TLR family. Immunol Lett 85(2):85–95

Cao J et al (1997) Characterization of colorectal-cancer-related cDNA clones obtained by subtractive hybridization screening. J Cancer Res Clin Oncol 123(8):447–451

Chen Z et al (2007) Immunoglobulin G expression in carcinomas and cancer cell lines. FASEB J 21(11):2931–2938

Connolly MK et al (2010) In hepatic fibrosis, liver sinusoidal endothelial cells acquire enhanced immunogenicity. J Immunol 185(4):2200–2208

Cordero-Espinoza L et al (2018) The balancing act of the liver: tissue regeneration versus fibrosis. J Clin Invest 128(1):85–96

Doherty DG et al (2000) Innate and adaptive lymphoid cells in the human liver. Immunol Rev 174:5–20

Erhardt A et al (2007) IL-10, regulatory T cells, and Kupffer cells mediate tolerance in concanavalin A-induced liver injury in mice. Hepatology 45(2):475–485

Fox-Robichaud A et al (2000) Molecular mechanisms of tumor necrosis factor alpha-stimulated leukocyte recruitment into the murine hepatic circulation. Hepatology 31(5):1123–1127

Gao B (2016) Basic liver immunology. Cell Mol Immunol 13(3):265–266

Geng LY et al (2007) Expression of SNC73, a transcript of the immunoglobulin alpha-1 gene, in human epithelial carcinomas. World J Gastroenterol 13(16):2305–2311

He B et al (2004) CpG DNA induces IgG class switch DNA recombination by activating human B cells through an innate pathway that requires TLR9 and cooperates with IL-10. J Immunol 173(7):4479–4491

Hemmi H et al (2000) A toll-like receptor recognizes bacterial DNA. Nature 408(6813):740–745

Hu J et al (2002) Expression of a novel immunoglobulin gene SNC73 in human cancer and its significance. Zhonghua Zhong Liu Za Zhi 24(1):38–40

Hu JB et al (2003) Expression of a novel immunoglobulin gene SNC73 in human cancer and non-cancerous tissues. World J Gastroenterol 9(5):1054–1057

Hu F et al (2012) Spontaneous production of immunoglobulin M in human epithelial cancer cells. PLoS One 7(12):e51423

Kalra A et al (2023) Physiology, liver. StatPearls Publishing, Treasure Island (FL)

Knolle PA et al (1998) IL-10 down-regulates T cell activation by antigen-presenting liver sinusoidal endothelial cells through decreased antigen uptake via the mannose receptor and lowered surface expression of accessory molecules. Clin Exp Immunol 114(3):427–433

Knolle PA et al (1999) Endotoxin down-regulates T cell activation by antigen-presenting liver sinusoidal endothelial cells. J Immunol 162(3):1401–1407

Krieg AM et al (1995) CpG motifs in bacterial DNA trigger direct B-cell activation. Nature 374(6522):546–549

Kubes P et al (2018) Immune responses in the liver. Annu Rev Immunol 36(1):247–277

Landers CD et al (2005) CpG oligodeoxynucleotides stimulate cord blood mononuclear cells to produce immunoglobulins. Clin Immunol 116(3):236–245

Lei Y et al (2014) Expression and distribution of immunoglobulin G in the normal liver, hepatocarcinoma and postpartial hepatectomy liver. Lab Investig 94(11):1283–1295

McDonald B et al (2008) Interaction of CD44 and hyaluronan is the dominant mechanism for neutrophil sequestration in inflamed liver sinusoids. J Exp Med 205(4):915–927

Oda M et al (2003) Regulatory mechanisms of hepatic microcirculation. Clin Hemorheol Microcirc 29(3–4):167–182

Onoe T et al (2005) Liver sinusoidal endothelial cells tolerize T cells across MHC barriers in mice. J Immunol 175(1):39–46

Qian S et al (2001) Hepatocyte induced apoptosis of activated T cells, a mechanism of liver transplant tolerance, is related to the expression of ICAM-1 and hepatic lectin. Transplant Proc 33(1–2):226

Qiu X et al (2003) Human epithelial cancers secrete immunoglobulin g with unidentified specificity to promote growth and survival of tumor cells. Cancer Res 63(19):6488–6495

Robinson MW et al (2016) Liver immunology and its role in inflammation and homeostasis. Cell Mol Immunol 13(3):267–276

Sana G et al (2014) Adult human hepatocytes promote CD4(+) T-cell hyporesponsiveness via interleukin-10-producing allogeneic dendritic cells. Cell Transplant 23(9):1127–1142

Shao W et al (2016a) Identification of liver epithelial cell-derived Ig expression in mu chain-deficient mice. Sci Rep 6:23669

Shao W et al (2016b) Epithelial cells are a source of natural IgM that contribute to innate immune responses. Int J Biochem Cell Biol 73:19–29

Shuai Z et al (2016) Adaptive immunity in the liver. Cell Mol Immunol 13(3):354–368

Tang JS et al (2018) Lung squamous cell carcinoma cells express non-canonically glycosylated IgG that activates integrin-FAK signaling. Cancer Lett 430:148–159

Tian Z et al (2016) Regional immunity in tissue homeostasis and diseases. Sci China Life Sci 59(12):205–209

Vernon H et al (2020) StatPearls. StatPearls Publishing LLC, Treasure Island (FL)

Wang ZH et al (2020) Cancer-derived sialylated IgG promotes tumor immune escape by binding to Siglecs on effector T cells. Cell Mol Immunol 17(11):1148–1162

Wong J et al (1997) A minimal role for selectins in the recruitment of leukocytes into the inflamed liver microvasculature. J Clin Invest 99(11):2782–2790

Wu R et al (2023) Immune checkpoint therapy-elicited sialylation of IgG antibodies impairs antitumorigenic type I interferon responses in hepatocellular carcinoma. Immunity 56(1):180–192

Yi CH et al (2015) Elevated core-fucosylated IgG is a new marker for hepatitis B virus-related hepatocellular carcinoma. Onco Targets Ther 4(12):e1011503

Yin S et al (2020) Hepatocyte-derived Ig kappa exerts a protective effect against ConA-induced acute liver injury. Int J Mol Sci 21(24):9379

Zheng S et al (2001) Structure and expression of colorectal cancer related immunoglobulin novel gene SNC73. Zhonghua Yi Xue Za Zhi 81(8):485–488

Zhou Z et al (2016) Hepatocytes: a key cell type for innate immunity. Cell Mol Immunol 13(3):301–315

Expression and Clinical Significance of Non B Cell-Derived Immunoglobulins in the Urinary System and Male Reproductive System

8

Zhenling Deng, Yue Wang, Caipeng Qin, Zhengzuo Sheng, Tao Xu, and Xiaoyan Qiu

Abstract

The urinary system comprises kidneys, ureters, bladder, and urethra with its primary function being excretion, referring to the physiological process of transporting substances that are harmful or surplus out of the body. The male reproductive system consists of gonads (testis), vas deferens, and accessory glands such as the prostate. According to classical immunology theory, the tissues and organs mentioned above are not thought to produce immunoglobulins (Igs), and any Ig present in the relevant tissues under physiological and pathological conditions is believed to be derived from B cells. For instance, most renal diseases are associated with uncontrolled inflammation caused by pathogenic Ig deposited in the kidney. Generally, these pathological Igs are presumed to be produced by B cells. Recent studies have demonstrated that renal parenchymal cells can produce and secrete Igs, including IgA and IgG. Glomerular mesangial cells can express and secrete IgA, which is associated with cell survival and adhesion. Likewise, human podocytes demonstrate the ability to produce and secrete IgG, which is related to cell survival and adhesion. Furthermore, renal tubular epithelial cells also express IgG, potentially involved in the epithelial-mesenchymal transition (EMT). More significantly, renal cell carcinoma, bladder cancer, and prostate cancer have been revealed to express high levels of IgG, which promotes tumour progression. Given the widespread Ig expression in the urinary and male reproductive systems, continued efforts to elucidate the roles of Igs in renal physiological and pathological processes are necessary.

Keywords

Immunoglobulin · Mesangial cell · Podocyte · Renal tubular epithelial cell · Renal cell carcinoma · Bladder cancer · Prostate cancer

Z. Deng (✉)
Department of Nephrology, Peking University Third Hospital, Beijing, China

Y. Wang
Peking University Third Hospital, Beijing, China

C. Qin · T. Xu
Department of Urology, Peking University People's Hospital, Beijing, China
e-mail: qincaipeng@pkuph.edu.cn; xutao@pkuph.edu.cn

Z. Sheng
Department of Thoracic Surgery, Fu Xing Hospital, Capital Medical University, Beijing, China

X. Qiu
Department of Immunology, School of Basic Medical Sciences, Peking University, Beijing, China
e-mail: qiuxy@bjmu.edu.cn

X. Qiu et al. (eds.), *Non B cell-Derived Immunoglobulins*, Advances in Experimental Medicine and Biology 1445, https://doi.org/10.1007/978-981-97-0511-5_8

8.1 Introduction

Kidneys are essential organs of the human body as they filter roughly 180 L of plasma daily to eliminate metabolic end-products and certain toxic wastes. Simultaneously, they significantly contribute to regulating water and salt metabolism and maintaining the acid–base equilibrium in bodily fluids. Moreover, kidneys also possess an endocrine function, producing renin, erythropoietin, active vitamin D3, prostaglandin, and kinin, among others. These functions of the kidneys ensure the stability of the body's internal environment, facilitating normal metabolic processes.

It is generally agreed that due to the critical role kidneys play in blood filtration, they often become sites for the deposition of antigen–antibody complexes. These complexes are produced when a systemic autoimmune reaction or response to microbes occurs. Complement activation by immunocomplexes seems to play an integral role in triggering uncontrolled inflammation caused by pathogenic Igs in kidneys (Yatim and Lakkis 2015). For instance, IgA nephropathy involves extensive IgA deposition in the mesangial region (Roberts 2014), and the pathogenesis of membranous nephropathy (MN) entails the subepithelial deposition of immune-complex (podocyte antigen and autoantibody IgG, especially IgG4) on the glomerular capillary wall (Couser 2017; Hoxha et al. 2022); multiple classes of Igs (IgG, IgA, IgM) and complement are commonly found as mesangial, subendothelial, and subepithelial deposits in lupus nephritis (LN) (George 2011). Immune complex and complement deposition in the tubular basement membrane (TBM) are significant features of IgG4-related tubulointerstitial nephritis (IgG4-TIN) (Cortazar and Stone 2015). Furthermore, renal amyloidosis is typically mediated by Ig light chain, which has been viewed as a plasma cell dyscrasia (Kidd and Carl 2016). Although it is challenging to comprehend how the glomerular basement membrane (GBM) allows such macromolecules to pass through, and why some nephropathy patients fail to respond to B cell depletion therapy, particularly in IgAN (Lafayette et al. 2017), this understanding forms the basis of the current classical theory "only B cells produce Igs". Therefore, scant attention has been paid to the question of whether local cells in kidneys can produce Ig and its potential physiological and pathological function.

In recent years, burgeoning evidence suggested that Ig can be produced by non B cells, such as epithelial cells, cells in immune-privileged sites (eyes, neurons, and spermatogenic cells), and cardiomyocytes (Niu et al. 2011a, b; Huang et al. 2008, 2009; Li et al. 2014). This fact was further bolstered by the results obtained on B cell-deficient μMT mice, in which different classes of Igs were identified in cardiomyocytes (Zhu et al. 2017) and liver epithelial cells (Shao et al. 2016). More significantly, cancer cells, including epithelial cancers (breast, colon, lung, prostate, and pancreatic cancer, etc.) (Zheng et al. 2001; Qiu et al. 2003; Jiang et al. 2014; Liu et al. 2012), mesenchymal tissue tumours (osteosarcoma and liposarcoma, etc.), and myeloid-origin-haematopoietic tumours [acute myeloid leukaemia (AML)] (Wu et al. 2020), could express Ig, especially IgG.

This chapter will discuss Ig expression in renal parenchymal cells, including glomerular mesangial cells, podocytes, and renal tubular epithelial cells, as well as in spermatogenic cells. Additionally, a unique sialylated IgG (SIA-IgG) has been discovered to be highly expressed in cancer cells of renal cell carcinoma (RCC), bladder cancer, and prostate cancer. The potential role of Igs from renal local cells in the pathogenesis of kidney diseases and cancer progression will also be summarised, providing further insights for basic and clinical studies.

8.2 Ig Expression in Renal Parenchymal Cells and Correlation Between Renal Ig Deposition and Diseases

8.2.1 Physiological and Pathological Significance of IgA Expression in Glomerular Mesangial Cell

8.2.1.1 Mesangial Cells Could Spontaneously Produce and Secrete Gd-IgA1

The functions of mesangial cells encompass the formation of capillary loops during development, interactions with other renal cells, regulation of capillary flow through concentrations, and removal of macromolecules. Beyond these actions, mesangial cells also function as immunologically active cells by producing numerous chemokines and cytokines. They play a prominent role in the development of inflammatory diseases and autoimmune disorders of the kidney. However, it remains unclear whether mesangial cells can produce Igs, especially IgA.

To ascertain if glomerular mesangial cells can express IgA, Deng et al. initially analysed the transcription of IgA in a human mesangial cell line (HMC, C2M12) and primary human renal mesangial cells (HRMCs) (Deng et al. 2018). They successfully amplified constant region transcripts of IgA, Igκ, and Igλ in both HMC and HRMCs. Furthermore, in line with previous reports suggesting the absence of diversity in the variable region of non-B cell-derived Ig, glomerular mesangial cell-derived IgA, especially the Vκ-Jκ rearrangement, exhibited restricted and conservative Vκ-Jκ patterns, such as Vκ3–20/Jκ1 predominantly in HMC and Vκ1–12/Jκ4 in HRMCs. As anticipated, IgA expression in the mesangial cells was found in the cytoplasm, membranes, and secretion (Fig. 8.1). In addition, IgA1 expression fluctuated dynamically with the cell cycle, with the highest level observed during the G2/M phase. The endogenous pro-inflammatory factor angiotensin II (Ang II) significantly upregulated IgA protein expression.

The immortalised mesangial cell lines provide a tool for studying mesangial cells in vitro. However, these cells do not retain every original characteristics of mesangial cells. Thus, it is necessary to detect Ig expression in primary mesangial cells. By using nested PCR and Sanger sequencing, it was confirmed the expression of Igs, including IgA, IgG, IgM, and IgE, in 15 single primary mesangial cells. These cells were identified by specific marker genes ITGA8 or/and ACTA2. Co-expression of IgA and IgG was detected in 1 out of the 15 single mesangial cells. A similar expression pattern was observed in single mesangial cells from C57BL/6 mice, and the detection of IgA and IgG transcripts in 1 out of the 14 sequenced single mesangial cells further confirmed the ability of mesangial cells to express Igs (Lu et al. 2017).

Given the absence of galactose in the pathogenic IgA1 (Gd-IgA1) leading to IgA nephropathy, Deng et al. further investigated whether mesangial cells produced IgA lacking galactose. To detect Gd-IgA1 in HRMCs, they procured a monoclonal antibody, KM55, which specifically recognises the Gd-IgA1, from Suzuki et al. (Yasutake et al. 2015). The results showed the presence of Gd-IgA1 in the cytoplasm of HRMCs and its secretion into cell culture supernatant. Moreover, the expression and secretion of Gd-IgA1 were significantly elevated in HRMCs stimulated with Staphylococcus aureus (SAC), staphylococcal enterotoxin B (SEB), and lipopolysaccharide (LPS). Furthermore, SEB and LPS promoted extracellular granular deposition of Gd-IgA1 in HRMCs, which is consistent with the renal pathology of IgAN (unpublished data). These findings suggested that IgA produced by mesangial cells might be a major source of leading to IgA nephropathy.

In addition, Deng et al. further addressed the potential function of IgA in mesangial cells and found that siRNA of IgA significantly reduced cell adhesion ability and enhanced early apoptosis, indicating that IgA expression in HMCs might play a crucial role in cell adhesion, growth, and survival of mesangial cells (Deng et al. 2018).

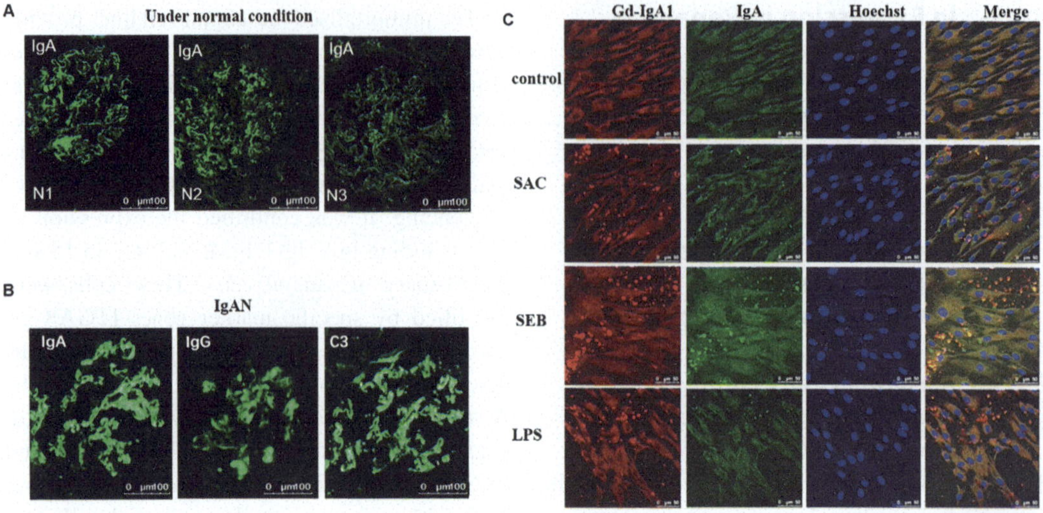

Fig. 8.1 The expression and location of IgA in the kidney and mesangial cells. (**a**) Positive staining of IgA in glomeruli from three normal kidneys (N1, N2, and N3 refer to the glomerulus from the normal renal cortex). (**b**) Renal biopsy findings in a patient with IgAN. Immunofluorescence image showing granular staining of IgA in glomerular mesangium and along the capillary walls, accompanied by positive staining of C3 and weak staining of IgG in the mesangium. (**c**) Under normal culture conditions, Gd-IgA1 can be expressed by HRMC, but Gd-IgA1 production increased significantly and presented granular deposition under the stimulation of pathogenic factors, such as SAC, SEB, and LPS

8.2.1.2 Possible Pathological Significance of IgA Expression in IgA Nephropathy

Immunoglobulin A nephropathy (IgAN) is the most common primary glomerulonephritis worldwide, with 30–40% of patients progressing to end-stage renal disease (ESRD) within 20–30 years after the first clinical presentation (Suzuki et al. 2011). IgAN is the most prevalent primary glomerulonephritis, characterised by mesangial cell proliferation, extracellular matrix (ECM) accumulation with mesangial expansion, and deposition (or possible in situ formation) of pathogenetic polymeric IgA1 immune complexes (occasionally with IgG and IgM) in the glomerular mesangium (Wyatt and Julian 2013). IgAN has been defined as an autoimmune disease, and its pathogenesis is believed to follow a multi-hit hypothesis. Four major steps are presumed to be required for the onset and progression of IgAN: (1) overproduction of galactose-deficient IgA1 (Gd-IgA1) (first hit), (2) generation of IgG or IgA autoantibodies specific for Gd-IgA1 (second hit), (3) subsequent immune complex formation (third hit), and (4) deposition of those complexes in glomeruli (fourth hit). Alternatively, it was speculated that the pathogenesis might involve an initial in situ formation of immune deposits, but the supporting evidence is lacking.

The exact pathogenesis of IgAN is not well defined, but it is widely accepted that Gd-IgA1 is one of the key effector culprit molecules. IgG and/or IgA autoantibodies recognise Gd-IgA1 to form immune complexes (containing Gd-IgA1) which deposit in the glomeruli, activating mesangial cells and leading to the local overproduction of cytokines, chemokines, and complements, ultimately resulting in podocyte and tubulointerstitial injury via humoral crosstalk.

IgA1 galactose deficiency is related to decreased expression and activity of core 1 β-1,3-galactosyltransferase (C1GalT1) which adds galactose to GalNAc (Ju et al. 2002), elevated expression and activity of α-*N*-acetylgalactosaminide α-2,6-sialyltransferase 2 (ST6GalNAc-II) which adds sialic acid to GalNAc (Raska et al. 2007). Reduced expression or activity of C1GalT1 usually occurs in peripheral blood mononuclear cells (PBMCs) of IgAN patients. A positive correlation between the serum level of Gd-IgA1 and IgAN progression has also been shown (Zhao et al. 2012). It has been suggested that the serum levels of Gd-IgA1 could be used as a biomarker to predict the progression of IgAN. In addition, the serum level of the IgG autoantibodies is associated with disease activity and disease progression (Berthoux et al. 2012), further supporting the crucial role of the autoantibodies in the pathogenesis of IgAN. Recent clinical studies also suggest that using elevated levels of circulating antiglycan antibodies as a disease biomarker outperforms serum Gd-IgA1 levels (Yanagawa et al. 2014).

In a study involving 2249 cases of IgAN, all cases (100%) had positive staining for IgA, with 43% for IgG and 54% for IgM, and in most cases, the tissue sample showed stronger staining for the lambda light chain than for the kappa light chain (Magistroni et al. 2015). A small cohort study confirmed that mesangial IgG co-deposits were present in patients with IgAN, which was associated with more severe clinical features, including more capillary-wall IgA deposits and persistent urinary abnormalities. Multivariate analysis indicated that mesangial IgG co-deposit was an independent determinant of disease progression (Wada et al. 2013).

Despite its prevalence, the precise pathogenesis of IgAN remains unclear. However, current biochemical and genetic data suggest that the overproduction of Gd-IgA1 has been identified as one of the key effector molecules in IgAN pathogenesis. However, the vast majority of patients with IgAN fail to respond to B cell-depleting therapy using rituximab (Lafayette et al. 2017). Adversely, deleting B cells increases the risk of infection, which may aggravate disease progression. The discovery that mesangial cells have the capacity to synthesise and secrete IgA, especially the Gd-IgA1, introduces mesangial cell-derived Igs into the pathogenesis of IgAN, and suggests that IgA deposited in the mesangium can be synthesised in situ (Fig. 8.1). Bacterial components, such as SAC, SEB, and LPS, stimulate mesangial cells to induce the excessive production and granular deposition of Gd-IgA1 in the extracellular matrix of mesangial cells. Moreover, SAC can increase mesangial cell proliferation, and LPS can promote the production of inflammatory cytokines (IL-6, MCP-1, and IL-8). These results indicate that excessive IgA production in pathological conditions may be involved in the pathogenesis of IgAN. It is reasonable to speculate that a small amount of IgA produced by mesangial cells maintains the physiological functions of the cells under physiological conditions, but with the stimulation of pathogenic factors, excessive Gd-IgA1 deposition in the mesangial area promotes the proliferation of mesangial cells and the secretion of proinflammatory cytokines and chemokines, leading to podocyte and tubulointerstitial injury via humoral crosstalk. This hypothesis may explain the failure of the rituximab-mediated B cell depletion therapy in most patients with IgAN, and the relatively high proportion (4–16%) of IgA deposits observed in autopsies. Therefore, revealing the regulatory relationships between mucosal infections and the production of IgA in mesangial cells will aid in developing more targeted approaches to block IgA overproduction and effectively treat IgAN.

To sum up, it is necessary to further explore and clarify the role of Gd-IgA1 produced by mesangial cells in IgA nephropathy, and clarify the new strategy of taking Gd-IgA1 produced by mesangial cells as a new target.

8.2.2 Physiological and Pathological Significance of Non-B-Igs Expression in Podocytes

8.2.2.1 Non B-Igs in Podocytes

Jing et al. were the first to have demonstrated the expression of Igs, particularly IgG4, Igκ, and Igλ (Fig. 8.2), in the cytoplasm and on the cell membrane of podocytes (Jing et al. 2018). They noted stronger positive staining of IgG, the transcripts of IgG, Igκ, and Igλ, along with the IgG subclasses (including IgG1, IgG3, and IgG4) are found in podocytes. Moreover, several predominant VHDJH or Vκ-Jκ rearrangement patterns showed in podocyte-derived IgG, such as VH3–

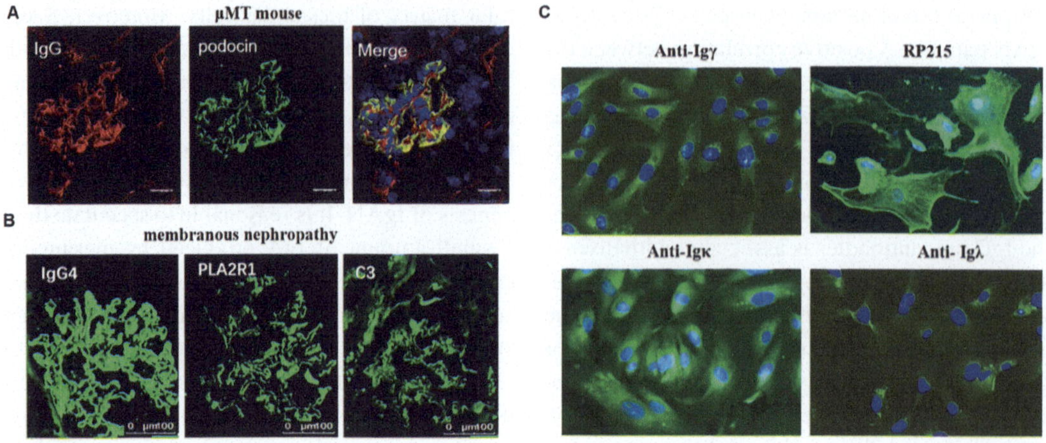

Fig. 8.2 The expression and location of IgG in the kidney and podocytes. (**a**) In B cell-deficient µMT mouse, positive staining of IgG was detected in podocytes. The podocin is a specific marker for podocytes. (**c**) Renal biopsy findings in a patient with MN. Immunofluorescence image showing strong staining of IgG4, accompanied by positive staining of PLA2R1 and weak staining of C3 along the capillary walls. (**c**) Positive staining of IgG heavy chains (including Ig γ4) and light chains (Ig-κ and Igλ) was detected in a human podocyte cell line (× 200)

23/D3–22/JH4, VH3–23/D3–16/JH4, and VH3–23/D4–17/JH6, as well as Vκ1–39/Jκ1.

Interestingly, SAC significantly upregulated the expression level of IgG and altered the conservative rearrangements in the VHDJH pattern in podocytes; the VH3–23/D1–7/JH6, VH3–23/D2–8/JH6, VH3–30/D3–16/JH6, and Vκ1–12/Jκ4 rearrangement patterns were frequently found in SAC-treated podocytes. These findings suggest that pathogens or their metabolites can change the rearrangement pattern and specificity of the IgG variable region. However, the specific mechanism and significance need to be elucidated.

The establishment of immortalised podocyte cell lines has provided opportunities for in vitro podocyte research. However, these cells do not retain some of the original characteristics of podocytes. Deng et al. (2021a) further explored if the Ig transcription occurs in human primary podocytes. They sorted a single podocyte from four cases of IgAN, three cases of MN, one case of ischaemic nephropathy, and one healthy donor. A total of 235 single cells were used for podocyte identification and Igs transcript amplification using nested PCR combined with Sanger sequencing. They identified 48 single cells as podocytes based on the expression of specific marker genes, NPHS1 or/and NPHS2, and the CD19 transcript was used to exclude the possibility of B cell contamination. Interestingly, they found that all five classes of Ig heavy chain transcripts, including IgG4, exist in a single podocyte.

Further analysis revealed that all five classes of podocyte-derived Igs displayed a classic VHDJH rearrangement pattern, with typical N nucleotide and P nucleotide additions at the VH–DH junctions and DH–JH junctions. The rate of nonfunctional VHDJH recombination (4.2%) was similar to that observed in B cells. Interestingly, podocytes showed a higher frequency usage of the VH1 gene family, such as IGHV1–18, IGHV1–24, and IGHV1–8, compared to B cells. Every single podocyte displayed a unique VHDJH recombination, which also showed diversity within an individual. Surprisingly, some podocytes shared the same VHDJH pattern, either between different podocytes or across individuals, which was considered impossible, particularly given the small number of B cells involved. Additionally, 50% of single podocytes expressed more than one functional VHDJH pattern, with a higher frequency of multi-clonal VHDJH patterns shown in the podocytes from patients.

In addition, Deng et al. (2021a) also found that podocyte-derived Ig gene rearrangements showed clear somatic hypermutation (SHM) in the variable gene segments, and SHM primarily occurred at the hotspot motif (RGYW/WRCY, W = A/T, R = A/G, Y = C/T), which is regarded as classic antigen-induced SHM. The SHM and hotspot mutation frequency of IgG in patients were significantly higher compared to the healthy control group. Furthermore, higher SHM and hotspot mutation frequency in IgM were exclusively observed in patients with IgAN.

In a preliminary functional study conducted by Jing et al. they found that siRNA targeting Igγ reduced cell volume, viability, and adhesive capability. These results indicate that podocyte-derived IgG is associated with cellular bioactivities, including survival, proliferation, and adhesion (Jing et al. 2018).

8.2.2.2 IgG Produced by Podocytes Is a Potential Cause Leading to Membranous Nephropathy

Primary membranous nephropathy (pMN) is an immune-mediated disease that results from the deposition of IgG (predominantly IgG4) in the subepithelial and on the outer surface of the glomerular capillary wall (Lai et al. 2015). IgG has been found to be associated with the MN glomerular lesion since the 1960s, but the autoimmune nature of the antibody specificity was not determined until 2009, when Beck identified a target autoantigen, M-type phospholipase A2 receptor (PLA2R), in human MN. About 70% of adult patients with primary MN (pMN) exhibited IgG4 autoantibody against PLA2R expressed in podocytes (Beck Jr et al. 2009). In 2014, Tomas et al. reported a second IgG4 autoantibody specific to thrombospondin type 1 domain-containing 7A (THSD7A), another podocyte membrane antigen (Tomas et al. 2014). Subsequently, circulating autoantibodies against THSD7A have been observed in a small percentage of patients with pMN (5–10%), whose PLA2R antibodies test negative. The autoimmune response against either PLA2R or THSD7A has been identified in most MN patients, and autoantibodies against either antigen are MN-specific because these antibodies cannot be detected in other proteinuric nephropathies. Furthermore, several studies have demonstrated that both autoantibody titres are positively correlated with disease activity and patient outcomes (Beck Jr et al. 2011; Bech et al. 2014; Hoxha et al. 2014; Ruggenenti et al. 2015), and low autoantibody levels at diagnosis predict spontaneous remission (Hofstra et al. 2012; Timmermans et al. 2015), whereas high baseline anti-PLA2R antibody levels predict a high risk of relapse and progressive loss of kidney function (Kanigicherla et al. 2013). Testing for antibodies against M-PLA2R and THSD7A has provided a new approach for diagnosis, immunological activity surveillance, and prognosis in patients with primary MN (Angioi et al. 2018).

To date, various medications used to treat MN, such as alkylating agents (cyclophosphamide or chlorambucil), steroids, calcineurin inhibitors, and so forth (Ruggenenti et al. 2017), have adverse side effects, causing myelotoxicity, infections, cancer, or nephrotoxicity. In addition, most patients become dependent on treatment and relapse after treatment withdrawal. Therefore, there is a significant need for improved therapeutics.

The identification of nephrogenic autoantibodies targeting podocyte antigens, including PLA2R and THSD7A, provides a clear pathophysiological basis for specifically targeting B cells to inhibit the production of autoantibodies and the subepithelial deposition of immune complexes. About two-thirds of patients with pMN achieve remission after receiving the RTX, an anti-CD20 monoclonal antibody, intervention. This targeted therapy reduces autoantibodies production and potentially prevents complement activation and kidney injury. Following a single dose of RTX, MN patients have relatively stable plasma IgG levels, even in the absence of circulating $CD19^+$ and $CD20^+$ B cells. However, since RTX treatment does not benefit about one-third of pMN patients, as membranous nephropathy is caused by the destruction of glomerular basement membrane caused by the deposition of IgG4 between podocyte and glomerular basement membrane, and podocyte itself produces IgG4, so it is reasonable to speculate that non-B-Igs may be involved to some extent in the pathogenesis of MN.

It is necessary to further explore and clarify the role of IgG4 produced by podocyte in MN, and clarify the new strategy of taking IgG4 produced by podocyte as a new target.

8.2.3 IgG Produced by Renal Tubular Epithelial Cells May be Involved in Renal Fibrosis

Proximal tubular epithelial cells (PTECs) are the predominant cell type in the kidney and are responsible for functions such as reabsorption and endocrine functions. They play a crucial role in renal repair and the progression of chronic kidney disease (Schlondorff 2008). Moreover, PTECs have immune functions, as they are capable of expressing toll-like receptors (TLR1, 2, 3, 4, 9) (Donadio et al. 2014), as well as antigen-presenting molecules, including MHCII, CD74, CD80, and CD86, both in a normal or diseased state (Breda et al. 2019). These innate immune properties of PTECs enable them to respond to a wide range of stimuli, subsequently producing and releasing bioactive mediators such as proinflammatory cytokines, chemokines, and complement components to induce interstitial inflammation and fibrosis.

Tubular basement membrane (TBM) immune deposits (TBMID) have been observed in various kidney diseases, such as lupus nephritis, membranous nephropathy, IgG4-associated nephritis, and other autoimmune diseases (Bracamonte et al. 2007). The predominant type of deposition on TBM is IgG as evidenced by its presence in about 1% (73/7177) renal allograft biopsy specimens. No monoclonal IgA or IgM TBMID was found (Sawada et al. 2019). In a retrospective study of renal transplants, IgG deposits were observed in 43% (41/95) of TBM cases (Orfila et al. 1991).

Proximal tubular epithelial cells (PTECs) expressed neonatal Fc receptor (FcRn), which retains the ability to bind IgG in a pH-dependent way and mediates transcytosis of IgG (Kobayashi et al. 2002). However, it has not been reported whether PTECs themselves express Igs.

To determine if renal tubular epithelial cells can also express Ig, Deng et al. (2021b) analysed IgG and Ig light chains (Igκ and Igλ) presented in human renal tissue using immunohistochemistry. Positive staining of IgG heavy chain, Igκ and Igλ were detected not only in the proximal tubular epithelial cells but also in epithelial cells of distal convoluted tubule. Unlike using a commercial anti-IgG antibody that showed a wider recognition in glomeruli and tubules, the use of RP215, a specific antibody for non-B-IgG, demonstrated the specificity for IgG expression in renal tubular epithelial cells (Fig. 8.3).

To eliminate the interference of the residual blood in the kidney cortex and obtain direct evidence of IgG expression in PTECs, Deng et al. (2020) used specific antibodies for PTEC markers, CD10 and CD13, to sort single PTECs from two healthy donors using FACS. Among the sorted 111 PTECs, only IgG heavy chain was successfully amplified in 82% of single PTECs (91/111) single PTECs, whereas none of the Igα, μ, ε, or δ chains were detected. Sanger sequencing results showed functional and classic VHDJH rearrangement with nucleotide additions at the junctions and somatic hypermutations in the VH region of PTECs-derived IgG. Interestingly, most single PTECs displayed unique VHDJH rearrangements, which also showed relative diversity within individuals. However, PTEC-derived IgG repertoires exhibited more conservative VHDJH rearrangements, with VH1–24/DH2–15/JH4 being the predominant pattern in 12.6% (14/111) PTECs. Moreover, ten pairs of single PTECs shared the same VHDJH rearrangements with identical junctions across different individuals, and so do five pairs of single PTECs within the same individual. These results collectively demonstrate limited diversity of PTEC-derived IgG repertoires.

In addition, an immortalised proximal tubular epithelial cell line, HK-2, was selected to further confirm the IgG expression (Deng et al. 2021b). Immunofluorescence and Western blotting analysis demonstrated the expression of Igγ, Igκ, and

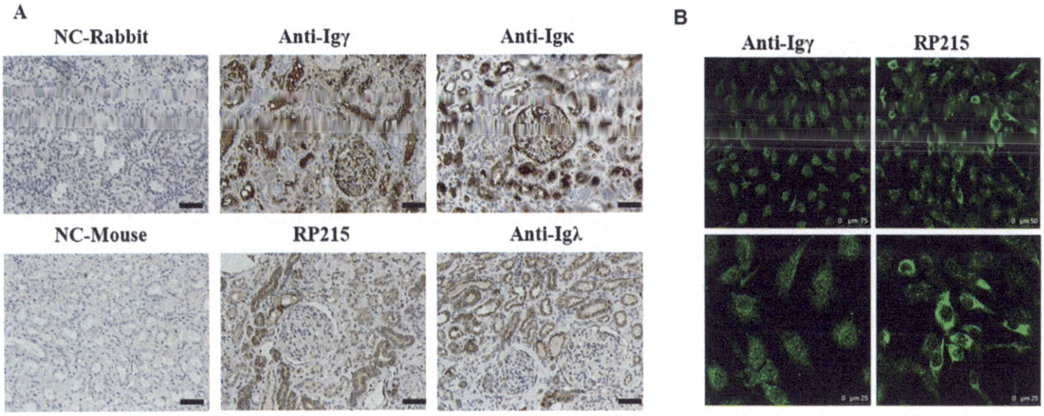

Fig. 8.3 The expression and location of IgG in the kidney and PTECs (Deng et al. 2021b). (**a**) Positive staining of Igγ, Igκ, and Igλ in renal tubular epithelial cells from a normal kidney. The red and blue arrows indicate proximal tubular epithelial cells and distal convoluted tubular cells, respectively. (**b**) Positive staining of IgG heavy chain was found in PTEC cell line HK-2. Scale bar, 75 μm (upper panel), 25 μm (lower panel)

Igλ in HK-2, with particularly stronger staining observed on the cell membrane by RP215 (Fig. 8.3). Similarly, Igγ in HK-2 also displayed a conservative V(D)J rearrangement pattern. Furthermore, RAG1/2 and AID, which are necessary for V(D)J rearrangement and somatic hypermutation in B cells, were detected in HK-2, indicating that similar mechanisms of Ig synthesis as in B cells may occur in HK-2.

Interstitial nephritis and fibrosis are common kidney diseases (Kodner and Kudrimoti 2003). PTECs play an important role in restoring tissue integrity or promoting the progression of chronic kidney disease (Matic et al. 2020). Transforming growth factor β1 (TGF-β1) is a master profibrotic mediator that contributes to the pathogenesis of renal tubular epithelial-mesenchymal transition (EMT) and renal fibrosis. During EMT, renal tubular epithelial cells lose their epithelial characteristics (such as a decrease in E-cadherin) and upregulated α-SMA expression, which mediates further renal fibrosis. Preliminary studies on the biological activity of IgG in PTECs have shown that TGF-β1 could upregulate IgG expression in PTECs, suggesting that the IgG may potentially play a role in EMT and contribute to further renal fibrosis (Deng et al. 2021b). However, the physiological and pathological significance of IgG expressed in PTECs requires further investigation.

8.2.4 Non B-Igs May be Involved in Other Diseases Caused by Pathogenic Ig Deposition in the Kidneys

8.2.4.1 Non B-Igs May be Involved in Lupus Nephritis

Lupus nephritis (LN) is the most common cause of serious kidney injury in systemic lupus erythematosus (SLE), affecting approximately 50% of all lupus patients. Roughly 10–20% of patients with LN may progress to end-stage renal disease (ESRD) (Almaani et al. 2017). Although the precise mechanism of LN progression remains elusive, a substantial body of evidence suggests that the production of anti-dsDNA autoantibodies and the formation of glomerular immune complex deposits are crucial initial events in the pathogenesis of LN. Immune complexes can deposit in multiple kidney sites, such as the mesangium, subendothelial region, and/or subepithelial space near the glomerular basement membrane of the kidney, which can instigate an inflammatory response, activate the complement pathway, resulting in the infiltration of neutrophils and the release of proinflammatory cytokines, and ultimately lead to chronic scarring and ESRD (Jorge et al. 2019). In addition to anti-dsDNA immune complex deposits, IgG, IgA, IgM, and complement are commonly found as mesangial, subendothelial, and subepithelial deposits. Tubulointerstitial lesions, including immune complex depositions (ICDs) in the tubular basement membrane, as demonstrated by granular staining for IgG, are among the pathological features of LN (George 2011).

Given that B cells are implicated in the pathogenesis of LN through the production of autoantibodies, a series of clinical trials involving B-cell depletion therapies with monoclonal antibodies have been conducted. However, the EXPLORER and LUNAR studies demonstrated that the addition of rituximab to immunosuppressant and corticosteroid therapy failed to benefit patients with class III or IV lupus nephritis (Rovin et al. 2012; Merrill et al. 2010). Similarly, the BELONG trial showed that the addition of ocrelizumab to standard therapy did not improve clinical outcomes (Mysler 2013). The failure of these trials indicates that B cells alone may not be the sole source of autoantibodies in LN, and Ig produced by local renal parenchymal cells may be involved in its pathogenesis and progression.

8.2.4.2 Igs in Renal Amyloidosis

Amyloidosis is a rare disease caused by extracellular deposition of protein in an insoluble β-pleated physical conformation. The kidney is a common organ of amyloid deposition. Ig-derived amyloidosis is one of the most prevalent types of renal amyloidosis and can be induced by Ig light chains, heavy chains, or combinations (Kidd and Carl 2016).

Immunoglobulin light chain amyloidosis (AL) is a dominant type of renal amyloidosis, caused by the insoluble amyloid fibrillar aggregates of Ig light chains, leading to structural and functional damage in multiple organs. Somatic mutations in the variable region of light chains reduce the fold stability, favouring the production of variable light chain domains that can cause amyloidosis. However, the exact product of amyloid formation and the precise molecular mechanisms underlying the targeting of the organ by amyloid are unknown. Several studies have shown that certain structural features associated with the IGLV gene and gene family increase the risk of specific organ involvement. For example, IGLV6–57 has been linked to renal involvement and is present more frequently in AL systemic amyloidosis than in the normal B cell repertoire. Mesangial cells have a tendency to form amyloid fibrils when treated with IGLV6–57 (Merlini et al. 2018).

AL amyloidosis is classified as a plasma cell dyscrasia in which clonal plasma cells produce a monoclonal Ig protein (M protein). The M protein light chains form insoluble fibrils, which subsequently deposit into the extracellular matrix of the kidney, leading to massive proteinuria and renal dysfunction. Currently, it is not clear whether the Igs in renal amyloidosis originate from B cells or if Igs produced by renal parenchyma cells are involved.

8.2.5 Prospect

The pivotal role of the kidney in haemofiltration renders it susceptible to toxins or metabolites present in the blood. Furthermore, the kidney often serves as the site for the accumulation of antigen–antibody complexes, which subsequently activate complement to cause renal pathology. Most kidney diseases are related to Igs, such as IgAN, MN, IgM nephropathy, IgG4-RKD, renal amyloidosis, and so forth. Igs deposits in the kidney directly or indirectly induce renal injury (Tecklenborg et al. 2018). B cell depletion utilising monoclonal antibodies targeting CD20 are not satisfactory (Rovin et al. 2012; Merrill et al. 2010; Mysler 2013).

Igs are found to be widely expressed in various renal parenchymal cells, such as mesangial cells, podocytes, endothelial cells (unpublished data), and renal tubular epithelial cells (Fig. 8.4). These non B-Igs are involved in cell proliferation, survival, and adhesion, and are upregulated in response to pathogenic factors such as SAC, Ang II, or TGF-β1. This discovery suggests that these locally produced Igs have significantly pathological role in the pathogenesis of renal diseases and update our current understanding of the mechanism underlying kidney diseases pathogenesis. More importantly, once it is confirmed that these locally generated Igs are the pathogenic factors of various kidney diseases, it will

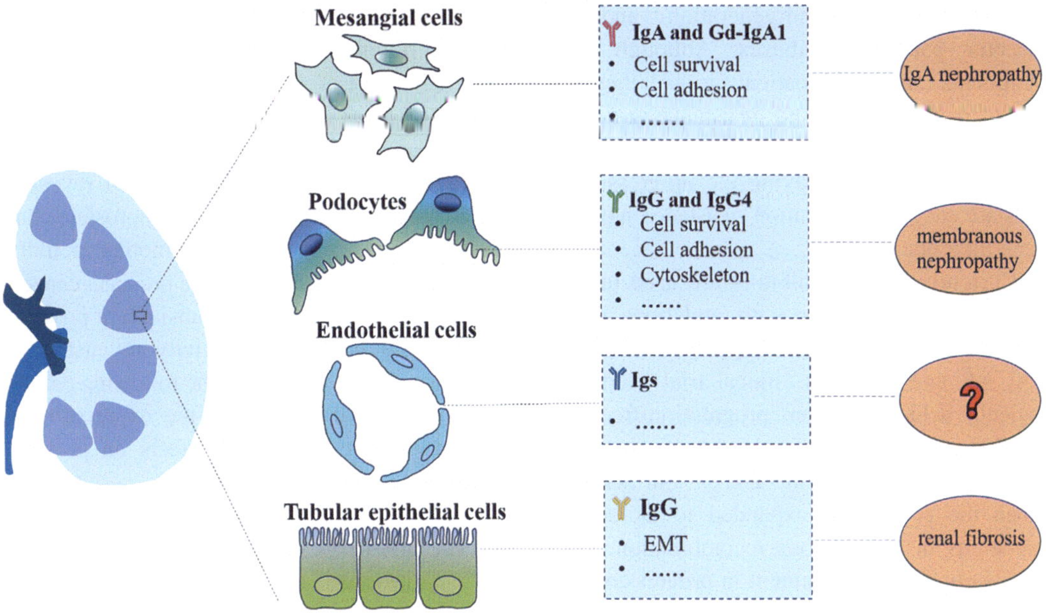

Fig. 8.4 Renal parenchyma cells express immunoglobulins and may be involved in renal diseases

revolutionise the treatment strategies for these diseases.

Current experimental evidence suggests that Igs can be expressed by mesangial cells, podocytes, endothelial cells (unpublished data), and renal tubular epithelial cells. These locally produced Igs may be involved in the pathogenesis of renal diseases, such as IgAN, MN and renal fibrosis.

8.3 Expression and Clinical Significance of Non B Cell-Derived Immunoglobulins in Urinary System Tumours

It has been introduced in other chapters of this book that non B cell-derived immunoglobulins are highly expressed in tumours of various epithelial origins, play a growth factor-like role, participate in the maintenance of tumour stemness phenotypes, and participate in the process of tumour immune escape. Similarly, the expression of non B cell-derived immunoglobulins is also detected in tumours of the urinary system, such as prostate cancer, renal cancer, and bladder cancer, and plays an important role in the development of tumours.

8.3.1 Prostate Cancer and IgG

Prostate cancer ranks as the most commonly diagnosed malignant tumour in males. According to statistics, in 2021, there were 268,490 new cases of prostate cancer in the United States, accounting for 27% of all newly diagnosed malignant tumours in men (Siegel et al. 2022; Miller et al. 2022).

Androgen promotes the metabolism and biosynthesis of prostate cells by binding to the androgen receptor (AR), which is essential for maintaining cell growth. As a result, hormone therapy, which either inhibits androgen secretion or blocks androgen binding to its receptors, is currently the standard treatment for patients with locally advanced and metastatic prostate cancer. However, prostate cancer patients receiving endocrine therapy eventually develop resistance to the treatment and progress to castration-resistant prostate cancer (CRPC). Existing treatments for CRPC include chemotherapy with docetaxel and second-generation androgen receptor pathway inhibitors. Although these treatments can prolong patient survival, they cannot provide a cure, and resistance gradually emerges (Yuan et al. 2014), Currently, the molecular mechanisms underlying the progression of prostate cancer to castration resistance have not been fully elucidated.

Currently, PARP inhibitors are used to treat metastatic CRPC cases with mutations in the homologous recombination repair genes BRCA2, BRCA1, and ATM. In clinical trials, although patients achieved better progression-free survival, there was no significant benefit in overall survival (Agarwal et al. 2022). Furthermore, when the scope was expanded to include 15 mutations in homologous recombination repair genes, no clear improvement in progression-free survival was observed (Stopsack 2021). In the TRITON2 clinical trial, for patients with BRCA1 or BRCA2 mutations, the PARP inhibitor demonstrated an objective response rate of 43.5% in terms of imaging and 54.8% in terms of PSA, which greatly limited its application to the population. Therefore, there is a need to explore new drug treatment targets to expand the potential patient population.

Due to the low immunogenicity of prostate cancer, the limited number of tumour-associated antigens and neoantigens in the microenvironment (Cha et al. 2020), and the downregulation of PD-L1 expression in many advanced cases, the current clinical trials involving immune checkpoint inhibitors (ICI) in prostate cancer have been unsuccessful. The outcomes of these clinical trials have shown that patients do not benefit from the treatment. Therefore, the identification of novel immune checkpoints may help overcome the challenge of applying existing ICIs in prostate cancer (Nicholson and Fong 2020).

Current studies have shown that SIA-IgG is widely expressed in various epithelial malignancies such as lung cancer, oesophagus cancer, liver cancer, pancreatic cancer, colorectal cancer, breast cancer, cervical cancer, endometrial carcinoma of the uterus, kidney cancer, bladder cancer, and prostate cancer. It shows growth factor-like effects on tumour cells, promotes cell migration, and is related to tumour differentiation and poor prognosis (Qiu et al. 2003; Chen and Gu 2007; Zhang et al. 2012; Cui et al. 2020; Yang et al. 2013; Sheng et al. 2016a, b; Qin et al. 2020). In addition, SIA-IgG has been found to play a role in tumour immune escape. Qin et al. revealed that SIA-IgG is highly expressed in prostate cancer tissue, exhibits tumour-promoting activity and immune escape effects in prostate cancer, and acts as an initiator in the transition of prostate cancer from hormone sensitivity to castration resistance (Qin et al. 2020). The following part of this chapter will summarise the expression of SIA-IgG and its biological effects on prostate cancer.

8.3.1.1 SIA-IgG Expression in Prostate Cancer

Several research groups have investigated prostate cancer cell lines and have found ectopic expression of IgG in prostate cancer tissue. It is expressed in both cell membrane and cytoplasm, but mainly in cytoplasm. Moreover, IgG expression is significantly lower in the benign prostate hyperplasia tissues than in prostate cancer tissues, Qin et al. downloaded 492 cases of prostate cancer and normal prostate tissues (52 cases) from the TCGA database and added the transcriptome data of 100 cases of normal prostate tissues from the GTEx database. IgG four subtypes (IGHG1–4) mRNA expression levels in cancer tissues are higher than those in normal tissues, and Igγ expression is positively correlated to histological grade and Gleason score (Liu et al. 2012; Pan et al. 2013; Xu et al. 2016). However, previous studies have been limited by the inadequacy of the antibodies used to distinguish between B cell-derived IgG and tumour-derived IgG (SIA-IgG), and cannot obtain reliable protein-level experimental results. In contrast, Tang et al. have identified a tumour-derived IgG called SIA-IgG, which possesses a non-classical glycosylation site that can be recognised by its specific antibody, RP215, distinguishing it from B cell-derived IgG (Tang et al. 2018). With the help of RP215, Qin et al. further confirm the high expression of SIA-IgG in prostate cancer tissues at the protein level, and the expression level of

SIA-IgG is positively correlated with the Gleason score and clinical stage of the patients, and for the first time in the single-cell transcriptomics data of prostate cancer, the expression of IgG is detected in three epithelial cells: basal epithelial cells, luminal epithelial cells, and neuroendocrine epithelial cells, at the same time it is confirmed that SIA-IgG produced by tumour cells can be secreted to the outside of prostate cancer cells to play a corresponding tumour-promoting effect.

8.3.1.2 SIA-IgG Promote the Development of Prostate Cancer

In 2012, Gu et al. found that the ectopic expression of IgG in prostate cancer tissue was associated with tumour cell differentiation, and clinical staging and grading. They proposed IgG as a biomarker for prostate cancer and a potential target for drug therapy (Liu et al. 2012). Subsequently, other researchers reported that the knockdown of IGHG1 using siRNA in prostate cancer cell lines (LNCaP and PC3) could inhibit cells from entering the division phase, restrain cell proliferation, migration, and invasion, and promote apoptosis (Pan et al. 2013; Xu et al. 2016). Qin et al. identified the SIA-IgG expressed by prostate cancer cells, which promotes tumour cell proliferation, invasion, and migration, and the expression level of SIA-IgG is negatively correlated with AR: the expression of SIA-IgG in castration-resistant cell lines with AR low expression is much higher than that in hormone-sensitive prostate cancer cell lines. At the tissue level, it is confirmed in prostate cancer tissue samples receiving neoadjuvant endocrine therapy that after treatment, the androgen signalling pathway is inhibited and SIA-IgG is highly expressed. Mechanism studies have shown that AR negatively regulates SOX2 transcription, while SIA-IgG is positively regulated by SOX2. After the AR signalling pathway is inhibited, the negative regulation of SOX2 and downstream SIA-IgG is released, which further leads to upregulation of SIA-IgG expression, promoting the activation of downstream tumour-related pathways such as AKT, MEK/ERK Simultaneously start the EMT process (Fig. 8.5) (Qin et al. 2020), and finally leads to the occurrence of castration-resistant prostate cancer. In addition, preliminary experimental results show that the killing T cells around prostate cancer cells with high IgG expression decreased, and the communication with T cell killing-related cells is weakened, suggesting that IgG may be involved in the process of immune escape of prostate cancer. In vivo experiments confirmed that a monoclonal antibody targeting SIA-IgG can inhibit tumour growth, suggesting that it is a potential drug target for prostate cancer (Sung et al. 2021).

8.3.2 Renal Cancer and IgG

Renal cancer is one of the most common malignancies in the genitourinary system, with an estimated 431,288 new cases and 179,368 deaths worldwide in 2020 (Sung et al. 2021). Targeted therapy plays a crucial role in the clinical management of RCC, especially when chemotherapy and radiotherapy demonstrate limited efficacy (Gray and Harris 2019). However, long-term exposure to these drugs, either an anti-angiogenesis inhibitor or an mTOR inhibitor, often leads to the development of drug resistance, highlighting the need for more specific markers and effective targets for RCC.

Sheng et al. have found that SIA-IgG is expressed in RCC cell lines (CAKI-I, 786–0, and ACHN), which is present in the cell cytoplasm, filopodia structure, and extracellular area. And the positive staining of SIA-IgG was frequently detected to locate in the cytoplasm of cancer cells in cRCC tissues (66/75, 88%) and in the adjacent normal kidney tissues (63/75, 84%) which is significantly associated with histological grades, tumour size, and the recurrence rate. Moreover, high SIA-IgG levels tended to be poor differentiation and advanced stage, which is not conducive to the survival of RCC patients. Functionally, knockdown SIA-IgG would lead to the remarkable inhibition of renal cancer cell proliferation ability, cell survival, cell migration, and invasion ability, which supposed that SIA-IgG is an important tumour-promoting factor, which may serve

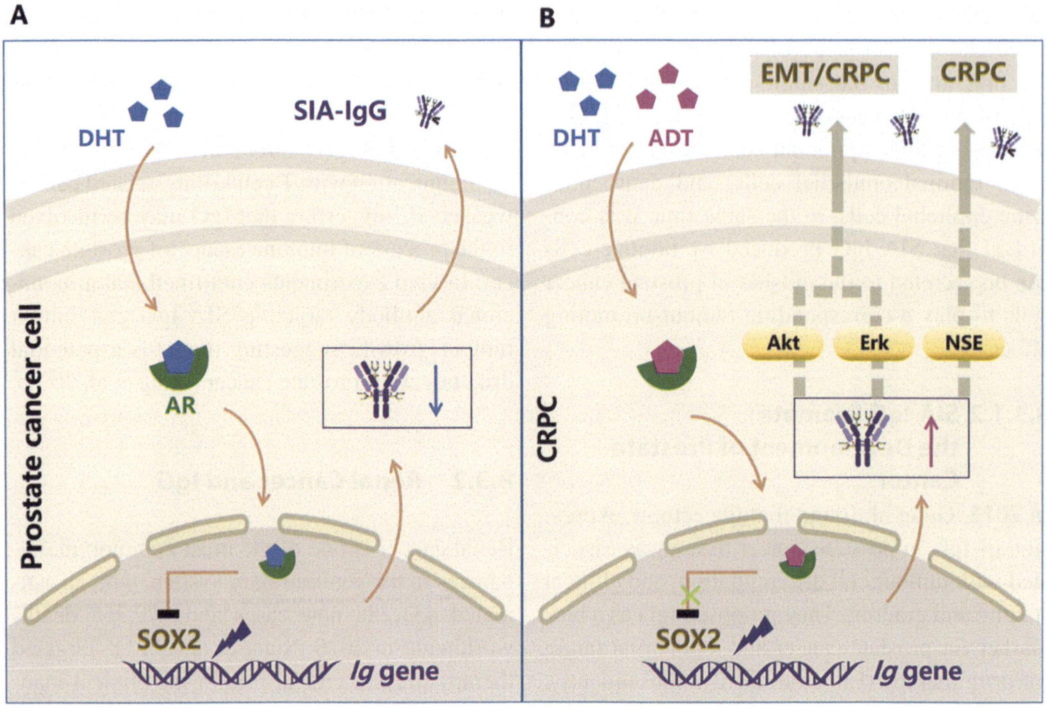

Fig. 8.5 Proposed model of SIA-IgG-mediated CRPC. In hormone-sensitive prostate cancer cells, SIA-IgG is inhibited by DHT (dihydrotestosterone) activated androgen signalling through SOX2 (**a**). After receiving ADT (androgen deprivation therapy), the androgen signalling pathway is inhibited, resulting in the upregulation of SOX2 expression, which further induces the upregulation of SIA-IgG expression and activates downstream Akt, Erk, NSE, signalling, leads to the transformation of prostate cancer into CRPC (castration against prostate cancer) (**b**)

as a potential therapeutic target of RCC treatment (Sheng et al. 2016a).

8.3.3 Bladder Cancer and IgG

In 2013, Liang et al. first identified the IgG expression in bladder cancer cell lines (Liang et al. 2013). In 2021, Sheng et al. identified the existence of IgG in bladder cancer cell lines (EJ, 5637, and BIU87) and bladder cancer tissues at mRNA and protein levels, Moreover, SIA-IgG expression is significantly correlated with a high histological grade, tumour size, and disease recurrence. Functional test results show that SIA-IgG can promote the proliferation, invasion, and migration of bladder cancer cells (Sheng et al. 2016b).

Furthermore, among the different bladder cancer cell lines the bladder cancer-VHDJH rearrangements had distinct usage and features that differed from those of B-cell-derived IgG, as well as a low frequency of mutation. For example, the usage of VH3 and IgJH4 was frequent in the bladder cancer cells and the same mutations are frequently present among different VHDJH rearrangements derived from bladder cancer cells. Since V(D)J rearrangement is a critical process for Ig production and diversity to recognise and neutralise highly varied antigens, IgG transcripts with unique patterns of VHDJH rearrangements determine the distinct biological activity in bladder cancer cells (Sheng et al. 2016b). These results suggested that IgG expression might be used as a prognostic marker and a potential therapy target in bladder cancer.

8.3.4 Prospects

SIA-IgG is highly expressed in tumours of the urinary system, and is involved in the progression of tumours, and mediates the occurrence of castration resistance of tumour cells in prostate cancer, which provides a theoretical reference for early drug combination therapy in this field in the future. For renal cancer that is insensitive to radiotherapy and chemotherapy, it provides a potential drug treatment target, which may break through the dilemma of existing treatment methods. Research on artificial cells designed for this target in bladder cancer, such as CAR-T cells, is underway. In future, it may be applied through local administration of the bladder to benefit patients.

References

Agarwal N et al (2022) Talapro-3: a phase 3, double-blind, randomized study of enzalutamide (ENZA) plus talazoparib (TALA) versus placebo plus enza in patients with DDR gene mutated metastatic castration-sensitive prostate cancer (mCSPC). Am Soc Clin Oncol 33:S273

Almaani S, Meara A, Rovin BH (2017) Update on lupus nephritis. Clin J Am Soc Nephrol 12(5):825–835

Angioi A et al (2018) Treatment of primary membranous nephropathy: where are we now? J Nephrol 31(4):489–502

Bech AP et al (2014) Association of anti-PLA$_2$R antibodies with outcomes after immunosuppressive therapy in idiopathic membranous nephropathy. Clin J Am Soc Nephrol 9(8):1386–1392

Beck LH Jr et al (2009) M-type phospholipase A2 receptor as target antigen in idiopathic membranous nephropathy. N Engl J Med 361(1):11–21

Beck LH Jr et al (2011) Rituximab-induced depletion of anti-PLA2R autoantibodies predicts response in membranous nephropathy. J Am Soc Nephrol 22(8):1543–1550

Berthoux F et al (2012) Autoantibodies targeting galactose-deficient IgA1 associate with progression of IgA nephropathy. J Am Soc Nephrol 23(9): 1579–1587

Bracamonte E et al (2007) Tubular basement membrane immune deposits in association with BK polyomavirus nephropathy. Am J Transplant 7(6):1552–1560

Breda PC et al (2019) Renal proximal tubular epithelial cells exert immunomodulatory function by driving inflammatory CD4(+) T cell responses. Am J Physiol Ren Physiol 317(1):F77–F89

Cha HR, Lee JH, Ponnazhagan S (2020) Revisiting immunotherapy: a focus on prostate cancer. Cancer Res 80(8):1615–1623

Chen Z, Gu J (2007) Immunoglobulin G expression in carcinomas and cancer cell lines. FASEB J 21(11):2931–2938

Cortazar FB, Stone JH (2015) IgG4-related disease and the kidney. Nat Rev Nephrol 11(10):599 609

Couser WG (2017) Primary membranous nephropathy. Clin J Am Soc Nephrol 12(6):983–997

Cui M et al (2020) High expression of cancer-derived glycosylated immunoglobulin G predicts poor prognosis in pancreatic ductal adenocarcinoma. J Cancer 11(8):2213–2221

Deng H et al (2018) Expression of immunoglobulin A in human mesangial cells and its effects on cell apoptosis and adhesion. Mol Med Rep 17(4):5272–5282

Deng Z et al (2020) Single-cell RNA sequencing confirms IgG transcription and limited diversity of V_HDJ_H rearrangements in proximal tubular epithelial cells. Sci Rep 10(1):19657

Deng Z et al (2021a) Single-cell sequencing confirms transcripts and V_HDJ_H rearrangements of immunoglobulin genes in human podocytes. Genes (Basel) 12(4):472

Deng Z et al (2021b) Expression of immunoglobulin G in human proximal tubular epithelial cells. Mol Med Rep 23(5):327

Donadio ME et al (2014) Toll-like receptors, immunoproteasome and regulatory T cells in children with Henoch–Schonlein purpura and primary IgA nephropathy. Pediatr Nephrol 29(9):1545–1551

George CT (2011) Systemic lupus erythematosus. N Engl J Med 365(22):2110–2121

Gray RE, Harris GT (2019) Renal cell carcinoma: diagnosis and management. Am Fam Physician 99(3):179–184

Hofstra JM et al (2012) Antiphospholipase A2 receptor antibody titer and subclass in idiopathic membranous nephropathy. J Am Soc Nephrol 23(10):1735–1743

Hoxha E et al (2014) Phospholipase A2 receptor autoantibodies and clinical outcome in patients with primary membranous nephropathy. J Am Soc Nephrol 25(6):1357–1366

Hoxha E, Reinhard L, Stahl RAK (2022) Membranous nephropathy: new pathogenic mechanisms and their clinical implications. Nat Rev Nephrol 18(7):466–478

Huang J et al (2008) Expression of immunoglobulin gene with classical V–(D)–J rearrangement in mouse brain neurons. Int J Biochem Cell Biol 40(8):1604–1615

Huang J et al (2009) Expression of immunoglobulin gene with classical V–(D)–J rearrangement in mouse testis and epididymis. J Histochem Cytochem 57(4):339–349

Jiang C et al (2014) Immunoglobulin G expression in lung cancer and its effects on metastasis. PLoS One 9(5):e97359

Jing Z et al (2018) Expression of immunoglobulin G in human podocytes, and its role in cell viability and adhesion. Int J Mol Med 41(6):3296–3306

Jorge A et al (2019) Renal transplantation and survival among patients with lupus nephritis: a cohort study. Ann Intern Med 170(4):240–247

Ju T et al (2002) Cloning and expression of human core 1 beta1,3-galactosyltransferase. J Biol Chem 277(1):178–186

Kanigicherla D et al (2013) Anti-PLA2R antibodies measured by ELISA predict long-term outcome in a prevalent population of patients with idiopathic membranous nephropathy. Kidney Int 83(5):940–948

Kidd J, Carl DE (2016) Renal amyloidosis. Curr Probl Cancer 40(5–6):209–219

Kobayashi N et al (2002) FcRn-mediated transcytosis of immunoglobulin G in human renal proximal tubular epithelial cells. Am J Physiol Ren Physiol 282(2):F358–F365

Kodner CM, Kudrimoti A (2003) Diagnosis and management of acute interstitial nephritis. Am Fam Physician 67(12):2527–2534

Lafayette RA et al (2017) A randomized, controlled trial of rituximab in IgA nephropathy with proteinuria and renal dysfunction. J Am Soc Nephrol 28(4):1306–1313

Lai WL et al (2015) Membranous nephropathy: a review on the pathogenesis, diagnosis, and treatment. J Formos Med Assoc 114(2):102–111

Li J et al (2014) Two ultrastructural distribution patterns of immunoglobulin G in human placenta and functional implications. Biol Reprod 91(5):128

Liang PY et al (2013) Overexpression of immunoglobulin G prompts cell proliferation and inhibits cell apoptosis in human urothelial carcinoma. Tumour Biol 34(3):1783–1791

Liu Y et al (2012) IgG gene expression and its possible significance in prostate cancers. Prostate 72(6):690–701

Lu Y et al (2017) Single-cell RNA-sequence analysis of mouse glomerular mesangial cells uncovers mesangial cell essential genes. Kidney Int 92(2):504–513

Magistroni R et al (2015) New developments in the genetics, pathogenesis, and therapy of IgA nephropathy. Kidney Int 88(5):974–989

Matic M et al (2020) Renal proximal tubular epithelial cells: review of isolation, characterization, and culturing techniques. Mol Biol Rep 47(12):9865–9882

Merlini G et al (2018) Systemic immunoglobulin light chain amyloidosis. Nat Rev Dis Primers 4(1):38

Merrill JT et al (2010) Efficacy and safety of rituximab in moderately-to-severely active systemic lupus erythematosus: the randomized, double-blind, phase II/III systemic lupus erythematosus evaluation of rituximab trial. Arthritis Rheum 62(1):222–233

Miller KD et al (2022) Cancer treatment and survivorship statistics, 2022. CA Cancer J Clin 72(5):409–436

Mysler EF (2013) Efficacy and safety of ocrelizumab in active proliferative lupus nephritis: results from a randomized, double-blind, phase III study. Arthritis Rheum 65(9):2368–2379

Nicholson LT, Fong L (2020) Immune checkpoint inhibition in prostate cancer. Trends Cancer 6(3):174–177

Niu N et al (2011a) Expression and distribution of immunoglobulin G and its receptors in an immune privileged site: the eye. Cell Mol Life Sci 68(14):2481–2492

Niu N et al (2011b) Expression and distribution of immunoglobulin G and its receptors in the human nervous system. Int J Biochem Cell Biol 43(4):556–563

Orfila C et al (1991) Immunofluorescent deposits on the tubular basement membrane in human renal transplant. Nephron 57(2):149–155

Pan B et al (2013) Suppression of IGHG1 gene expression by siRNA leads to growth inhibition and apoptosis induction in human prostate cancer cell. Mol Biol Rep 40(1):27–33

Qin C et al (2020) Cancer-driven IgG promotes the development of prostate cancer though the SOX2-CIgG pathway. Prostate 80(13):1134–1144

Qiu X et al (2003) Human epithelial cancers secrete immunoglobulin G with unidentified specificity to promote growth and survival of tumor cells. Cancer Res 63(19):6488–6495

Raska M et al (2007) Identification and characterization of CMP-NeuAc:GalNAc-IgA1 alpha2,6-sialyltransferase in IgA1-producing cells. J Mol Biol 369(1):69–78

Roberts IS (2014) Pathology of IgA nephropathy. Nat Rev Nephrol 10(8):445–454

Rovin BH et al (2012) Efficacy and safety of rituximab in patients with active proliferative lupus nephritis: the lupus nephritis assessment with rituximab study. Arthritis Rheum 64(4):1215–1226

Ruggenenti P et al (2015) Anti-phospholipase A2 receptor antibody titer predicts post-rituximab outcome of membranous nephropathy. J Am Soc Nephrol 26(10):2545–2558

Ruggenenti P, Fervenza FC, Remuzzi G (2017) Treatment of membranous nephropathy: time for a paradigm shift. Nat Rev Nephrol 13(9):563–579

Sawada A et al (2019) Monoclonal immunoglobulin G deposits on tubular basement membrane in renal allograft: is this significant for chronic allograft injury? Nephrol Dial Transplant 34(4):711–717

Schlondorff DO (2008) Overview of factors contributing to the pathophysiology of progressive renal disease. Kidney Int 74(7):860–866

Shao W et al (2016) Identification of liver epithelial cell-derived Ig expression in mu chain-deficient mice. Sci Rep 6:23669

Sheng Z et al (2016a) IgG is involved in the migration and invasion of clear cell renal cell carcinoma. J Clin Pathol 69(6):497–504

Sheng Z et al (2016b) Involvement of cancer-derived IgG in the proliferation, migration and invasion of bladder cancer cells. Oncol Lett 12(6):5113–5121

Siegel RL et al (2022) Cancer statistics, 2022. CA Cancer J Clin 72(1):7–33

Stopsack KH (2021) Efficacy of PARP inhibition in metastatic castration-resistant prostate cancer is very different with non-BRCA DNA repair alterations: reconstructing prespecified endpoints for cohort B from the phase 3 PROfound trial of Olaparib. Eur Urol 79(4):442–445

Sung H et al (2021) Global cancer statistics 2020: GLOBOCAN estimates of incidence and mortality worldwide for 36 cancers in 185 countries. CA Cancer J Clin 71(3):209–249

Suzuki H et al (2011) The pathophysiology of IgA nephropathy. J Am Soc Nephrol 22(10):1795–1803

Tang J et al (2018) Lung squamous cell carcinoma cells express non-canonically glycosylated IgG that activates integrin-FAK signaling. Cancer Lett 430:148–159

Tecklenborg J, Clayton D, Siebert S, Coley SM (2018) The role of the immune system in kidney disease. Clin Exp Immunol 192(2):142–150

Timmermans SA et al (2015) Anti-PLA2R antibodies as a prognostic factor in PLA2R-related membranous nephropathy. Am J Nephrol 42(1):70–77

Tomas NM et al (2014) Thrombospondin type-1 domain-containing 7A in idiopathic membranous nephropathy. N Engl J Med 371(24):2277–2287

Wada Y et al (2013) Clinical significance of IgG deposition in the glomerular mesangial area in patients with IgA nephropathy. Clin Exp Nephrol 17(1):73–82

Wu L et al (2020) High levels of immunoglobulin expression predict shorter overall survival in patients with acute myeloid leukemia. Eur J Haematol 105(4):449–459

Wyatt RJ, Julian BA (2013) IgA nephropathy. N Engl J Med 368(25):2402–2414

Xu Y et al (2016) IgG silencing induces apoptosis and suppresses proliferation, migration and invasion in LNCaP prostate cancer cells. Cell Mol Biol Lett 21:27

Yanagawa H et al (2014) A panel of serum biomarkers differentiates IgA nephropathy from other renal diseases. PLoS One 9(5):e98081

Yang B et al (2013) Correlation of immunoglobulin G expression and histological subtype and stage in breast cancer. PLoS One 8(3):e58706

Yasutake J et al (2015) Novel lectin-independent approach to detect galactose-deficient IgA1 in IgA nephropathy. Nephrol Dial Transplant 30(8):1315–1321

Yatim KM, Lakkis FG (2015) A brief journey through the immune system. Clin J Am Soc Nephrol 10(7):1274–1281

Yuan X et al (2014) Androgen receptor functions in castration-resistant prostate cancer and mechanisms of resistance to new agents targeting the androgen axis. Oncogene 33(22):2815–2825

Zhang L et al (2012) Expression of immunoglobulin G in esophageal squamous cell carcinomas and its association with tumor grade and Ki67. Hum Pathol 43(3):423–434

Zhao N et al (2012) The level of galactose-deficient IgA1 in the sera of patients with IgA nephropathy is associated with disease progression. Kidney Int 82(7):790–796

Zheng S, Cao J, Geng L (2001) Structure and expression of colorectal cancer related immunoglobulin novel gene SNC73. Zhonghua Yi Xue Za Zhi 81(8):485–488

Zhu Z et al (2017) Immunoglobulin M, a novel molecule of myocardial cells of mice. Int J Biochem Cell Biol 88:172–180

Functions and Clinical Significance of Myocardial Cell-Derived Immunoglobulins

9

Zhu Zhu, Meng Zhang, and Xiaoyan Qiu

Abstract

Immunoglobulins (Igs) have been widely accepted to be exclusively expressed by B cells. Nonetheless, this theory is challenged by mounting evidence which suggests that Igs can also be generated by non B cells (non B-Ig), including cardiomyocytes (CM). Non B-Ig exhibits unique physical and chemical characteristics, unique variable region sequences and functions, which diverge from those of B-Ig. For instance, non B-Ig demonstrates hydrophobicity, limited diversity in the variable region, and extracellular matrix protein activity. Likewise, cardiomyocytes can express different classes of Igs, including IgM, IgG, and free Igκ light chains (cardiomyocyte derived-Igs, CM-Igs). In particular, CM-Igs can be secreted into the extracellular space in various cardiovascular diseases, such as myocardial ischaemia and myocardial fibrosis where they might be involved in complement activation and direct damage to cardiomyocytes. Nevertheless, the precise pathological activity of CM-Igs remains unclear. Recently, Zhu et al. focused on studying the sequence characteristics and functions of CM-Igκ; they discovered that the CM-Igκ exhibits a unique VJ recombination pattern, high hydrophobicity, and is principally located on the intercalated discs and cross striations of the cardiomyocytes. Interestingly, loss of Igκ in cardiomyocytes results in structural disorders in intercalated discs and dysfunction in myocardial contraction and conduction. Mechanically, Igκ promotes the stabilisation of plectin, a cytoskeleton cross-linker protein that connects desmin to desomsome, to maintain the normal structure of the intercalated disc. This finding indicates that CM-Igκ plays an integral role in maintaining cytoskeleton structure. Consequently, it is imperative to reveal the physiological functions and mechanisms of pathological injury associated with CM-Igs.

Keywords

Immunoglobulin · Antibody · IgM · Igκ · FLC · Cardiomyocyte · Myocardial infarction · Intercalated disc · Viral myocarditis · Heart failure

Z. Zhu (✉) · X. Qiu
Department of Immunology, School of Basic Medical Sciences, Peking University, Beijing, China
e-mail: zhuzhu@bjmu.edu.cn; qiuxy@bjmu.edu.cn

M. Zhang
Department of Cardiology, Aerospace Center Hospital, Beijing, China

X. Qiu et al. (eds.), *Non B cell-Derived Immunoglobulins*, Advances in Experimental Medicine and Biology 1445, https://doi.org/10.1007/978-981-97-0511-5_9

9.1 General Considerations

Cardiac muscle is a type of involuntary striated muscle that contributes to the heart's pumping function via its systole and diastole (Garcia-Dorado 2004). It was formerly believed that Igs was an immune molecule not expressed in cardiomyocytes and therefore had no role in cardiomyocyte function.

However, during various cardiovascular diseases, including acute myocardial infarction (AMI), heart failure (HF), and myocarditis, Ig levels are frequently increased in the local area of injured tissue or peripheral blood, and they can mediate heart tissue injury through complement activation and other mechanisms. For example, IgM is often found to be upregulated and deposited in cardiac tissue during myocardial ischaemia/reperfusion injury, leading to complement activation (Krijnen et al. 2005). Free Ig light chain (FLC) is significantly upregulated in heart tissue during viral myocarditis, contributing to improved heart function by inhibiting virus replication and increasing IL-10 expression (Matsumori et al. 2010). FLC is also found to be elevated in the peripheral blood of HF patients and associated with a poor prognosis (Jackson et al. 2015). It has been considered that these Igs are autoantibodies produced by B cells. However, Zhu et al. have proven that cardiomyocytes themselves can produce Igs, including different types of Igs and FLC (Zhu et al. 2017). These CM-Igs showed unique physiological and pathological functions which will be discussed later in this chapter.

9.2 Elevated Ig Level and the Implications to Cardiovascular Diseases

9.2.1 IgM and Myocardial Infarction

Acute myocardial ischaemia-reperfusion injury (I/R) is a predominant cause of mortality in contemporary society. Although the restoration of blood flow (reperfusion) is critical for survival, it can also induce an acute inflammatory response known as I/R. The hallmark of reperfusion injury, which includes the activation of innate and adaptive immune responses, is amongst the pathological mechanisms underlying myocardial inflammation and inflammatory cell infiltration (Garcia-Dorado 2004; Frangogiannis et al. 2002; Eltzschig et al. 2004). Krijnen analysed 59 patients with myocardial infarction and discovered that IgM was deposited in the area of myocardial infarction, which was colocalised with complement C3 and C-reactive protein (CRP). These IgM deposits were mainly situated on the myocardial cell membrane, in the cytoplasm, and within the myocardial cross striation (Krijnen et al. 2005). Compared to their wild-type littermates, IgM-deficient mice developed significantly less I/R injury which was restored by supplementation with pooled serum or mouse IgM (Zhang et al. 2006).

These IgM have been considered as autoantibodies, which are from the circulation after ischemia/reperfusion. Is it self-expressed by myocardial cells? Is the pathological damage of myocardial cells caused by the release of Igs to the outside under cell stress conditions such as ischemia/hypoxia? These need to be answered.

9.2.2 IgG and Dilated Cardiomyopathy

IgG is the most abundant immunoglobulin in serum. It is not only a critical molecule in the anti-infection immunity process, but also a common autoantibody in various diseases, particularly viral myocarditis and dilated cardiomyopathy (DCM). IgG autoantibodies are more pathogenic than IgM autoantibodies. It is a consensus that there are multiple antibodies against myocardial cell components in the local and circulatory areas of various cardiomyopathy. These IgG autoantibodies include anti-myosin autoantibodies (Caforio et al. 1992), anti-beta-1 adrenergic receptor autoantibodies (Limas et al. 1990; Wallukat et al. 1991), anti-cTnI receptor autoantibodies (Okazaki et al. 2003), and anti-Na-K-ATPase autoantibodies (Baba et al. 2002). The production of these autoantibodies is believed to be triggered by the exposure of some autoantigens during myocardial injury, which is recognised by IgG autoantibodies produced by B cells. This recognition activates complements and immune cells such as T cells, leading to tissue damage. However, it has also been found that IgG can directly interact with IgG Fc receptor on the myocardial cell membrane, resulting in direct myocardial cell damage and activation of complements and immune cells (Staudt et al. 2007).

Similarly, these IgGs have been considered as autoantibodies, which are from B cells; it remains unclear if these autoantibodies are secreted from myocardial cells.

9.2.3 FLC and Amyloid Cardiomyopathy

Free light chain (FLC) amyloidosis is a protein misfolding disease caused by the secretion and deposition of misfolded monoclonal Ig light chains in multiple tissues, including the heart. Approximately half of patients with systemic amyloidosis experience cardiac dysfunction, manifested by impaired diastolic filling and restrictive cardiomyopathy, which rapidly progresses to congestive heart failure with a poor 5-year survival rate. Although the disease has been known since the mid-nineteenth century, it remains unclear how amyloid deposition leads to rapid cardiac dysfunction (Klein et al. 1989; Skinner et al. 1996; Dubrey et al. 1997). A recent study found that circulating FLC isolated from amyloidosis patients with heart failure can cause direct cardiomyocyte injury rather than contribute to fibrosis formation. More importantly, in these patients, only FLC derived from severe heart disease can cause ventricular dysfunction in vitro. However, the source of FLC and the specific mechanism through which FLC leads to direct myocardial damage remain unclear (Liao et al. 2001). For a long time, it was believed that systemic amyloidosis is secondary to multiple myeloma cells, and its pathogenic FLC is produced by myeloma, but in fact, only 12–15% of patients with myocardial amyloidosis are accompanied by, whereas most patients are exempt from multiple myeloma. Consequently, the origin of these abnormally elevated light chains deposited in heart tissue is unclear.

9.2.4 FLC and Heart Failure

Heart failure (HF) is a complex clinical syndrome characterised by symptoms such as breathlessness, fatigue, and peripheral oedema. Structural or functional cardiac abnormalities can lead to impaired ventricular filling or ejection of blood into the systemic circulation, resulting in elevated intracardiac pressures or reduced cardiac output. Heart failure has a high morbidity and mortality rate and remains a prevalent disorder worldwide. The causes of heart failure are numerous, and among them, inflammation is now believed to play an important role in its pathophysiology, including the activation of innate immune responses following tissue injury, elevated inflammatory cytokines, and oxidative stress (Elster et al. 1956; Bouras et al. 2014; van Kimmenade et al. 2012; Hartupee et al. 2013).

It was documented that increased FLC was found in HF. Colette et al. recently developed a novel assay for measuring kappa and lambda immunoglobulin FLCs together, known as combined free light chain (cFLC). She found that almost one-half of the 628 patients recently hospitalised with decompensated HF had elevated cFLC concentrations. Moreover, patients with elevated cFLC concentrations had an increased risk of death, suggesting that cFLC could be used as a novel prognostic indicator for heart failure. However, the cause of elevated cFLCs in patients with HF is unknown (Jackson et al. 2015).

Based on the above phenomenon and the fact that myocardial cells can produce FLC, we can assume that FLC secreted by myocardial cells can be deposited in the heart, and then activate the Mast cell, thereby promoting local inflammation. The FLC has a pro-inflammatory effect by activating mast cells, which play a central role in the inflammatory process, including producing mediators such as cytokines, causing apoptosis of cardiac myocytes, and promoting proliferation of non-myocardial cells. Compared with normal hearts, increased mast cell density was also observed in hearts with left ventricular systolic dysfunction (Redegeld et al. 2002; Patella et al. 1998; Hara et al. 1999).

9.2.5 FLC and Viral Myocarditis

Myocarditis is characterised by infiltration of mononuclear cells into the myocardium, and it presently stands as one of the most significant clinical challenges in cardiology. Mild myocarditis may lead to difficulties in breathing, while severe myocarditis can progress to acute heart failure and even sudden death (Cooper 2009; Sagar et al. 2012). The usual cause of myocarditis is pathogenic infection from agents such as viruses, bacteria, parasites, and fungi. However, autoimmune disorders and drug-induced hypersensitivity can also trigger myocarditis. Among these causes, viruses including enteroviruses, adenoviruses, influenza viruses, cytomegaloviruses, parvoviruses, herpes viruses, and the human immunodeficiency virus are primarily culpable (Dennert et al. 2008).

FLC has been demonstrated to have protective effects on viral myocarditis. In 2010, Akira Matsumori et al. found that FLC production was significantly enhanced during EMC virus infection in mice. Moreover, FLC showed protective effects on viral myocarditis. When the FLC antagonist F991 was administered during the initial phase of viral infection, it worsened the pathology of viral myocarditis, suggesting that FLC plays a protective role in myocarditis. Conversely, administering FLC supplements at the onset of viral infection reduced necrosis and notably improved the survival rate. The study suggests that FLC may inhibit viral replication, as demonstrated in an in vitro plaque assay. Indeed, the viral load in heart tissue was significantly reduced in animals treated with FLC. This effect is likely achieved through direct antiviral activity and an anti-inflammatory effect brought about by increased IL-10 expression. Moreover, the gene expressions of interferon-α and -γ tended to increase in the hearts of mice treated with FLC, suggesting that these effects may also contribute to the inhibition of viral replication and improvement of myocardial lesions (Matsumori et al. 2010).

The precise cause of elevated FLC in patients with cardiovascular disease, as mentioned earlier, remains unknown. In cardiac amyloidosis, the abnormally elevated FLC is thought to be produced by dysplastic plasma cells, but not all patients have multiple myeloma (MM). Furthermore, most patients with HF do not have MM. In addition, FLC levels remain elevated for at least several months after HF exacerbation (Shantsila et al. 2014). It is worth noting that activated B cells are upregulated during episodes of acute heart failure and then return to basal levels once it resolves (Burmeister et al. 2014; TorreAmione et al. 1996; Cordero-Reyes et al. 2014). The elevated FLC level may be more than just a result of general immune system activation.

9.3 Igs Can Be Expressed by Cardiomyocytes

Igs have also been detected in normal non B cells, such as epithelial cells, skin epidermal cells, and neurons. Increasing evidence shows that cardiomyocytes can also produce Igs. In 2009, Mehta et al. detected Igh and Igl transcripts in murine heart tissue and the cardiac cell line-HL-1 (Kang et al. 2009). Additionally, other groups found Ig transcripts in mice heart tissue using cDNA microarray analysis (Ashton et al. 2006; Vasu et al. 2007). However, these studies only detected Ig transcripts instead of Ig protein in cardiomyocytes.

Zhu et al. have demonstrated that cardiomyocytes can produce various Ig transcripts and proteins. To rule out the possibility of contamination from B cell-derived Igs, they used B cell-deficient mice-μMT mice, myocardial cell line-HL-1, and human-induced pluripotent stem cell-cardiomyocyte (iPS-CM). First, they found IgM to be expressed in mouse cardiomyocytes and located on the intercalated disc and cell membrane. Moreover, it was found that the V(D)J pattern of IgM heavy chain derived from myocardial cells displayed unique characteristics, such as IGHV1-39/D1-1/J4 was frequently found in HL-1 cells. Subsequently, to further identify if the use of the unique VDJ rearrangements pattern has a cell lineage specificity, Zhu et al. compared the unique VDJ rearrangements pattern from the cardiomyocyte with the VDJ rearrangement pattern expressed by fibroblasts from the same cardiac tissue, and found that the V(D)J pattern of IgM heavy chains between myocardial cells (IGHV1-59/D1-1/J2) and fibroblasts (IGHV1-82/D1-1/J4) were different, suggesting that V(D)J pattern correlates with cell type, which may play an important role in specific lineages in a manner that differs from classical antibody activity (Zhu et al. 2017). Subsequently, they also found Igκ expression in human and mouse cardiomyocytes. Similar to IgM, Igκ was also located on the intercalated disc and cross striation (Fig. 9.1), indicating its potential involvement in myocardial conduction and contraction. Single-cell sequencing technology was then used to analyse the variable region sequence of Igκ generated by cardiomyocytes. The results showed that the VκJκ usage pattern in Igκ produced by cardiomyocytes is also highly conserved, which is different from the pattern of diversity observed in B cells. IGKV17-121 (28.3%, 95/336), IGKV9-120 (25%, 84/336), and IGKV14-100 (15.5%, 52/336) were the dominant Vκ usages in cardiomyocytes of adult mice. Notably, IGKV9-120 was also the dominant Vκ usage in hepatocytes and other non B cells (Shao et al. 2016). However, these Vκ sequences, which are predominant in cardiomyocytes, accounted for less than 3% in B cells. This limited diversity suggests that cardiomyocyte-derived Igκ may play physiological roles other than antibody activities.

9.4 The Physiological Function of Igκ in Cardiomyocytes

The heart is an organ in which individual cardiomyocytes must adhere to and communicate with neighbour cells along longitudinally oriented fibres to transmit electrical impulses rapidly and drive muscle contraction. To achieve this goal, cardiomyocytes are connected end-to-end by a specialised structure known as the intercalated disc (ID). The ID is a highly organised structure arranged both transversely and longitudinally in

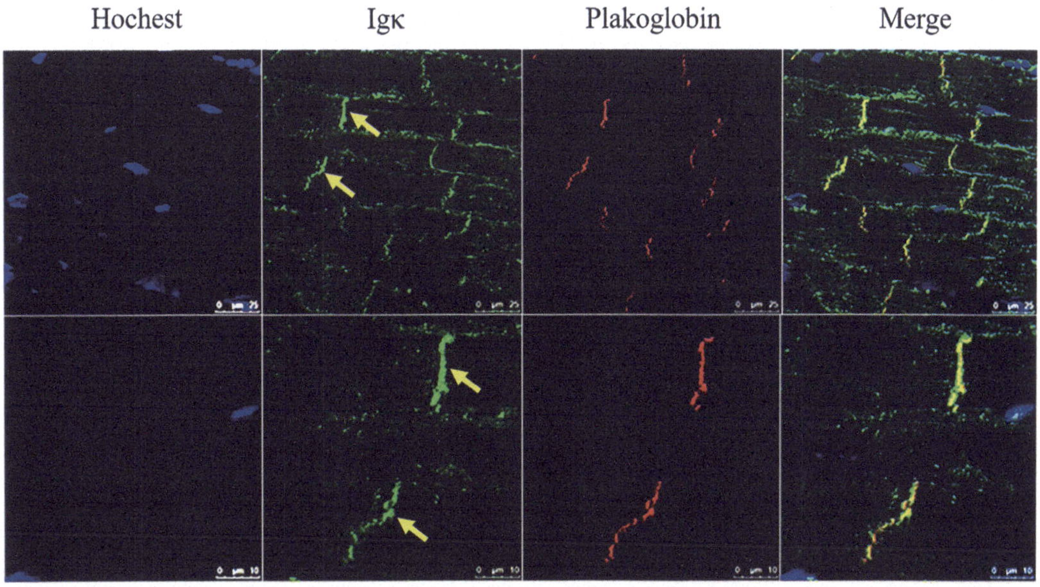

Fig. 9.1 Cardiomyocyte-derived Igκ colocalised with Plakoglobin on the intercalated disc. Positive Igκ staining (green) was detected on the intercalated disc (yellow arrow) of mice heart tissue and colocalised with plakoglobin, an ICD marker (red). Nuclei were detected with Hoechst (blue). Bar = 25 μm

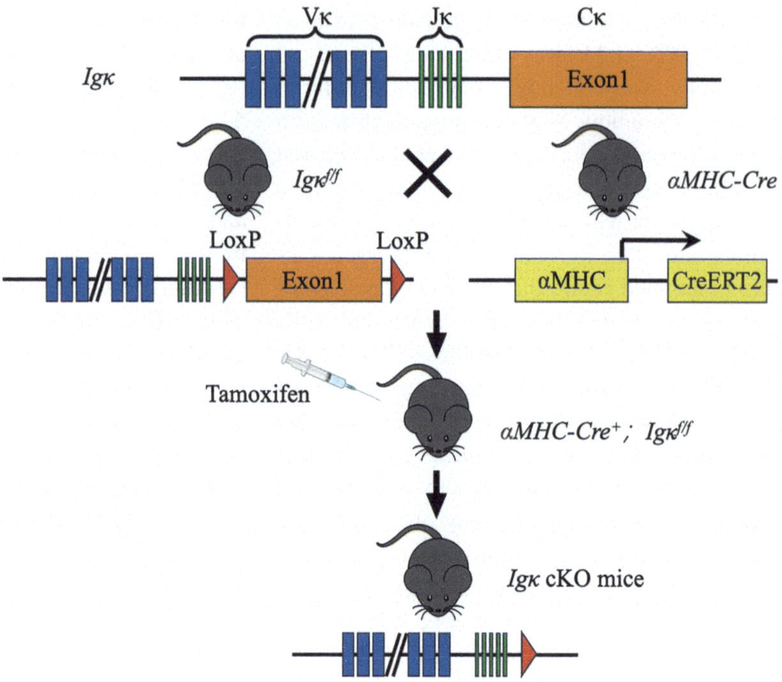

Fig. 9.2 Scheme for generation of cardiac conditional Igκ knock-out mouse (cKO). LoxP sites were inserted to flank exons in the *Igκ* gene constant region. Mice homozygous for LoxP insertion (f/f) were crossed with mice expressing Cre recombinase under the control of the αMHC promoter. After tamoxifen administration, Cre-mediated excision of the *Igκ* constant region resulted in a defect of *Igκ* gene structure and dysfunction

a staircase-like fashion with "steps" and "risers" (Hoyt et al. 1989; Shimada et al. 2004; Tandler et al. 2006). The ID is mainly composed of adhesion junctions (AJs), desmosomes, and gap junctions (GJs). Desmosomes and AJs can enhance the stability of ID by anchoring the actin cytoskeleton and intermediate filaments (IFs) to ID, thereby providing mechanical connection between adjacent cardiomyocytes. GJs facilitate the propagation of electrical stimuli throughout heart muscle cells by connecting the cytoplasm of neighbouring cardiomyocytes metabolically and electrotonically. AJs, GJs, and desmosomes collectively integrate the electrical and mechanical functions of the ID. In addition, desmosomes and AJs synergistically enhance mechanical coupling as an area composita. In the area composita, desmoplakin, a key component of desmosomes, is considered to associate with the intermediate filament desmin, while α-catenin, β-catenin, and vinculin in AJs are considered to associate with actin (Manring et al. 2018; Zhao et al. 2019; Gard et al. 2005; Li et al. 2005, 2008; Yang et al. 2006).

To investigate the function of Igκ produced by myocardial cells, an inducible mouse that specifically knocks out *Igk* in cardiomyocytes was established by Zhu et al. Tamoxifen was used to induce *Igk* knockout specifically in cardiomyocytes of 6–8-week-old mice (Fig. 9.2).

Using Igκ cKO mice, Zhu et al. examined the role of Igκ in myocardial contraction and electrical conduction. The results showed that Igκ cKO mice displayed a lower survival rate and reduced blood pressure. Echocardiography analysis demonstrated decreased contractility with lower ejection fraction (EF%) and fractional shortening (FS%) on day 3 after Igκ was knocked out. Furthermore, a contractile shortening assay was performed, and the results showed that myocytes from Igκ cKO mice exhibited a more attenuated contractile response than wild-type mice, indicating that Igκ deficiency impaired contractility capability. Moreover, the ID structure was disorganised and expanded with a decrease in the expression of N-cadherin, desmoplakin, and Cx43.

How does Igκ knockout affect the localisation and expression of ID-related proteins? Zhu et al. found that Igκ interacts with plectin, a large cytolinker protein which plays an important role in binding and promoting the assembly of actin and intermediate filaments (Favre et al. 2011; Eisenberg et al. 2013; Wiche et al. 2015; Matsubara et al. 2017). It has been reported that plectin interacts with desmoplakin on IDs and connects intermediate filaments/desmin to desmosomes (Castanon et al. 2013; Pieperhoff et al. 2012). Loss of plectin leads to disassembly of desmin and F-actin, and disruption of AJs and GJs, which results in abnormal electrical coupling of cardiomyocytes and reduced conduction velocity.

Igκ interacts with plectin, primarily depending on its variable region, and enhances the stability of plectin on IDs via the proteasome. However, the precise mechanism by which Igκ regulates plectin and its role in ID assembly still requires further investigation (Fig. 9.3).

9.5 Significance and Prospect

Although little attention has been paid to the essential physiological role of Ig molecule as structural proteins in the myocardium in cardiovascular research, as described in this chapter, growing evidence has proved that cardiomyocytes can produce various types of Igs, including IgM, IgG, IgE, and free Igκ. It has been demonstrated that cardiomyocytes-derived Igκ is located on the intercalated disc of the myocardium and interacts with plectin, a protein that plays an important role in microfilaments and intermediate filaments assembly and maintains the typical structure of the intercalated disc. This suggests that the Ig molecule is an unidentified myocardial structural protein and plays an indispensable physiological role. However, the physiological functions of other types of Igs produced by cardiomyocytes, such as IgM, IgG, and IgE, have not been revealed. It is believed that these Ig molecules also play essential roles in the physiological function of cardiomyocytes and heart.

In addition, as mentioned earlier, elevated levels of cardiomyocyte-derived Igs have been found in various heart diseases and have been associated with pathological damage to the myocardial tissue, which indicates that the Igs produced by cardiomyocytes might play a more important role in the pathogenesis of heart diseases than autoantibodies produced by B cells in the traditional theory. The diagnosis and treatment of related heart diseases will benefit if this hypothesis is further supported by more experimental evidence.

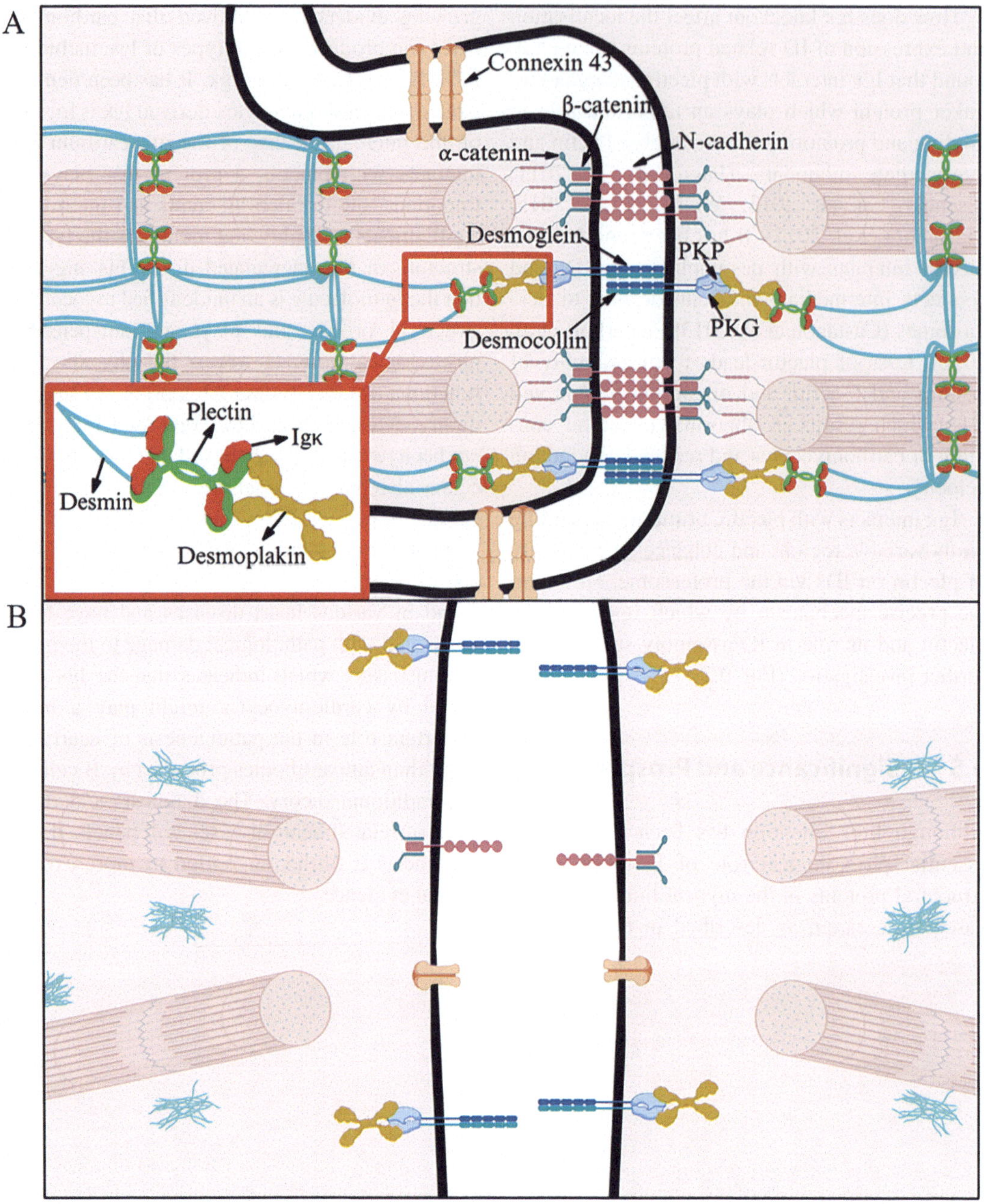

Fig. 9.3 The mechanism of Igκ in maintaining the stability of the intercalated disc. (**a**) The mechanism of Igκ to maintain the stability of the intercalated discs by binding with plectin. (**b**) The putative mechanism of Igκ gene knockout that leads to ID dysfunction

References

Ashton KJ et al (2006) Age-associated shifts in cardiac gene transcription and transcriptional responses to ischemic stress. Exp Gerontol 41(2):189–204

Baba A et al (2002) Autoantibodies produced against sarcolemmal Na-K-ATPase: possible upstream targets of arrhythmias and sudden death in patients with dilated cardiomyopathy. J Am Coll Cardiol 40(6):1153–1159

Bouras G et al (2014) Inflammation and chronic heart failure: from biomarkers to novel anti-inflammatory therapeutic strategies. Med Chem 10(7):682–699

Burmeister A et al (2014) The relationship between high-sensitivity CRP and polyclonal free light chains as markers of inflammation in chronic disease. Int J Lab Hematol 36(4):415–424

Caforio ALP et al (1992) Identification of alpha-cardiac and beta-cardiac myosin heavy-chain isoforms as major autoantigens in dilated cardiomyopathy. Circulation 85(5):1734–1742

Castanon MJ et al (2013) Plectin-intermediate filament partnership in skin, skeletal muscle, and peripheral nerve. Histochem Cell Biol 140(1):33–53

Cooper LT (2009) Myocarditis. N Engl J Med 361(4):423–424

Cordero-Reyes AM et al (2014) Stimulated B-cell products induce fibroblast production of promoters of negative remodeling-a novel concept of the role of B-cells in heart failure. J Card Fail 20(8):S24

Dennert R et al (2008) Heymans, acute viral myocarditis. Eur Heart J 29(17):2073–2082

Dubrey SW et al (1997) Familial and primary (AL) cardiac amyloidosis: echocardiographically similar diseases with distinctly different clinical outcomes. Heart 78(1):74–82

Eisenberg JL et al (2013) Plectin-containing, centrally localized focal adhesions exert traction forces in primary lung epithelial cells. J Cell Sci 126(16):3746–3755

Elster SK et al (1956) A study of C-reactive protein in the serum of patients with congestive heart failure. Am Heart J 51(4):533–541

Eltzschig HK et al (2004) Vascular ischaemia and reperfusion injury. Br Med Bull 70:71–86

Favre B et al (2011) Plectin interacts with the rod domain of type III intermediate filament proteins desmin and vimentin. Eur J Cell Biol 90(5):390–400

Frangogiannis NG et al (2002) The inflammatory response in myocardial infarction. Cardiovasc Res 53(1):31–47

Garcia-Dorado D (2004) Myocardial reperfusion injury: a new view. Cardiovasc Res 61(3):363–364

Gard JJ et al (2005) Remodeling of gap junctions and slow conduction in a mouse model of desmin-related cardiomyopathy. Cardiovasc Res 67(3):539–547

Hara M et al (1999) Mast cells cause apoptosis of cardiomyocytes and proliferation of other intramyocardial cells in vitro. Circulation 100(13):1443–1449

Hartupee J et al (2013) Positioning of inflammatory biomarkers in the heart failure landscape. J Cardiovasc Transl Res 6(4):485–492

Hoyt RH et al (1989) Saffitz, distribution and 3-dimensional structure of intercellular-junctions in canine myocardium. Circ Res 64(3):563–574

Jackson CE et al (2015) Combined free light chains are novel predictors of prognosis in heart failure. Jacc-Heart Failure 3(8):618–625

Kang BY et al (2009) LOX-1 dependent overexpression of immunoglobulins in cardiomyocytes in response to angiotensin II. Biochem Biophys Res Commun 379(2):395–399

Klein AL et al (1989) Doppler characterization of left-ventricular diastolic function in cardiac amyloidosis. J Am Coll Cardiol 13(5):1017–1026

Krijnen PA et al (2005) IgM colocalises with complement and C reactive protein in infarcted human myocardium. J Clin Pathol 58(4):382–388

Li J et al (2005) Cardiac-specific loss of N-cadherin leads to alteration in connexins with conduction slowing and arrhythmogenesis. Circ Res 97(5):474–481

Li JF et al (2008) N-cadherin haploinsufficiency affects cardiac gap junctions and arrhythmic susceptibility. Eur J Cell Biol 87(7):451–451

Liao RL et al (2001) Infusion of light chains from patients with cardiac amyloidosis causes diastolic dysfunction in isolated mouse hearts. Circulation 104(14):1594–1597

Limas CJ et al (1990) Anti-beta-receptor antibodies in human dilated cardiomyopathy and correlation with Hla-Dr antigens. Am J Cardiol 65(7):483–487

Manring HR et al (2018) At the heart of inter- and intracellular signaling: the intercalated disc. Biophys Rev 10(4):961–971

Matsubara T et al (2017) Regulation of osteoclast differentiation and actin ring formation by the cytolinker protein plectin. Biochem Biophys Res Commun 489(4):472–476

Matsumori A et al (2010) Effects of free immunoglobulin light chains on viral myocarditis. Circ Res 106(9):1533–1540

Okazaki T et al (2003) Autoantibodies against cardiac troponin I are responsible for dilated cardiomyopathy in PD-1-deficient mice. Nat Med 9(12):1477–1483

Patella V et al (1998) Stem cell factor in mast cells and increased mast cell density in idiopathic and ischemic cardiomyopathy. Circulation 97(10):971–978

Pieperhoff S et al (2012) The plaque protein myozap identified as a novel major component of adhering junctions in endothelia of the blood and the lymph vascular systems. J Cell Mol Med 16(8):1709–1719

Redegeld FA et al (2002) Immunoglobulin-free light chains elicit immediate hypersensitivity-like responses. Nat Med 8(7):694–701

Sagar S et al (2012) Myocarditis. Lancet 379(9817):738–747

Shantsila E et al (2014) Free light chains in patients with acute heart failure secondary to atherosclerotic coronary artery disease. Am J Cardiol 114(8):1243–1248

Shao WW et al (2016) Identification of liver epithelial cell-derived Ig expression in mu chain-deficient mice. Sci Rep 6:23669

Shimada T et al (2004) Cytoarchitecture and intercalated disks of the working myocardium and the conduction system in the mammalian heart. Anat Rec A Discov Mol Cell Evol Biol 280(2):940–951

Skinner M et al (1996) Treatment of 100 patients with primary amyloidosis: a randomized trial of melphalan, prednisone and colchicine versus colchicine only. Am J Med 100(3):290–298

Staudt A et al (2007) Fc receptors IIa on cardiomyocytes and their potential functional relevance in dilated cardiomyopathy. Eur Heart J 28:794–794

Tandler B et al (2006) High resolution scanning electron microscopy of the intracellular surface of intercalated disks in human heart. Tissue Cell 38(6):417–420

TorreAmione G et al (1996) Proinflammatory cytokine levels in patients with depressed left ventricular ejection fraction: a report from the studies of left ventricular dysfunction (SOLVD). J Am Coll Cardiol 27(5):1201–1206

van Kimmenade RR et al (2012) Emerging biomarkers in heart failure. Clin Chem 58(1):127–138

Vasu VT et al (2007) Genome-wide screening of alpha-tocopherol sensitive genes in heart tissue from alpha-tocopherol transfer protein null mice (ATTP(−/−)). FEBS Lett 581(8):1572–1578

Wallukat G et al (1991) Autoantibodies against the beta-adrenergic-receptor in human myocarditis and dilated cardiomyopathy—beta-adrenergic agonism without desensitization. Eur Heart J 12:178–181

Wiche G et al (2015) Networking and anchoring through plectin: a key to IF functionality and mechanotransduction. Curr Opin Cell Biol 32:21–29

Yang Z et al (2006) Desmosomal dysfunction due to mutations in desmoplakin causes arrhythmogenic right ventricular dysplasia/cardiomyopathy. Circ Res 99(6):646–655

Zhang M et al (2006) The role of natural IgM in myocardial ischemia-reperfusion injury. J Mol Cell Cardiol 41(1):62–67

Zhao G et al (2019) Intercalated discs: cellular adhesion and signaling in heart health and diseases. Heart Fail Rev 24(1):115–132

Zhu Z et al (2017) Immunoglobulin M, a novel molecule of myocardial cells of mice. Int J Biochem Cell Biol 88:172–180

Cancer-Derived Immunoglobulin G and Pancreatic Cancer

10

Ming Cui and Xiaoyan Qiu

Abstract

Immunoglobulin (Ig) is traditionally believed to be produced solely by B cells. Nonetheless, mounting evidence has demonstrated that various types of Igs are extensively expressed in many cell types. Among them, IgG is found to be highly expressed in cancer cells and is thus labeled as cancer-derived IgG. Cancer-derived IgG shares identical fundamental structures with B cell-derived IgG, but displays several unique characteristics, including restricted variable region sequences and unique glycosylation modifications for those expressed by epithelial cancers. Cancer-derived IgG plays multiple crucial roles in carcinogenesis, including facilitating cancer invasion and metastasis, enhancing cancer stemness, contributing to chemoresistance, and remodeling the tumour microenvironment. Recent studies have discovered that cancer-derived sialylated IgG (SIA-IgG) is extensively expressed in pancreatic cancer cells and is predominantly located in the cytoplasm and on the cell membrane. Cancer-derived IgG expressed by pancreatic cancer presents a restrictive variable region sequence and contains a unique sialylation site of the Fab region. Functionally, cancer-derived IgG participates in pancreatic cancer progression via different mechanisms, such as promoting proliferation, facilitating migration and invasion, resisting apoptosis, inducing inflammation, and modulating the tumour microenvironment. SIA-IgG has shown potential as a clinical biomarker. The expression of SIA-IgG is associated with poor tumour differentiation, metastasis, and chemoresistance in pancreatic cancer. High expression of SIA-IgG can serve as an independent prognostic factor for pancreatic cancer. Additionally, SIA-IgG expression elevated with malignant progression for the precursor lesions of pancreatic cancer. These findings present a prospect of applying cancer-derived IgG as a novel diagnostic and therapeutic target in the management of pancreatic cancer, and aiding in overcoming the challenge in the treatment of this stubborn malignancy.

Keywords

Pancreatic cancer · Immunoglobulin · SIA-IgG · Diagnosis · Prognosis

M. Cui (✉)
Department of General Surgery, Peking Union Medical College Hospital, Chinese Academy of Medical Sciences, Peking Union Medical College, Beijing, China
e-mail: cuiming@pumch.cn

X. Qiu
Department of Immunology, School of Basic Medical Sciences, Peking University, Beijing, China
e-mail: qiuxy@bjmu.edu.cn

X. Qiu et al. (eds.), *Non B cell-Derived Immunoglobulins*, Advances in Experimental Medicine and Biology 1445, https://doi.org/10.1007/978-981-97-0511-5_10

10.1 The Physiological Function of the Pancreas

The pancreas is a crucial organ that plays a substantial role in digestion. The pancreas performs two main functions including exocrine function and endocrine function. With respect to exocrine function, pancreatic acinar cells and ductal cells collaborate to produce enzymes such as lipase, protease, and amylase that break down fats, proteins, and sugars. Regarding the endocrine function, several clusters of pancreatic endocrine cells produce hormones such as insulin, glucagon, gastrin, and somatostatin that assist in regulating blood sugar levels and stimulating stomach acids and appetite.

Previous studies have identified IgG expression in both normal pancreatic acinar cells and islet cells (Wan et al. 2015; Cui et al. 2020a). Thus, it is evident that IgG is broadly expressed in both exocrine and endocrine glands of the normal pancreas. Notably, IgG expressed in the normal pancreas does not present unique sialylation specifically identified in epithelial cancer-derived IgG, which is revealed by using antibodies with different specificities (Cui et al. 2020a).

In addition, the association between IgG expression in the pancreas and the diabetic condition has been preliminarily investigated (Wan et al. 2015). However, relevant studies are still limited to elucidate the structural characteristics and physiological functions of IgG expressed by the normal pancreas, and more uncharted territories need to be further explored in the future.

10.2 The Expression of IgG in Pancreatic Cancer

Pancreatic cancer continues to be one of the most lethal solid malignancies (Mizrahi et al. 2020). The incidence of pancreatic cancer has been increasing annually, whilst the 5-year survival rate stays at roughly 10% (Siegel et al. 2022; Tempero et al. 2017). Extensive experimental and epidemiological data demonstrate that pancreatic cancer is a genetic disease caused by the accumulation of genetic alterations in several specific genes, including *KRAS*, *CDKN2A*, *TP53*, and *SMAD4* (Iacobuzio-Donahue 2012; Storz and Crawford 2020). Given the high malignancy and metastatic potency of pancreatic cancer, it should be viewed as a systemic disease comprising primary and metastatic tumours.

Surgery remains the only curative method for pancreatic cancer, crucial to remove the primary lesion. Nevertheless, effective systemic treatment is equally vital to improve the clinical outcomes for pancreatic cancer patients (Tempero et al. 2017). Given the current challenges in pancreatic cancer management, novel biomarkers and therapeutic targets are urgently required to enhance early detection and systemic treatment strategies for pancreatic cancer patients.

The existence of an "Ig-like" protein in multiple types of cancer cells was initially identified in 1996 (Qiu and Yang 1996). Subsequently, several research groups have studied this molecule and proved the protein is Ig expressed by cancer cells, thus naming it cancer-derived Ig (Qiu et al. 2003; Cui et al. 2021). Cancer-derived Igs have been detected in various types of epithelial cancers, soft tissue tumours, and haematopoietic malignancies (Chen and Gu 2007; Chen et al. 2010; Zheng et al. 2007; Qiu et al. 2013). Among the above various types of cancer-derived Igs, IgG is significantly elevated and exhibits several unique characteristics.

V(D)J rearrangement is a critical process for Ig production and diversity to recognise and neutralise highly varied antigens (Chi et al. 2020). However, studies have shown that the V(D)J transcripts of cancer-derived Ig present restricted and unique V(D)J recombination patterns even among different tumour samples (Zheng et al. 2009), suggesting that cancer-derived Ig exhibits similar biological activity in different types of cancer cells. Interestingly, cancer-derived IgG expressed by epithelial cancer cells displays unique sialylation at asparagine 162 (Asn162) of the Fab region, which cannot be detected in B cell-derived IgG (Tang et al. 2018). Thus, this type of cancer-derived sialylated IgG (SIA-IgG) can be distinguished from B cell-derived IgG using a specific antibody recognising this aberrant sialylated epitope.

Regarding the mechanism of action, cancer-derived IgG acts as a growth factor-like molecule that directly promotes the malignant behaviours of cancer cells via activating multiple pathways, such as integrin/FAK-Src, MEK/ERK/c-Myc, and reactive oxygen species (ROS) (Tang et al. 2018; Wang et al. 2013; Chu et al. 2019). Moreover, cancer-derived IgG can also modulate the tumour microenvironment and promote tumour immune escape. Sialic acid-binding immunoglobulin-type lectins (Siglecs), crucial membrane receptors of immune cells, have been proven to be the receptors for SIA-IgG. Binding to Siglecs, SIA-IgG can significantly inhibit T cell proliferation and reduce effector T cell frequencies (Wang et al. 2020).

Conventionally, the relationship between Ig and pancreas has been previously studied in autoimmune pancreatitis (Silva et al. 2011; Maneva-Radicheva et al. 2014). In recent years, the existence of the heavy-chain constant region of Ig transcripts was also detected in the human pancreatic cancer cell line HC48 (Kimoto 1998). The expression of IgG at the protein level in pancreatic cancer was first described in 2011 (Li et al. 2011). In this study, IgG expression in pancreatic cancer tissues was identified by comparative proteomic analysis aimed at identifying differentially expressed proteins between pancreatic cancer tissues and adjacent noncancerous tissues. The existence of IgG was further confirmed in human pancreatic cancer cell lines SW1990 and HC48 using an anti-human IgG1 antibody (Li et al. 2011). In 2015, Wan et al. expanded the detection of Ig in pancreatic cancer and associated diabetes. They detected the expression of IgG heavy (γ) and light chains (λ and κ) in 50 cases of human pancreatic cancer by immunohistochemistry and in situ hybridisation. The adjacent pancreatic islets also showed strong positive signals of Ig expression. Furthermore, the expression of Ig was confirmed in four human pancreatic cancer cell lines AsPC-1, BxPC-3, MIA PaCa-2, and PANC-1 by Western blotting, immunofluorescence, and RT-PCR. The expression of RAG1, RAG2, and AID, which are essential enzymes for IgG functional recombination, was also detected in the above pancreatic cancer cell lines (Wan et al. 2015).

Previous studies have shown that SIA-IgG is the main functional component of cancer-derived IgG in epithelial cancer cells (Tang et al. 2018; Wang et al. 2020). Using a specific monoclonal antibody that specifically recognises SIA-IgG, the SIA-IgG was found strongly expressed in human pancreatic cancer cell lines BxPC-3 and T3M4, and moderately expressed in human pancreatic cancer cell lines AsPC-1, CFPAC-1, PANC-1, and HPAF. Immunofluorescence analysis showed that SIA-IgG was mainly located in the cytoplasm and on the cell membrane of pancreatic cancer cells. The expression of SIA-IgG has also been investigated in human pancreatic cancer tissues by immunohistochemistry. Both the monoclonal antibody RP215 and commercial anti-human IgG antibody were used in the immunohistochemical staining. The results showed that RP215 specifically detected SIA-IgG in pancreatic cancer cells, whilst the commercial antibody extensively stained IgG in lymphocytes, mesenchymal cells, normal pancreatic cells, and cancer cells (Fig. 10.1). SIA-IgG was detected in nearly 90% of PDAC tissues, with high expression observed in about 30% of PDAC tissues. The expression of SIA-IgG in PDAC tissues was significantly higher than that in peritumoral tissues (Cui et al. 2020a).

One of the most critical characteristics of cancer-derived IgG is its unique and conservative variable region. The rearranged transcripts of the variable region sequence of Igγ in human pancreatic cancer cell lines BxPC-3 and PANC1 were further amplified and analysed. The recombination of the $V_H\gamma D\gamma J_H\gamma$ variable region in BxPC-3 and PANC-1 cells showed a restricted pattern of V5-51/D3-16/J4, which is in accordance with that found in several other epithelial-derived cancers, such as lung cancer (Cui et al. 2020a; Zheng et al. 2009).

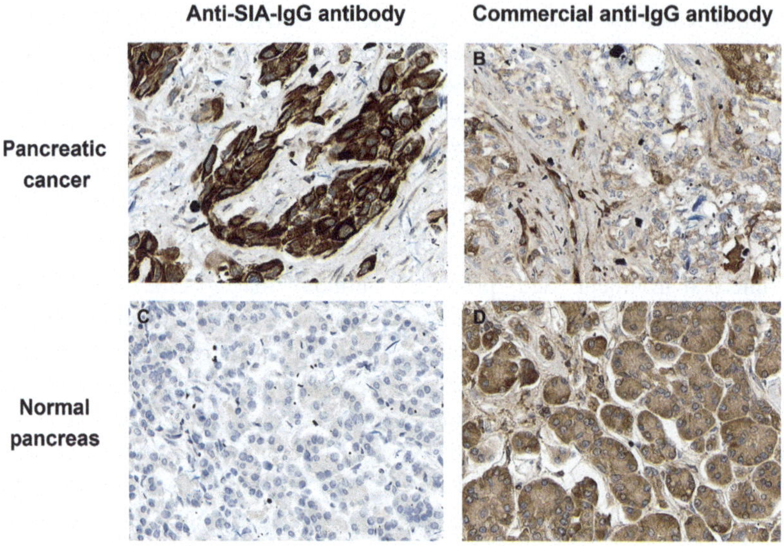

Fig. 10.1 Immunohistochemical staining of IgG by anti-SIA-IgG antibody and commercial anti-IgG antibody in pancreatic cancer and normal pancreas tissue

10.3 The Biological Activity of Cancer-Derived IgG in Pancreatic Cancer

Studies have investigated the functions of cancer-derived IgG in the tumorigenesis of pancreatic cancer. Silencing cancer-derived IgG significantly suppressed the capacity for proliferation, migration, and invasion of human pancreatic cancer cell lines BxPC-3 and T3M4 (Cui et al. 2020a). Similarly, Wan et al. discovered that siRNAs and antibodies targeting IgG triggered apoptosis of pancreatic cancer cell lines BxPC-3 and PANC-1 (Wan et al. 2015). Furthermore, Li et al. reported that the presence of IGHG1 in human pancreatic cancer downregulated the cytotoxic activity of natural killer (NK) cells and facilitated immune evasion mechanisms (Li et al. 2011). Chen et al. reported that IgG was the primary functional component of cancer cell-derived debris that promotes metastasis of pancreatic cancer by inducing inflammation via tumour-associated macrophages (Chen et al. 2019). According to data from the TCGA database, the density of tumour-promoting M2-type macrophages positively correlated with IgG expression. SIA-IgG was further demonstrated to have the potency to polarise macrophages from M0 type to tumour-promoting M2 type, thus it might mediate tumour immune escape of pancreatic cancer. The above findings suggest that cancer-derived IgG plays crucial roles in the tumorigenesis of pancreatic cancer through different mechanisms. More research is required to clarify the pathophysiological functions of cancer-derived IgG in pancreatic cancer (Fig. 10.2).

10.4 The Clinical Significance of Cancer-Derived IgG in Pancreatic Cancer

SIA-IgG can serve as an independent prognostic marker for human pancreatic cancer. Further analysis indicated that high SIA-IgG expression was associated with poor tumour differentiation, metastasis, and chemoresistance in pancreatic cancer patients during follow-up. More importantly, high SIA-IgG expression was significantly associated with worse overall survival (OS) and disease-free survival (DFS) in pancreatic cancer patients. The accuracy of an established prognosis prediction model was improved when incorporating SIA-IgG expression into 1-year OS and 2-year OS. Patients with high SIA-IgG expres-

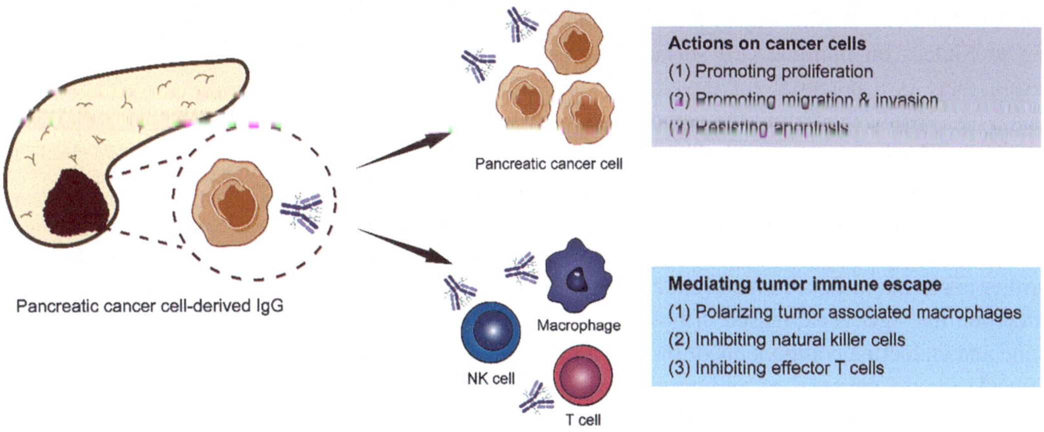

Fig. 10.2 The biological activity of cancer-derived IgG in the tumorigenesis of pancreatic cancer

sion showed a lower survival rate after receiving postoperative adjuvant chemotherapy, suggesting that this subgroup of patients may not benefit from traditional chemotherapy (Cui et al. 2020a). Additionally, a recent study revealed that SIA-IgG expression in treatment-naïve fine needle aspirate biopsy specimens could serve as a predictive factor for the pathological response to neoadjuvant therapy in pancreatic cancer (Cui et al. 2023). Therefore, antibodies targeting SIA-IgG could be promising for this subgroup of patients characterised by high SIA-IgG expression. Given the high complexity and heterogeneity of pancreatic cancer, SIA-IgG could be extremely important in identifying and categorising different subgroups of pancreatic cancer patients who might benefit from personalised therapy targeting SIA-IgG. In another study, SIA-IgG expression was detected in 88 cases of intraductal papillary mucinous neoplasm (IPMN) of the pancreas, which is considered a precursor lesion of pancreatic cancer. SIA-IgG was expressed in IPMN lesions, and its expression increased with the advancing malignant progression of IPMN, highlighting its essential role in the initial tumorigenesis of pancreatic cancer. Receiver operating characteristic curves analysis demonstrated that SIA-IgG showed favourable value in distinguishing high-risk IPMNs from low-risk IPMNs and could serve as a novel diagnostic biomarker to guide clinical decisions (Cui et al. 2020b).

10.5 Prospective

SIA-IgG holds promise as a novel therapeutic target for pancreatic cancer. Pancreatic cancer is currently one of the most malignant neoplasms. While surgery is a direct method to treat primary lesions of pancreatic cancer, effective systemic treatments are urgently needed to control the recurrence or metastasis of pancreatic cancer. Although chemotherapy such as gemcitabine, FOLFIRINOX, and nab-paclitaxel have shown therapeutic effects in the systemic control of pancreatic cancer, the overall outcome of pancreatic cancer patients was far from satisfactory. In recent years, immunotherapy such as antibody targeting PD-L1 or CTLA-4 has shown promise in treating various malignancies. However, its effectiveness in pancreatic cancer has been limited (Royal et al. 2010; Brahmer et al. 2012). A possible explanation is the distinct microenvironment of pancreatic cancer (Ho et al. 2020; Balachandran et al. 2019). For instance, pancreatic cancer is considered a type of "cold tumour" with scarce infiltration of T lymphocytes, which barely responds to antibodies targeting PD-L1 or CTLA-4. Exploring novel therapeutic targets that conform with pancreatic cancer characteristics may offer hope for this stubborn malignancy. Given its distinctive sialic acid modification, SIA-IgG exhibits the capacity to enhance the stemness (self-renewal and metastatic potential) of cancer stem cells, while also bolstering the tumour's immune evasion capabilities (via binding to Siglecs on immune cell surfaces). As a result, there is a potential for developing anticancer therapeutics targeting the sialic acid epitope of SIA-IgG, which could exert an anti-pancreatic cancer effect through dual mechanisms. Last but not least, the physiological functions of IgG in the normal pancreas are worth further exploration. Based on current evidence, it is certain that IgG is expressed in both the exocrine and endocrine components of the pancreas. Considering the pancreas as an important digestive and endocrine organ, it can be speculated that IgG expressed by normal pancreatic cells may participate in multiple physiological processes, such as the digestion process, glucose regulation, and even intestinal microecology.

References

Balachandran VP, Beatty GL, Dougan SK (2019) Broadening the impact of immunotherapy to pancreatic cancer: challenges and opportunities. Gastroenterology 156(7):2056–2072

Brahmer JR et al (2012) Safety and activity of anti-PD-L1 antibody in patients with advanced cancer. N Engl J Med 366(26):2455–2465

Chen Z, Gu J (2007) Immunoglobulin G expression in carcinomas and cancer cell lines. FASEB J 21(11):2931–2938

Chen Z et al (2010) Immunoglobulin G is present in a wide variety of soft tissue tumors and correlates well with proliferation markers and tumor grades. Cancer 116(8):1953–1963

Chen Q et al (2019) Tumour cell-derived debris and IgG synergistically promote metastasis of pancreatic cancer by inducing inflammation via tumour-associated macrophages. Br J Cancer 121(9):786–795

Chi X, Li Y, Qiu X (2020) V(D)J recombination, somatic hypermutation and class switch recombination of immunoglobulins: mechanism and regulation. Immunology 160(3):233–247

Chu J et al (2019) IGHG1 regulates prostate cancer growth via the MEK/ERK/c-Myc pathway. Biomed Res Int 2019:1–10

Cui M et al (2020a) High expression of cancer-derived glycosylated immunoglobulin G predicts poor prognosis in pancreatic ductal adenocarcinoma. J Cancer 11(8):2213–2221

Cui M et al (2020b) Expression of cancer-derived immunoglobulin G during malignant progression in intraductal papillary mucinous neoplasms: a pilot study. Pancreas 49(6):e51–e54

Cui M et al (2021) Immunoglobulin expression in cancer cells and its critical roles in tumorigenesis. Front Immunol 12:613530

Cui M et al (2023) Cancer-cell-derived sialylated IgG as a novel biomarker for predicting poor pathological response to neoadjuvant therapy and prognosis in pancreatic cancer. Int J Surg 109(2):99–106

Ho WJ, Jaffee EM, Zheng L (2020) The tumour microenvironment in pancreatic cancer—clinical challenges and opportunities. Nat Rev Clin Oncol 17(9):527–540

Iacobuzio-Donahue CA (2012) Genetic evolution of pancreatic cancer: lessons learnt from the pancreatic cancer genome sequencing project. Gut 61(7):1085–1094

Kimoto Y (1998) Expression of heavy-chain constant region of immunoglobulin and T-cell receptor gene transcripts in human non-hematopoietic tumor cell lines. Genes Chromosom Cancer 22(1):83–86

Li X et al (2011) The presence of IGHG1 in human pancreatic carcinomas is associated with immune evasion mechanisms. Pancreas 40(5):753–761

Maneva-Radicheva L et al (2014) Autoimmune diabetes is suppressed by treatment with recombinant human tissue Kallikrein-1. PLoS One 9(9):e107213

Mizrahi JD et al (2020) Pancreatic cancer. Lancet 395(10242):2008–2020

Qiu X, Yang G (1996) Existance of Ig-like protein in malignant tumor cells. J Norman Bethune Univ Med Sci 22(6):572–575

Qiu X et al (2003) Human epithelial cancers secrete immunoglobulin g with unidentified specificity to promote growth and survival of tumor cells. Cancer Res 63(19):6488–6495

Qiu X et al (2013) Immunoglobulin gamma heavy chain gene with somatic hypermutation is frequently expressed in acute myeloid leukemia. Leukemia 27(1):92–99

Royal RE et al (2010) Phase 2 trial of single agent Ipilimumab (anti-CTLA-4) for locally advanced or metastatic pancreatic adenocarcinoma. J Immunother 33(8):828–833

Siegel RL et al (2022) Cancer statistics, 2022. CA Cancer J Clin 72(1):7–33

Silva DG et al (2011) Anti-islet autoantibodies trigger autoimmune diabetes in the presence of an increased frequency of islet-reactive CD4 T cells. Diabetes 60(8):2102–2111

Storz P, Crawford HC (2020) Carcinogenesis of pancreatic ductal adenocarcinoma. Gastroenterology 158(8):2072–2081

Tang J et al (2018) Lung squamous cell carcinoma cells express non-canonically glycosylated IgG that activates integrin-FAK signaling. Cancer Lett 430:148–159

Tempero MA et al (2017) Pancreatic adenocarcinoma, version 2.2017, NCCN Clinical Practice Guidelines in Oncology. J Natl Compr Cancer Netw 15(8):1028–1061

Wan X et al (2015) Pancreatic expression of immunoglobulin G in human pancreatic cancer and associated diabetes. Pancreas 44(8):1304–1313

Wang J et al (2013) Cancer-derived immunoglobulin G promotes tumor cell growth and proliferation through inducing production of reactive oxygen species. Cell Death Dis 4(12):e945

Wang Z et al (2020) Cancer derived sialylated IgG promotes tumor immune escape by binding to Siglecs on effector T cells. Cell Mol Immunol 17(11):1148–1162

Zheng H et al (2007) Expression and secretion of immunoglobulin alpha heavy chain with diverse VDJ recombinations by human epithelial cancer cells. Mol Immunol 44(9):2221–2227

Zheng J et al (2009) Immunoglobulin gene transcripts have distinct VHDJH recombination characteristics in human epithelial cancer cells. J Biol Chem 284(20):13610–13619

Non B Cell-Derived Immunoglobulins in Intestinal Tract

11

Zihan Geng, Lina Wu, Qianqian Wang, Junfan Ma, and Zhan Shi

Abstract

Intestinal epithelium constitutes a barrier to the unrestricted movement of pathogens, and other detrimental substances from the external world (gut lumen) into the interstitial environment. Intestinal epithelial cells obstruct harmful substances passing through the epithelium as a physical and chemical barrier; Moreover, the epithelial cells can express Toll-like receptors (TLRs) and cytokines to exert innate immune function. In addition, high levels of immunoglobulin A (IgA) and other antibodies exist in the intestinal mucosa, maintaining intestinal immune homeostasis in conjunction with intestinal probiotics. Traditionally, these antibodies have been deemed to be secreted by submucosal plasma cells. Nonetheless, in recent years, it has been demonstrated that intestinal epithelial cells produce a substantial amount of Igs, especially IgA or free Ig light chains, which are involved in intestinal immune homeostasis and the survival of normal epithelial cells. Furthermore, mounting evidence affirms that many human carcinoma cells, including colorectal cancer (CRC), can overexpress Igs, particularly IgG. Cancer-derived Igs exhibit a unique V(D)J rearrangement pattern distinct from B cell-derived Ig; moreover, this cancer cell-derived IgG also has a unique sialic acid modification on the 162 site of CH1 domain (SIA-IgG). The SIA-IgG plays a crucial role in promoting cancer initiation, progression, metastasis, and tumour immune escape. Simultaneously, CRC cells can also express free Ig light chains, which promote colitis, colitis-associated colon carcinogenesis, and CRC progression. Therefore, Igs expressed by CRC cells could be a potential target for diagnosing and preventing the transformation of inflammation into cancer, as well as treating CRC.

Z. Geng (✉)
Institute of Chinese Materia Medica, China Academy of Chinese Medical Sciences, Beijing, China
e-mail: zhgeng@icmm.ac.cn

L. Wu
Central Laboratory, Key Laboratory of Carcinogenesis and Translational Research (Ministry of Education/Beijing), Peking University Cancer Hospital & Institute, Beijing, China
e-mail: lnwu@bjmu.edu.cn

Q. Wang
School of Food and Drug, Shenzhen Polytechnic University, Shenzhen, China
e-mail: wangqq@szpt.edu.cn

J. Ma
Department of Clinical Research, Sinocelltech Group Limited, Beijing, China
e-mail: junfan_ma@sinocelltech.com

Z. Shi
Department of Ultrasound in Medicine, The Second Affiliated Hospital of Zhejiang University School of Medicine, Hangzhou, China
e-mail: shizhan@zju.edu.cn

X. Qiu et al. (eds.), *Non B cell-Derived Immunoglobulins*, Advances in Experimental Medicine and Biology 1445, https://doi.org/10.1007/978-981-97-0511-5_11

Keywords

Non B cell-derived IgG · IgA · Free Ig light chain · Ig repertoire · Colorectal cancer · Colitis · Inflammatory

11.1 General Consideration

The intestinal tract is among the most complex organs within the human body. The functions of the intestinal tract encompass food and water absorption, serving as a barrier, and immune regulation. The gastrointestinal mucosal surface is covered by a single layer of intestinal epithelial cells (IECs), extending up to 100 m^2, and functions as a selective barrier to block the passage of potentially harmful substances into the mucosal immune system (Artis 2008). This is achieved by the intercellular junctions among adjacent IECs that segregate the inner and the outer environments. IECs also secrete a broad array of inflammatory cytokines and provide constant activation of the mucosal immune system to the epithelial barrier (Xavier et al. 2007; Philip et al. 2004; Moossavi et al. 2012). Additionally, high levels of IgA, IgG and IgM exist in intestinal mucosa, maintaining intestinal immune homeostasis in conjunction with intestinal probiotics.

Disruptions in the epithelial barrier function often result in inflammatory bowel disease (IBD) and colorectal cancer (CRC), among other conditions. IBD is a chronic inflammatory disease of the intestinal tract, which clinically includes Crohn's disease, ulcerative colitis, and other conditions (Guan 2019; Szigethy et al. 2010). Moreover, chronic inflammation of the gut mucosa in IBD patients may enhance the risk for developing CRC due to the prolonged duration and extent of inflammation (Gillen et al. 1994). CRC is the third leading type of cancer for estimated new cancer cases and deaths in both males and females (Siegel et al. 2023), where inflammation is one of the most crucial factors contributing to cancer progression. Antibodies produced by IECs have been deemed to be secreted by submucosal plasma cells. Nonetheless, in recent years, it has been demonstrated that IECs produce a substantial amount of Igs, especially IgA or free Ig light chains, which are involved in intestinal immune homeostasis and the survival of normal epithelial cells. Under pathological conditions, the free Igκ light chains can promote inflammation, inflammation-cancer transformation, and CRC progression. This chapter aims to offer a comprehensive review of recent literature on the expression, functions, and clinical significance of non B cell-derived Ig in the intestinal tract.

11.2 Normal Intestinal Epithelial Cells Express Igs

The intestinal tract is considered the largest immune organ in the human body, and the intestinal mucosa harbours the largest population of plasma cells in the body, which produce an immense amount of IgA daily. IgA plays a pivotal role in safeguarding the mucosal epithelium from pathogens, toxins, and food antigens (Ag), shaping the intestinal microbiota, and regulating host-commensal homeostasis (Gommerman et al. 2014). It has been believed that IgA and other classes of Igs in the intestinal tract are produced by submucosal plasma cells and secreted through the mucosal epithelium (Castro-Dopico et al. 2019). However, in addition to the submucosal plasma cells generally thought to produce Igs into the intestine, there is evidence showing that Igs are also expressed in IECs.

11.2.1 Colon Epithelial Cells Express Ig Heavy Chains

The Ig gene has been first detected in human normal IECs by Sanger or next-generation sequencing. Zheng et al. at the beginning of the twenty-first century confirmed the expression of the IgA gene in IECs (Hu et al. 2002, 2003; Geng et al. 2007). In 2019, Geng et al. also reported the expression of Ig heavy chain mRNA in sorted IECs (Geng et al. 2019). At the same time, a positive signal was detected in human normal colon tissues using non B-IgG-specific monoclonal antibody (Jiang et al. 2019). An Ig repertoire is the sum of all functionally diverse Igs in the body at a particular time. As described in previous chapters, the basic structure of Ig is a tetramer consisting of two identical heavy chains and two identical light chains linked by disulphide bonds. The sequences of the region near the C-terminal are relatively conserved, called the constant region, which primarily determines the subclasses of Ig (IgA[α], IgD[δ], IgE[ε], IgG[γ], and IgM[μ]). On the other hand, the sequences of the region near the N-terminal vary greatly, named the variable region, which primarily recognises antigens. The variable region of the Ig gene is composed of variables (V), diversity (D), and joining (J) gene segments, and the primary diversification of Ig is mainly achieved through V(D)J recombination. After encountering antigens, the variable region of Ig undergoes somatic hypermutation (SHM) that introduces a great quantity of point mutations into the V region, thereby enhancing the affinity of Igs to specific antigens (Teng et al. 2007; Jung et al. 2006).

All five classes of Igs are found in human IECs, with IgA being the predominant class (Geng et al. 2019), which is consistent with the presence of large amounts of IgA in the intestinal mucosa. Gene sequencing reveals that Ig heavy chains expressed by IECs also undergo V(D)J recombination. However, unlike B cell-derived Igs (B-Igs), which display great diversity (Mourad 2020), non B cell-derived Ig expressed by IECs has restricted V(D)J recombination patterns. For example, non B cell-derived Ig heavy chains in IECs prefer to use VH3-74 and JH4 over other VH and JH segments. Compared to the Ig heavy chain expressed in CRC cells, Ig heavy chains in IECs have lower SHM rates, and the same mutation sites are usually shared across different patient samples (Geng et al. 2019). The expression of all five classes of Igs was also detected in single-cell RNA-seq datasets of human IECs (Single Cell Portal: SCP259, https://portals.broadinstitute.org/single_cell) (Fig. 11.1).

11.2.2 Colon Epithelial Cells Express Ig Light Chains

Generally, an Ig molecule is composed of two identical Ig heavy chains and two identical light chains. The Igκ light chain, but not Igλ light chain, has been frequently detected in IECs at both mRNA (transcriptional) and protein (translational) levels. Shi et al. carried out single-cell sequencing in mouse IECs, and revealed the existence of Igκ transcript in single IEC. Igκ expression profile in human IEC subpopulations was also analysed using single-cell RNA-seq datasets (Single Cell Portal: SCP259, https://portals.broadinstitute.org/single_cell). Igκ is highly expressed in Transit Amplifying (TA) cells and stem cells, suggesting that Igκ may participate in the proliferation and regeneration of IECs (Fig. 11.1). Interestingly, many Igκ exhibit a free light chain (FLC) form and exist as monomers, dimers, or polymers (Hopper et al. 1986; Shapiro et al. 1966; Ma et al. 2017). Clearly, the Igκ showed all forms of monomers, dimers, or polymers in $CD45^{-}EpCAM^{+}$ IECs digested from mouse colon tissues. As expected, sequence analysis of IEC-derived Igκ showed limited diversity and biased rearrangement patterns. For instance, the Igκ from human IECs usually showed Vκ4-1/Jκ3 rearrangement patterns, and the Igκ from mouse IECs usually showed Vκ9-120/Jκ1 and Vκ17-121/Jκ2 (unpublished data), suggesting that IEC-derived Igκ might perform unique biological functions in IECs.

11.3 IEC-Derived Free Igκ Can Promote Colitis and the Process from Inflammation to Cancer

11.3.1 IEC-Derived Free Igκ Can Promote Colitis

FLCs, ordinarily present in normal serum, urine, and cerebrospinal fluid, have traditionally been viewed as bystanders without any function (Bradwell 2005; Katzmann et al. 2005). However, they have been identified as biomarkers for many

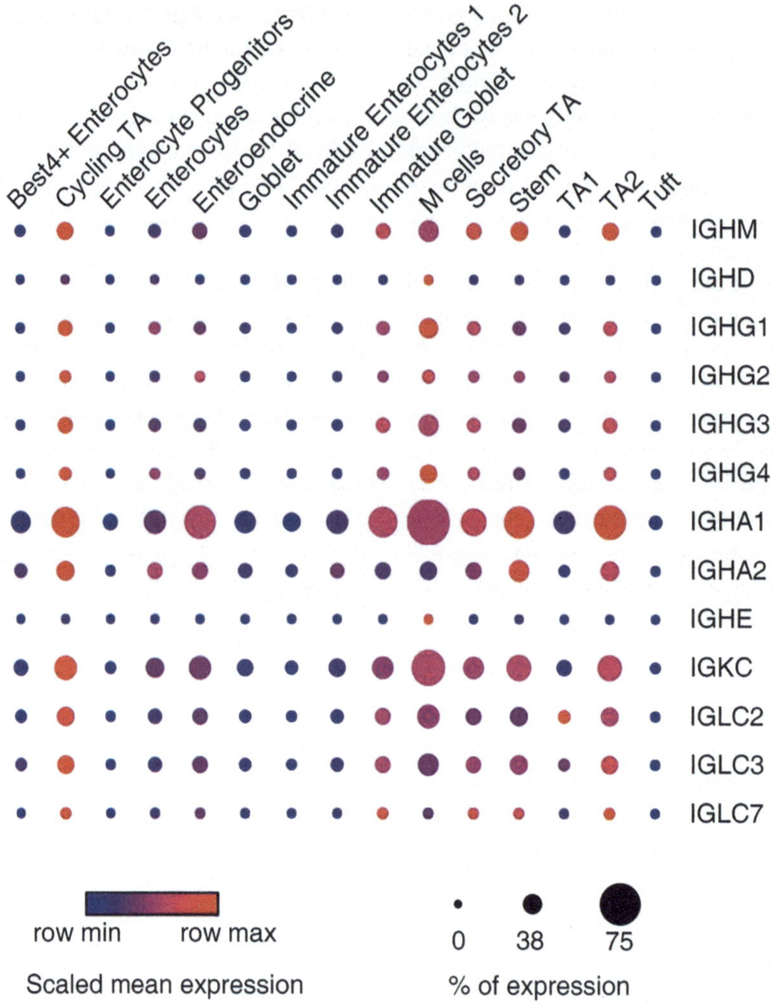

Fig. 11.1 Expression of Ig heavy chains and Ig light chains in human IECs from single-cell RNA-seq datasets (Single Cell Portal https://portals.broadinstitute.org/single_cell: SCP259, referring to https://doi.org/10.1016/j.cell.2019.06.029), including 365,492 cells from 18 ulcerative colitis patients and 12 healthy individuals. The colours represent the level of gene expression, with blue indicating low expression and red indicating high expression. Each dot represents the percentage of cells in each cell type that express the gene. Enterocytes are a group of cells with the ability to absorb water, sugar and peptide. According to the different stages of development, it can be divided into enterocyte progenitors, immature enterocytes and enterocytes. Bestrophin 4 positive enterocytes (BEST4⁺ enterocytes) are a group of mature absorptive IECs that play a role in biosynthesis that helps goblet cells produce mucus and acids (Elmentaite et al. 2021; Burclaff et al. 2022). Transit amplifying cells (TA) are a group of non-differentiated proliferating cells in transition between stem cells and differentiated IECs (Cancedda et al. 2023), classified as TA1 and TA2 according to different cell cycle phases (Fujii et al. 2018). Goblet cells are a group of mucosal epithelial cells that synthesise and secrete mucin, forming a mucosal barrier to protect intestinal epithelium (Johansson et al. 2013). M cells exist in the lymphoid follicular epithelium, and their main function is to take up and transport antigens (Miller et al. 2007). Tuft cells play a key role in immunity against parasitic infections (Howitt et al. 2016). Enteroendocrine cells spread as single cells throughout the intestinal tract, secrete motilin and other hormones, which coordinate the innate immune response of the host gut to symbiotic flora (Watnick et al. 2020)

inflammatory diseases, such as autoimmune diseases, diabetes mellitus, and central nervous system (CNS) inflammation (Nakano et al. 2011; van der Heijden et al. 2006). This view has only been revised as the functions of FLCs under various pathophysiological conditions have gradually been appreciated. Importantly, it has been reported that FLCs play a key role in promoting hypersensitivity-like IBD (Rijnierse et al. 2010).

To address if the non B-FLCs participate in colitis by using B cell-deficient mice, colitis model induced by Dextran Sulphate Sodium Salt (DSS) was utilised (Dharmani et al. 2011), and the level of FLCs was found to be significantly elevated in colitis (unpublished data). Ma et al. found that F991, a FLC-specific inhibitor, can markedly reduce infiltration of neutrophils and mast cell and injury of intestinal epithelium (Ma et al. 2017). Shi et al. further determined whether the pro-inflammatory FLCs mainly originate from IECs. By establishing mice that specifically knockout Igκ in IECs, they found that knockout of Igκ in IECs significantly delayed time of colitis onset, and reduced the severity of the lesion (unpublished data). These evidences indicated that IEC-derived FLCs participate in colitis.

11.3.2 IEC-Derived Free Igκ Can Promote the Process of Inflammation to Cancer

CRC is a prime example of a tumour strongly associated with prolonged inflammation. The process from inflammation to cancer involves, as the initial step, a breakdown of the epithelial barrier, followed by chronic inflammatory responses triggered by microbial activity or by bacterial strains that exist symbiotically with the host, and finally leading to the transformation from colitis to CRC (Kiesslich et al. 2012).

Ma et al. further analysed if the non B-FLCs can promote inflammatory-cancer transformation using AOM-DSS model in B cell-deficient mice, and demonstrated that non B-FLCs also promoted inflammatory-cancer transformation. Similarly, the F991 significantly inhibited tumorigenesis, reduced the infiltration of neutrophils and mast cell, and increased the number of CD103$^+$ dendritic cells (DCs), which maintained colonic lamina propria tolerance to commensal bacteria and food antigens (Hovav 2014), in colon tissues. Mechanically, the activity of non B-FLCs was promoting the formation of inflammasomes via regulating the activation of caspase-1, and promoting the maturation and secretion of IL-1β and IL-18 precursors (Ma et al. 2017). Subsequently, Shi et al. also found that the knockout of Igκ in IECs could significantly inhibit the tumorigenesis of CRC (unpublished data). These evidences indicated that IEC-derived FLCs participated in the transformation from colitis to CRC.

11.4 Expression and Characteristics of Non B Cell-Derived Igs in CRC

Over the past 30 years, growing reports have indicated that Igs are expressed in many types of cancer cells (Qiu et al. 2003; Zhu et al. 2008; Zheng et al. 2007; Chen et al. 2007; Babbage et al. 2006; Li et al. 2004; Qiu et al. 1996), including lung cancer, oesophageal cancer, and breast cancer. Accumulating evidence has also confirmed the expression of Igs in CRC cells, and its cancer-promoting effects (Geng et al. 2019; Qiu et al. 2003; Niu et al. 2012; Zheng et al. 2009).

11.4.1 Expression of Non B Cell-Derived Igs in CRC Cells

In 1996, Qiu et al. found and first reported the presence of IgG-like protein in CRC cells using immunohistochemical (IHC) staining (Qiu et al. 1996). Based on this result, Qiu et al. further carried out a series of experiments to identify if the non B cancer cells can produce Ig. They isolated cancer cells from breast cancer tissues and CRC tissues, purified the IgG-like protein and identified its nature by N-terminal sequencing. The results revealed that the IgG-like protein found in these cancer cells (including CRC cells) was indeed the classical IgG. They then applied laser cell microdissection (LCM) technology to capture individual cancer cells from tumour tissues. IgG and IgM transcripts were found to be present in these cancer cells (including CRC cells) as well as CRC cell line HT-29, proving evidences that CRC cells can produce Ig. In 2003, Qiu et al. for the first time reported that non B cancer cells can produce Ig (Qiu et al. 2003). Subsequently, IgG expression was also confirmed in many CRC cell lines HCT116, HCT-8, LOVO, SW1116, and SW480 (Jiang et al. 2019; Qiu et al. 2003; Niu et al. 2012; Zheng et al. 2009; Kimoto 1998). In addition, IgG-positive staining was stronger in poorly differentiated adenocarcinoma than that in well or moderately differentiated adenocarcinomas. Similarly, IgG expression was higher in cancer nests infiltrating into the deeper muscular layer than in the non-infiltrating cancer cells located in the mucosa (Niu et al. 2012).

Expression of IgA has also been found in CRC cells. In 2001, Zheng et al. discovered a novel Ig gene, SNC73, by subtractive hybridisation in CRC (Zheng et al. 2001). Open reading frame prediction indicated that SNC73 encoded a peptide identical to the constant region of an IgA at the carboxyl-terminus. Then, IgA heavy chain was also detected in CRC cell lines SW480 and LOVO (Geng et al. 2007). However, the expression of IgA in CRC cells is much lower than that in IECs of surgically severed colon. This was confirmed by Geng et al. (as shown in Fig. 11.2) (Geng et al. 2019), suggesting that non B cell-derived IgA may function in protecting the mucosal epithelium, differing from non B cell-derived IgG that promotes tumorigenesis and development of CRC.

In addition to tetrameric Ig molecules, FLCs are also expressed by cancer epithelial cells, including breast cancer, lung squamous cell carcinoma, and CRC cells. The CRC cell-derived Igκ has been found in the cytoplasm, cell culture supernatant, and the extracellular matrix (ECM) protein components. Importantly, non-reduction electrophoresis and Western blotting have revealed that the Igκ exists as a free polymer in the culture supernatant and ECM extract. In particular, Wang et al. discovered a strong positive Igκ signal in CRC cells using Vκ4-1/Jκ3-FLC monoclonal antibodies, with lower expression found in para cancer tissues (Fig. 11.3) (Wang et al. 2022). Therefore, non B FLCs not only promote inflammation-cancer transformation but are also closely linked to tumorigenesis and development.

11.4.2 Characteristics of Ig Variable Region Expressed by CRC Cells

Most Ig transcripts of the variable regions in sorted CRC cells exhibit functional gene recombination. Among the sequences of CRC cell-derived Igγ and Igμ analysed by Zheng et al., 87.5% were identified as functional and 12.3% as nonfunctional (Zheng et al. 2009). The rate of nonfunctional VHDJH rearrangement in non B

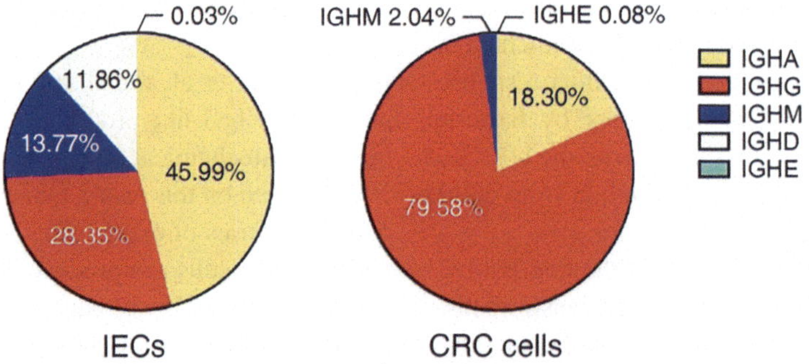

Fig. 11.2 Average proportions of Ig heavy chain classes in human IECs of surgically severed colon and CRC cells at transcript level. EpCAM⁺ epithelial cells were sorted from the intestinal cell suspension, and the V, D, J, and CH1 regions of Ig heavy chain were amplified and detected by next-generation sequencing. The sequence results were compared to the IMGT database to determine the classes of Ig heavy chains

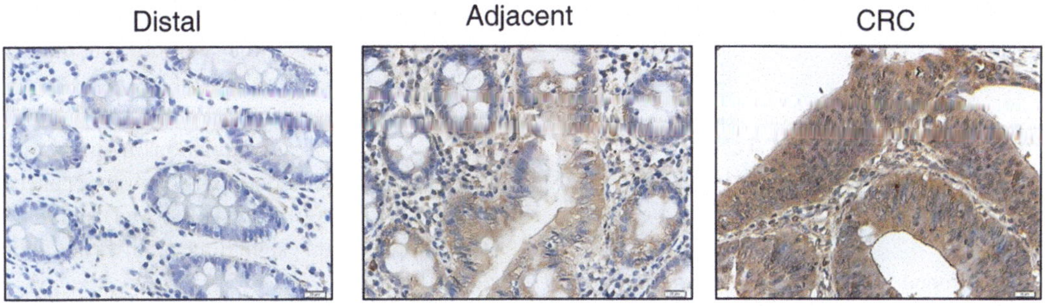

Fig. 11.3 IHC staining of human distal surgically severed tissue, adjacent tissue, and CRC tissues using a monoclonal antibody specifically recognises Vκ4-1/Jκ3-FLC (scale bars, 20 μm) (Wang et al. 2022)

cell-derived Igs is similar to that in B cell-derived Igs (Nzula et al. 2003). The usage analysis of Ig VH revealed that, like previous discoveries, several VHs, such as VH3-23, VH5-51, and VH6-1 are frequently detected in non B cell-derived Ig and are joined by diverse D and JH segments in the CRC cell lines HT-29 and SW480 (Jiang et al. 2019; Zheng et al. 2009). Moreover, VH5-51, VH3-23, and VH3-7 are predominantly used in the CRC cells from more than one patient with CRC (Geng et al. 2019; Zheng et al. 2009). By next-generation sequencing, it was demonstrated that Ig repertoires from CRC cells also exhibit very restricted and unique VHDJH patterns in each case examined, with a proportion of predominant rearrangement patterns even exceeding 80% (Geng et al. 2019). The rearrangement pattern such as VH5-51/D3-9/JH4 has been found to be used by different CRC cell lines or epithelial cancer cells of patients with CRC (Geng et al. 2019; Zheng et al. 2009), suggesting that these molecules may be potential diagnostic markers for CRC.

Wang et al. also analysed usage and VJ rearrangement patterns of Igκ expressed in CRC cells. Interestingly, the Vκ4−1/Jκ3 pattern expressed with absolute dominance in CRC (Wang et al. 2022), and this sequence is also frequently overexpressed in other tumour cells. But its unique expression regulation mechanism remains unclear and needs to be further explored.

In addition, theoretically, to screen for high-affinity antibodies, the V region of Ig in B cells undergoes SHM after antigen stimulation. Due to the fact that SHM occurs before Ig class switching (from expressing IgM to expressing IgG, IgA, or IgE), thereby, the SHM are often present on IgG, IgA, or IgE rather than IgM. Interestingly, SHM has been surprisingly observed in the VH and Vκ genes of CRC cell-derived IgG, and Igκ, but not IgM. More unexpectedly, the mutation pattern is also identical among VDJ or VJ rearrangements shared by different individuals (Geng et al. 2019; Zheng et al. 2009; Wang et al. 2022). This genetic characteristic cannot be explained with current knowledge.

11.4.3 The Glycosylation Characteristics of IgG Expressed by CRC Cells

It has been demonstrated that in addition to the characteristics of V(D)J recombination, epithelial cancer cell-derived IgG, including those produced by CRC cells, has a specific site of sialic acid modification. Apart from the classical sialylation site at the CH2 region of IgG (Asn297) (Kaneko et al. 2006), the IgG expressed in epithelial cancer cells has a novel N-glycosylation modification on 162 site of CH1 domain (called as SIA-IgG). Moreover, high sialylation modification was exhibited at the end of SIA-IgG glycan (Tang et al. 2018). Compared with commercial anti-IgG antibody, a monoclonal antibody named RP215 (described in Chap. 2), specifically recognises this N-glycan-related epitope (Jiang et al. 2019; Tang et al. 2018; Lee et al. 2008, 2009; Liao et al. 2015). It is worth noting that SIA-IgG is highly expressed in cancer cells, but not in B cells, and low or no expression in normal IECs.

11.5 The Biological Activity of Non B Cell-Derived Ig in CRC

As mentioned above, the V(D)J combination of CRC cell-derived Ig varies from that of IEC-derived Ig, suggesting that Ig expressed by CRC cells may hold unique pathological significance. Studies have revealed that CRC cell-derived Ig fosters the proliferation, migration, and invasion of CRC cells. IgG expressed by cancer cells typically carries a unique sialic acid modification at the Asn162 site, and this sialylated IgG is closely associated with tumour immune escape (Tang et al. 2018; Wang et al. 2020). Furthermore, FLCs in CRC also encourage tumour progression by interacting with Integrin β1 and activating the downstream FAK/Src pathway. Thus, non B cell-derived Ig produced by CRC cells may serve as a potential target for diagnosis and therapy of CRC.

11.5.1 Non B Cell-Derived IgG Directly Promotes Tumour Progression of CRC

Non B cell-derived IgG expressed by CRC cells has been proven to promote CRC progression, encompassing cell proliferation, infiltration, and metastasis. Significant inhibition of proliferation and higher rate of apoptosis in CRC cell lines were detected when a specific antibody against IgG and Igγ antisense vector was added to the cells (Qiu et al. 2003; Niu et al. 2012; Deng et al. 2006). The suppression of IgG expression also stymied tumour growth in vitro and in vivo. The transwell or invasion assay demonstrated that silencing non B cell-derived IgG expression diminished the migration and invasion ability of LOVO and SW480 cells (Jiang et al. 2019; Niu et al. 2012).

As described above, it was identified that IgG expressed in CRC cells exhibits a unique sialylation modification (SIA-IgG). The SIA-IgG is mainly expressed in cancer stem cells. SIA-IgG, depending on its sialylation modification, can promote progression of CRC or other cancer via binding to integrin α6β4 or c-Met, and activating downstream signalling pathway (Tang et al. 2018; Liao et al. 2015).

11.5.2 SIA-IgG Encourages Tumour Immune Escape

Glycosylation of IgG is a crucial mechanism for the diversification of antibody function, and different glycosylation modifications determine IgG activity. For example, IgG modified by N-acetyl galactose usually exhibits a pro-inflammatory effect, while IgG modified by salivary acid has an anti-inflammatory effect (Kaneko et al. 2006; Malhotra et al. 1995; Hodoniczky et al. 2005; Arnold et al. 2007). The terminal sialic acid linked to the Asn297 of IgG plays anti-inflammatory roles by binding to C-type lectins, such as dendritic cell-specific intercellular adhesion molecule-grabbing nonintegrin (DC-SIGN) on myeloid cells, and indirectly inhibits $CD4^+$ T cell activation (Schwab et al. 2013; Kaufman et al. 2015; Galeotti et al. 2017).

Wang et al. purified SIA-IgG from ovarian cancer tissue or ascites and found that SIA-IgG can directly inhibit the proliferation of $CD4^+$ and $CD8^+$ T cells, especially $CD8^+$ T cells, in vitro. It was confirmed in vitro and in vivo that the immunosuppressive effect of SIA-IgG is sialylation-dependent. SIA-IgG significantly promoted the progression of various epithelial cancer cells, including mouse melanoma and breast cancer cells, and reduced the proportion of $CD4^+$ and $CD8^+$ T cells in vivo. The tumour promotion and immunosuppressive effects of SIA-IgG mainly rely on its Fab fragments, where the Asn162 sialylation site is located, and thus could be blocked by RP215 (Wang et al. 2020).

It has been found that SIA-IgG inhibits T cells through Sialic acid-binding immunoglobulin-type lectins (Siglecs), which are frequently found on innate immune cells as well as B cells and T cells (Rodrigues et al. 2018; Macauley et al. 2014; Bandala-Sanchez et al. 2013; Ikehara et al. 2004; Nakamura et al. 1994; Stanczak et al. 2018). It has been reported that Siglec-10 expressed on T cells can bind α-2,3 sialic acid linkage with galactose in CD52, then recruit Src homology region 2 domain-containing phosphatase-1 (SHP1) to the immunoreceptor tyrosine-based inhibitory motif (ITIM) of Siglec-10 and interact with the TCR to suppress $CD4^+$ T cell signalling (Bandala-Sanchez et al. 2018). The Siglecs family may serve as a potential immune checkpoint molecule in cancer patients (Duan et al. 2020). Through GEO database mining, RT-PCR, and flow cytometry, Wang et al. found that $CD4^+$ and $CD8^+$ T cells from healthy donors and patients with cancer could express Siglecs (Wang et al. 2020), which contain ITIMs on their cytoplasmic tails (Macauley et al. 2014), at different levels. Both Siglec-7 and Siglec-10 blocking antibodies could alleviate the proliferation inhibition of $CD4^+$ and $CD8^+$ T cells induced by SIA-IgG, suggesting that SIA-IgG inhibits cell proliferation by binding to Siglec-7 or Siglec-10 on the surface of T cells. The upregulation of Siglecs expression in tumour-infiltrating T cells was also confirmed by microarray or RNA-Seq datasets in the GEO database analysis (Wang et al. 2020). Recently, Fan et al. compared the expression frequency of Siglecs on $CD4^+$ and $CD8^+$ T cells between the same patient's CRC tissue and its surgical severed intestinal mucosa. It was found that $CD8^+$ T cells, but not $CD4^+$ T cells, in cancer tissue showed high proportion of Siglec-7 expression. Moreover, the expression of Siglec-7 was markedly correlated with PD-1 expression (Fan et al. 2023).

As the Asn162-sialylated IgG is commonly found in various epithelial tumour tissues, including lung cancer, ovarian cancer, and CRC, the non B-IgG expressed by CRC cells may also have the same immunomodulatory effect. Thus, the role of cancer cell-derived sialylated IgG in tumour immune escape suggests a new avenue for tumour immunotherapy in CRC.

11.5.3 Non B Cell-Derived Free Igκ Fosters Progression of CRC by Activating Integrin β1

As mentioned earlier, free Igκ light chain with a unique Vκ4-1/Jκ3 rearrangement is generated in CRC cells. Employing a recombinant protein of free Igκ light chains with unique Vκ4-1/Jκ3 rearrangement, Wang et al. revealed that Vκ4-1/Jκ3-FLC notably fostered the proliferation, migration, and invasion of CRC cells in vitro, and growth of tumour in vivo was significantly inhibited by Vκ4-1/Jκ3-FLC antibody. Wang et al. also discovered that the binding of Vκ4-1/Jκ3-FLC to integrin β1 fostered its activation, leading to the subsequent activation of the focal adhesion kinase (FAK) signalling pathway, including the phosphorylation of Src, paxillin in the adhesion plaque complex, and the activation of ERK and AKT downstream. By using two monoclonal antibodies to inhibit Vκ4-1/Jκ3-FLC, they found that the activation of the integrin-FAK pathway and the ability of cell colony formation were dampened, indicating that Vκ4-1/Jκ3-FLC-specific monoclonal antibodies have the application value in anti-tumour therapy (Wang et al. 2022). The structure of the Ig light chain encompasses three complementarity-determining regions (CDRs) that are highly variable, and four relatively conserved framework regions (FWRs) (Solomon 1986). Wang et al. have discerned that integrin β1 is the primary interaction protein of Vκ4-1/Jκ3-FLC, and there is a high probability that the binding site is situated in the CDR2 region of Vκ4-1/Jκ3-FLC (Wang et al. 2022).

11.6 Clinical Significance of CRC-Derived Ig

CRC is classified into five stages from stage 0 (carcinoma in situ) to IV according to local invasion depth (T stage), lymph node involvement (N stage), and presence of distant metastases (M stage) (TNM system) (Brenner et al. 2014), which provide the basis for therapeutic decisions and are closely related to prognosis, especially the degree of tumour differentiation. Non B cell-derived Igs have been associated with tumour differentiation, especially TNM staging. IgG expression was higher in stages III-IV than in stages I-II. In terms of lymph node staging, high IgG expression was found in stages N1 and N2 of CRC. The relationship between non B cell-derived IgG and TNM stage indicates that IgG is more highly expressed in the patients with CRC with poor prognosis. Furthermore, patients with poorly differentiated CRC adenocarcinoma (67.3%) have a higher IgG expression level compared to well and moderately differentiated cases (Niu et al. 2012). Jiang et al. also demonstrated that high expression of non B cell-derived IgG in CRC tissue can predict worse overall survival by using Kaplan-Meier survival analysis (Jiang et al. 2019).

The proliferative activity of tumour cells is another critical prognostic marker in CRC. Cell cycle-specific markers, including proliferating cell nuclear antigen (PCNA), Cyclin D1, and NF-κB proteins, are used as standard markers of proliferation to assess the growth fraction of a cell population and the proliferative activity of cells (Inoue et al. 2014; Jurikova et al. 2016; Ioachim 2008). P53, an extensively researched tumour suppression gene, exhibits mutations in more than 50% of sporadic CRC cases. Other biomarkers, such as anti-apoptosis gene B-cell lymphoma 2 (Bcl-2) and invasion marker matrix metalloproteinases (MMP), are correlated with more aggressive tumour growth in CRC. The p53 negative/Bcl-2 positive phenotype is thought to be an independent indicator of prognosis in CRC (Lyall et al. 2006). Non B cell-derived IgG was found to co-localise with Cyclin D1, NF-κB, or PCNA in CRC tissues. Concurrently, IgG expression exhibited a significant positive correlation with the expressions of Cyclin D1, NF-κB, and PCNA, and a negative correlation with the expression of Bcl-2. Conversely, no significant correlation was found between IgG expression and the expressions of MMP-2 and p53. As mentioned above, non B cell-derived IgG can foster anchor-independent growth and infiltration. Niu et al. found that tumours can endogenously secrete IgG into circulation by using a severe combined immunodeficiency disease (SCID) mouse model without B cell contamination. This IgG can regulate the proliferation of cancer cells through autocrine signalling (Niu et al. 2012). High expression of non B cell-derived IgG predicts a poor prognosis, and targeting IgG may be a potential target in molecular targeted therapy for CRC.

In addition, FLCs exacerbate inflammatory damage and tumour load and thus decrease survival rates in murine models (Ma et al. 2017). The evidence about non B-FLC function indicates that blockage of FLC may be a potential therapeutic option in treating CRC, especially colitis-associated cancer.

11.7 Progress and Prospects

The expression of Igs in normal and malignant IECs is an important recent discovery. The above findings only reflect partial functional characteristics of Ig produced by IECs. The discovery that IECs can produce Ig will initiate a novel study field that reveals the physiological and pathological significance of Igs in maintaining the intestine.

11.7.1 Physiological and Pathological Significance of IEC-Derived IgA

IgA is the predominant Ig in the intestines under normal physiological conditions. IgA serves diverse functions, including neutralising toxins and viruses, blocking excessive live bacterial adherence or translocation, clearing unwanted macromolecular structures at the epithelial surface, and directed sampling of luminal antigen (Artis 2008; Gommerman et al. 2014; Castro-Dopico et al. 2019). Previously, it was believed that the IgA was only produced by plasma cells in the mucosa. The evidence in this chapter shows that not only plasma cells but also normal IECs can express IgA, which act as the first-line barrier in the intestinal tract together with plasma cells and gut microbiota. In addition, given the fact that high expression level of IgA in intestinal epithelial stem cells as well as CRC cells, it is hypothesised that IgA is also involved in the differentiation and development of IECs, and progression of CRC. The speculation needs further study.

11.7.2 Physiological and Pathological Significance of IEC-Derived IgG

The expression of IgG is higher in CRC than in normal cells. As the immune molecule present in tumours, non B cell-derived Ig plays vital biological activities in cell differentiation, infiltration, and metastasis of CRC, indicating its self-protecting role during cancer development. The mechanism of IEC-derived IgG in CRC progression needs to be explored. Jiang et al. found that CRC cells derived IgG directly interacted with E-cadherin and β-catenin (Jiang et al. 2019), indicating that IgG may activate β-catenin which maintains the stemness and malignancy of CRC cells. Tang et al. demonstrated that non B cell-derived IgG interacts with integrin α6β4 and activates the downstream oncogenic FAK and Src signalling pathways in the lung squamous cell carcinoma (Tang et al. 2018). Qin et al. found that non B cell-derived IgG promotes the development of prostate cancer through the SOX2 pathway (Qin et al. 2020). Understanding the exact mechanism of non B cell-derived IgG in the CRC will be of great significance for studying IgG as a target for cancer therapy. In addition, the physiological significance of IEC-derived IgG remains to be further revealed.

11.7.3 Physiological and Pathological Significance of IEC-Derived FLC

As mentioned in this chapter, the data provided by Ma et al. (Ma et al. 2017) and Wang et al. (Wang et al. 2022) suggest that FLCs promote tumour-associated inflammation, the transition from inflammation to CRC, and the tumorigenesis and development of CRC. Therefore, it is important to explore the application prospects of FLC as a new target for CRC-targeted therapy. Also, IECs spontaneously express Igκ FLC at a high level, the physiological significance of which still needs to be further revealed.

Apart from IgA, IgG and Igκ FLC, IgM, IgE, and a few Igλ FLC were also expressed in IECs. It is believed that these Igs also have important physiological and pathological significance.

References

Arnold JN et al (2007) The impact of glycosylation on the biological function and structure of human immunoglobulins. Annu Rev Immunol 25:21–50

Artis D (2008) Epithelial-cell recognition of commensal bacteria and maintenance of immune homeostasis in the gut. Nat Rev Immunol 8(6):411–420

Babbage G et al (2006) Immunoglobulin heavy chain locus events and expression of activation-induced cytidine deaminase in epithelial breast cancer cell lines. Cancer Res 66(8):3996–4000

Bandala-Sanchez E et al (2013) T cell regulation mediated by interaction of soluble CD52 with the inhibitory receptor Siglec-10. Nat Immunol 14(7):741–748

Bandala-Sanchez E et al (2018) CD52 glycan binds the proinflammatory B box of HMGB1 to engage the Siglec-10 receptor and suppress human T cell function. Proc Natl Acad Sci U S A 115(30):7783–7788

Bradwell AR (2005) Serum free light chain measurements move to center stage. Clin Chem 51(5):805–807

Brenner H et al (2014) Colorectal cancer. Lancet 383(9927):1490–1502

Burclaff J et al (2022) A roximal-to-distal survey of healthy adult human small intestine and colon epithelium by single-cell transcriptomics. Cell Mol Gastroenterol Hepatol 13(5):1554–1589

Cancedda R et al (2023) Transit Amplifying Cells (TACs): a still not fully understood cell population. Front Bioeng Biotechnol 11:1189225

Castro-Dopico T et al (2019) Clatworthy, IgG and Fcgamma receptors in intestinal immunity and inflammation. Front Immunol 10:805

Chen Z et al (2007) Immunoglobulin G expression in carcinomas and cancer cell lines. FASEB J 21(11):2931–2938

Deng Y et al (2006) [Immunoglobulin expression in colon cancer cell line HT-29 and its biological activities]. Chinese J Oncol 28(2): 88–91

Dharmani P et al (2011) Tumor necrosis factor-alpha and Muc2 mucin play major roles in disease onset and progression in dextran sodium sulphate-induced colitis. PLoS One 6(9):e25058

Duan S et al (2020) Siglecs as immune cell checkpoints in disease. Annu Rev Immunol 38:365–395

Elmentaite R et al (2021) Cells of the human intestinal tract mapped across space and time. Nature 597(7875):250–255

Fan T et al (2023) Sialylated IgG in epithelial cancers inhibits antitumor function of T cells via Siglec-7. Cancer Sci 114(2):370–383

Fujii M et al (2018) Human intestinal organoids maintain self-renewal capacity and cellular diversity in niche-inspired culture condition. Cell Stem Cell 23(6):787–793

Galeotti C et al (2017) IVIG-mediated effector functions in autoimmune and inflammatory diseases. Int Immunol 29(11):491–498

Geng LY et al (2007) Expression of SNC73, a transcript of the immunoglobulin alpha-1 gene, in human epithelial carcinomas. World J Gastroenterol 13(16):2305–2311

Geng ZH et al (2019) Human colorectal cancer cells frequently express IgG and display unique Ig repertoire. World J Gastrointest Oncol 11(3):195–207

Gillen CD et al (1994) Ulcerative colitis and Crohn's disease: a comparison of the colorectal cancer risk in extensive colitis. Gut 35(11):1590–1592

Gommerman JL et al (2014) Re-thinking the functions of IgA(+) plasma cells. Gut Microbes 5(5):652–662

Guan Q (2019) A comprehensive review and update on the pathogenesis of inflammatory bowel disease. J Immunol Res 2019:7247238

Hodoniczky J et al (2005) Control of recombinant monoclonal antibody effector functions by Fc N-glycan remodeling in vitro. Biotechnol Prog 21(6):1644–1652

Hopper JE et al (1986) Evidence by radioimmunoassay that mitogen-activated human blood mononuclear cells secrete significant amounts of light chain Ig unassociated with heavy chain. Cell Immunol 101(1):122–131

Hovav AH (2014) Dendritic cells of the oral mucosa. Mucosal Immunol 7(1):27–37

Howitt MR et al (2016) Tuft cells, taste-chemosensory cells, orchestrate parasite type 2 immunity in the gut. Science 351(6279):1329–1333

Hu J et al (2002) [Expression of a novel immunoglobulin gene SNC73 in human cancer and its significance]. Zhonghua Zhong Liu Za Zhi 24(1):38–40

Hu JB et al (2003) Expression of a novel immunoglobulin gene SNC73 in human cancer and non-cancerous tissues. World J Gastroenterol 9(5):1054–1057

Ikehara Y et al (2004) Paulson, negative regulation of T cell receptor signaling by Siglec-7 (p70/AIRM) and Siglec-9. J Biol Chem 279(41):43117–43125

Inoue A et al (2014) A small molecule inhibitor of monoubiquitinated Proliferating Cell Nuclear Antigen (PCNA) inhibits repair of interstrand DNA cross-link, enhances DNA double strand break, and sensitizes cancer cells to cisplatin. J Biol Chem 289(10):7109–7120

Ioachim E (2008) Expression patterns of cyclins D1, E and cyclin-dependent kinase inhibitors p21waf1/cip1, p27kip1 in colorectal carcinoma: correlation with other cell cycle regulators (pRb, p53 and Ki-67 and PCNA) and clinicopathological features. Int J Clin Pract 62(11):1736–1743

Jiang H et al (2019) Cancer IgG, a potential prognostic marker, promotes colorectal cancer progression. Chin J Cancer Res 31(3):499–510

Johansson ME et al (2013) Mucus and the goblet cell. Dig Dis 31(3–4):305–309

Jung D et al (2006) Mechanism and control of V(D) J recombination at the immunoglobulin heavy chain locus. Annu Rev Immunol 24:541–570

Jurikova M et al (2016) Ki67, PCNA, and MCM proteins: markers of proliferation in the diagnosis of breast cancer. Acta Histochem 118(5):544–552

Kaneko Y et al (2006) Ravetch, anti-inflammatory activity of immunoglobulin G resulting from Fc sialylation. Science 313(5787):670–673

Katzmann JA et al (2005) Diagnostic performance of quantitative kappa and lambda free light chain assays in clinical practice. Clin Chem 51(5):878–881

Kaufman GN et al (2015) Induction of regulatory T cells by intravenous immunoglobulin: a bridge between adaptive and innate immunity. Front Immunol 6:469

Kiesslich R et al (2012) Local barrier dysfunction identified by confocal laser endomicroscopy predicts relapse in inflammatory bowel disease. Gut 61(8):1146–1153

Kimoto Y (1998) Expression of heavy-chain constant region of immunoglobulin and T-cell receptor gene transcripts in human non-hematopoietic tumor cell lines. Genes Chromosomes Cancer 22(1):83–86

Lee G et al (2008) Molecular identity of a pan cancer marker, CA215. Cancer Biol Ther 7(12):2007–2014

Lee G et al (2009) Preclinical assessment of anti-cancer drugs by using RP215 monoclonal antibody. Cancer Biol Ther 8(2):161–166

Li M et al (2004) Expression of immunoglobulin kappa light chain constant region in abnormal human cervical epithelial cells. Int J Biochem Cell Biol 36(11):2250–2257

Liao Q et al (2015) Aberrant high expression of immunoglobulin G in epithelial stem/progenitor-like cells contributes to tumor initiation and metastasis. Oncotarget 6(37):40081–40094

Lyall MS et al (2006) Profiling markers of prognosis in colorectal cancer. Clin Cancer Res 12(4):1184–1191

Ma J et al (2017) Free immunoglobulin light chain (FLC) promotes murine colitis and colitis-associated colon carcinogenesis by activating the inflammasome. Sci Rep 7(1):5165

Macauley MS et al (2014) Paulson, Siglec-mediated regulation of immune cell function in disease. Nat Rev Immunol 14(10):653–666

Malhotra R et al (1995) Glycosylation changes of IgG associated with rheumatoid arthritis can activate complement via the mannose-binding protein. Nat Med 1(3):237–243

Miller H et al (2007) Intestinal M cells: the fallible sentinels? World J Gastroenterol 13(10):1477–1486
Moossavi S et al (2012) Inflammation in sporadic colorectal cancer. Arch Iran Med 15(3):166–170
Mourad A (2020) Immunogenetic aspect of B-cell antigen receptor diversity generation. In: Normal and malignant B-cell. InTechOpen, Rijeka
Nakamura Y et al (1994) Expression of CD33 antigen on normal human activated T lymphocytes. Blood 83(5):1442–1443
Nakano T et al (2011) Free immunoglobulin light chain: its biology and implications in diseases. Clin Chim Acta 412(11–12):843–849
Niu N et al (2012) IgG expression in human colorectal cancer and its relationship to cancer cell behaviors. PLoS One 7(11):e47362
Nzula S et al (2003) Antigen-driven clonal proliferation, somatic hypermutation, and selection of B lymphocytes infiltrating human ductal breast carcinomas. Cancer Res 63(12):3275–3280
Philip M et al (2004) Inflammation as a tumor promoter in cancer induction. Semin Cancer Biol 14(6):433–439
Qin C et al (2020) Cancer-driven IgG promotes the development of prostate cancer though the SOX2-CIgG pathway. Prostate 80(13):1134–1144
Qiu X et al (1996) Existance of Ig-like protein in malignant tumor cells. J Jilin Univ Med Ed 5:572–575
Qiu X et al (2003) Human epithelial cancers secrete immunoglobulin g with unidentified specificity to promote growth and survival of tumor cells. Cancer Res 63(19):6488–6495
Rijnierse A et al (2010) Ig-free light chains play a crucial role in murine mast cell-dependent colitis and are associated with human inflammatory bowel diseases. J Immunol 185(1):653–659
Rodrigues E et al (2018) Hypersialylation in cancer: modulation of inflammation and therapeutic opportunities. Cancers (Basel) 10(6):207
Schwab I et al (2013) Intravenous immunoglobulin therapy: how does IgG modulate the immune system? Nat Rev Immunol 13(3):176–189
Shapiro AL et al (1966) Synthesis of excess light chains of gamma globulin by rabbit lymph node cells. Nature 211(5046):243–245
Siegel RL et al (2023) Cancer statistics, 2023. CA Cancer J Clin 73(1):17–48
Solomon A (1986) Light chains of immunoglobulins: structural-genetic correlates. Blood 68(3):603–610
Stanczak MA et al (2018) Self-associated molecular patterns mediate cancer immune evasion by engaging Siglecs on T cells. J Clin Invest 128(11):4912–4923
Szigethy E et al (2010) Inflammatory bowel disease. Child Adolesc Psychiatr Clin N Am 19(2):301–318. ix
Tang J et al (2018) Lung squamous cell carcinoma cells express non-canonically glycosylated IgG that activates integrin-FAK signaling. Cancer Lett 430:148–159
Teng G et al (2007) Immunoglobulin somatic hypermutation. Annu Rev Genet 41:107–120
van der Heijden M et al (2006) Free immunoglobulin light chains as target in the treatment of chronic inflammatory diseases. Eur J Pharmacol 533(1–3):319–326
Wang Z et al (2020) Cancer-derived sialylated IgG promotes tumor immune escape by binding to Siglecs on effector T cells. Cell Mol Immunol 17(11):1148–1162
Wang Q et al (2022) A widely expressed free immunoglobulin kappa chain with a unique Vkappa4-1/Jkappa3 pattern promotes colon cancer invasion and metastasis by activating the integrin beta1/FAK pathway. Cancer Lett 540:215720
Watnick PI et al (2020) Microbial control of intestinal homeostasis via enteroendocrine cell innate immune signaling. Trends Microbiol 28(2):141–149
Xavier RJ et al (2007) Podolsky, unravelling the pathogenesis of inflammatory bowel disease. Nature 448(7152):427–434
Zheng S et al (2001) [Structure and expression of colorectal cancer related immunoglobulin novel gene SNC73]. Zhonghua Yi Xue Za Zhi 81(8):485–488
Zheng H et al (2007) Expression and secretion of immunoglobulin alpha heavy chain with diverse VDJ recombinations by human epithelial cancer cells. Mol Immunol 44(9):2221–2227
Zheng J et al (2009) Immunoglobulin gene transcripts have distinct VHDJH recombination characteristics in human epithelial cancer cells. J Biol Chem 284(20):13610–13619
Zhu X et al (2008) Immunoglobulin mRNA and protein expression in human oral epithelial tumor cells. Appl Immunohistochem Mol Morphol 16(3):232–238

12 Characteristics and Clinical Implications of Immunoglobulins Derived from Non B Cells in the Skin

Hui Dai, Dongyang Jiang, Wenjing Zhou, and Xiaoyan Qiu

Abstract

Skin is the most prominent tissue and organ, as well as the first line of defence, of the body. Because it is situated on the body's surface, it is constantly exposed to microbial, chemical, and physical factors such as mechanical stimulation. Therefore, skin has evolved substantial immune defences, regenerative ability, and anti-injury capacity. Epidermal cells produce antibacterial peptides that play a role in immune defence under physiological conditions. Additionally, IgG or IgA in the skin also participates in local anti-infective immunity. However, based on the classical theory of immunology, Ig can only be produced by B cells which should be derived from local B cells. This year, thanks to the discovery of Ig derived from non B cells (non B-Ig), Ig has also been found to be expressed in epidermal cells and contributes to immune defence. Epidermal cell-derived IgG and IgA have been demonstrated to have potential antibody activity by binding to pathogens. However, these epidermal cell-derived Igs show different microbial binding characteristics. For instance, IgG binds to *Staphylococcus aureus* and IgA binds to *Staphylococcus epidermidis*. Epidermal cells producing IgG and IgA may serve as an effective defense mechanism alongside B cells, providing a novel insight into skin immunity.

Keywords

Immunoglobulin · Keratinocytes · Epidermis · Ig-free light chain · VDJ rearrangement · Antimicrobial antibody · Psoriasis

H. Dai (✉)
Department of Immunology, School of Basic Medical Sciences, NHC Key Laboratory of Medical Immunology, Peking University, Beijing, China
e-mail: daihui@bjmu.edu.cn

D. Jiang
Department of Cardiology, Pan-vascular Research Institute, Shanghai Tenth People's Hospital, Tongji University School of Medicine, Shanghai, China
e-mail: jiangdy@tongji.edu.cn

W. Zhou
Institute of Thoracic Oncology, West China Hospital, Sichuan University, Chengdu, China
e-mail: zwj1993@wchscu.cn

X. Qiu
Department of Immunology, School of Basic Medical Sciences, Peking University, Beijing, China
e-mail: qiuxy@bjmu.edu.cn

12.1 General Considerations

Skin is an important barrier that defends against the invasion of pathogens. It employs many types of anti-infective defence mechanisms, including the water content, mechanical hardness, cuticle lipid, and antimicrobial peptides and defensins of

X. Qiu et al. (eds.), *Non B cell-Derived Immunoglobulins*, Advances in Experimental Medicine and Biology 1445, https://doi.org/10.1007/978-981-97-0511-5_12

the epidermis. Additionally, the growth of symbiotic bacteria on the skin surface, such as *Staphylococcus epidermidis*, can directly or indirectly protect the host from pathogenic bacteria, such as *Staphylococcus aureus* (Coruh et al. 1980). Despite the absence of blood vessels in the epidermis and the presence of a dense basement membrane that prevents the passage of macromolecules from the dermis to the epidermis, there is still a certain amount of immunoglobulin (Ig) in the epidermis under physiological conditions. However, for a long time, based on the classical theory of immunology, these antibodies were believed to be derived from B cells and to play a role in immune defence (Coruh et al. 1980). Under pathological conditions, such as inflammatory or autoimmune diseases, including psoriasis or bullous pemphigus, a large number of autoantibodies are found on the skin, causing pathological damage. These pathogenic Igs are also considered to be produced by B cells (Coruh et al. 1980).

Although it has been widely accepted that Ig is only produced by B cells, growing evidence has demonstrated that Igs, including different Ig classes such as IgM, IgG, and IgA, are also produced by non B cells (non B-Igs). These non B-Igs can exert both antibody and cell biological functions, such as promoting cell proliferation (Qiu et al. 2003, 2012, 2013; Huang et al. 2014), migration (Liu et al. 2015; Tang et al. 2018), and maintaining stem cell function (Huang et al. 2023). Many recent studies have shown that epidermal cells also produce Ig and play a crucial role in physiological and pathological processes (Jiang et al. 2015; Zhang et al. 2010).

12.2 IgG and IgA Expression in Epidermal Cells

Although a high level of human IgG has been well demonstrated in malignant squamous cells, including oesophageal squamous cell cancer, squamous cell lung cancer, oral squamous cell carcinoma, and normal oral squamous cells adjacent to the tumours, it remained unknown for long time whether squamous cells can produce Igs under steady-state conditions. Besides, the Ig deposition at the epidermal junction has also been observed in both murine lupus models and cutaneous lupus erythematosus (CLE) patients with serum positivity for anti-DNA IgG (Oren-Suissa et al. 2007; Dong et al. 2017). To identify if Ig was expressed in normal epidermal cells, Jiang et al. first analysed the expression of IgG and IgA in intact skin tissue by immunohistochemistry using commercial anti-human IgG or IgA, and found positive staining of commercial anti-human IgG or IgA in both the epidermis and dermis, with more pronounced staining in the dermis. Interestingly, the IgG staining was widely distributed in the dermis and epidermis, while IgA was mainly localised in basal cells of the epidermis. To further confirm the presence of IgG and IgA in epidermal cells, Jiang et al. isolated normal skin epidermal cells from circumcision samples. As expected, IgG and IgA were identified in the epidermal cells by immunofluorescence and Western blot analysis. In addition, Jiang et al. also identified IgG existence in the HaCaT cell line, an immortal human keratinocyte cell line (Jiang et al. 2015).

To further address the non B cell-derived Ig expression, public databases were used to analyse Ig expression in keratinocytes. Igκ transcripts were found in the RTS3b skin cancer cell line (GSE: 44327) (Jordanovski et al. 2013), human keratinocytes (GSE: 41905), follicle stem cells (GSE1096), and mouse epidermis (GSE: 3108) (Koster et al. 2006). IGLV6-57 of Igλ was detected in human keratinocytes by microarray (GSE: 41905). Additionally, Igκv28, Igκv4-59, Igκv10-96, Igκv9-120, and Igκv1-133 were frequently found in mouse neonatal skin-derived precursors (GSE: 18690) (Biernaskie et al. 2009). Igκ (IGVK4-101_IGKJ301_C) was found in human primary skin fibroblasts (GSE4914, 10123) and human keratinocytes (GSE: 32620) (Koster et al. 2006).

12.3 V(D)J Rearrangement Feature of Ig Gene Expressed in Epidermal Cells

B cells have been acknowledged as the only source of Ig. To discount the possibility of B lymphocyte contamination, Jiang et al. examined CD20 and CD19 transcripts, two B cell markers, to confirm the absence of B lymphocytes in the isolated epidermal cells. Subsequently, they demonstrated that the rearranged IgG and IgA transcripts exist in the epidermal cells from two donors by RT-PCR. Sequence analysis illustrated that epidermis-derived Igγ and Igα genes exhibited classical V(D)J rearrangements, with typical N nucleotide or P nucleotide insertions at the V-D or D-J junctions. Significantly, all rearrangements they discovered were functional. Furthermore, the transcripts of Igγ and Igα were also detected in the HaCaT keratinocyte cell line (Jiang et al. 2015).

Jiang et al. analysed VHDJH rearrangement features and usage of epidermal cell-derived Igγ and Igα genes, then found only one or two VHDJH patterns in epidermal cells of each individual, which was entirely different from the high diversity of B cell-derived rearrangements. For instance, the gene of epidermis-derived Igγ displayed only one VHDJH rearrangement pattern in each individual, including VH1-2/DH5-18/JH6 in individual No. 1 and VH2-5/DH2-21/JH4 in individual No. 2, respectively. The epidermis-derived Igα genes showed two V(D)J rearrangement patterns, with VH7-4-1/D6-13/JH6 and VH5-10-1/D2-2/JH3 being found in individual No. 1, as well as VH1-3/D4-17/JH1 and VH1-3/D4-17/JH4 being found in individual No. 2. Thus far, none of the epidermis-derived V(D)J rearrangement has been reported in B cell-derived rearrangement, either Igγ or Igα chains. Furthermore, the Igγ V(D)J rearrangements in healthy skin epithelial cells were also different from those observed in epithelial cancer cells (Jiang et al. 2015).

Additionally, Ig class switching is a well-established process to explain the appearance of IgG, IgA, or IgE after IgM during immune response in B cells. According to the class switching process, Igγ or Igα should share the same variable region with Igμ within the same cell population. However, no identical rearrangements were found between the VHγDγJHγ and VHαDαJHα in each sample. The above data suggest that epidermis-derived IgG and IgA may not arise by classical class switching; however, there is no reasonable explanation for the genetic mechanism by which epidermal cells produce IgG or IgA with different sequence of variable regions.

12.4 Keratinocyte-Derived IgG Has a Unique Sialylation Modification on the Asn162 Site of the Ig CH1 Domain

Recently, it has been discovered that epithelial cancer cells can express a unique form of IgG, known as sialylated IgG (SIA-IgG), which carries a sialylation modification on the Asn162 site of the Ig CH1 domain. Notably, a monoclonal antibody, RP215, has been found to recognise the unique sialylation modification epitope on the Asn162 site of the Igγ CH1 domain (Tang et al. 2018). Using RP215, Jiang et al. found that SIA-IgG staining was only present in epidermal cells, but not in the dermis. Moreover, intense staining was seen in the stratum spinosum and granulosum, while a weak signal was observed in the stratum basale and stratum corneum in both paraffin-embedded and frozen sections (Jiang et al. 2015).

12.5 IgG and IgA Secreted from Keratinocytes Serve as a Natural Antibody

It has been reported that human epithelial cells spontaneously produce natural antibody IgM to recognise microbial antigens (Hu et al. 2012). As the first protective barrier of the human body, the epidermis prevents the entry of a variety of pathogens, including bacteria and fungi. To address whether IgG or IgA secreted from keratinocytes serve as a natural antibody, Jiang et al. identified that IgG or IgA could be secreted into the extracellular space and culture supernatant of epidermal cells, and then purified the IgG or IgA. Subsequently, using *S. epidermidis*, *S. aureus*, *C. Albicans*, and *E. coli* they detected the microbial-binding ability of epidermis-derived IgG and IgA with pathogenic or commensal bacteria, and found that epidermis-derived Ig from keratinocytes of almost all of the skin samples recognised specific pathogens, with IgG and IgA exhibiting different reactions to different pathogens. For example, epidermis-derived IgG bound strongly with *S. aureus*, the main cause of skin infections, while epidermis-derived IgA bound both *S. aureus* and a commensal bacterium, *S. epidermidis* (Jiang et al. 2015). The potential correlation between the localisation or limited V(D) J rearrangement patterns of IgG and IgA and their microbial-binding patterns still needs to be investigated. These findings indicate that epidermal cells produce and secrete Ig to recognise and defend against pathogens, and that epidermis-derived Ig level increases upon microbe stimulation, potentially participating in immune responses to pathogens.

To rule out the possibility of B cell-derived Ig contamination, Jiang et al. examined the human keratinocyte cell line HaCaT for producing and microbe-binding activity of Ig, and they found that HaCaT cell line could also recognise several kinds of pathogens, especially *Staphylococcus aureus*. HaCaT cells spontaneously secreted IgG with anti-*S. aureus* activity, and IgA with anti-*E. coli* and anti-*S. aureus* activity, confirming that all IgG or IgA originated from human keratinocytes. The microbial-recognition profiles of HaCaT-derived Ig were consistent with epidermis-derived Ig. Interestingly, secreted Ig level elevated significantly following microbe stimulation and showed comparable microbial-binding profiles with those of Ig derived from epidermis. Thus, the microbial-recognising antibodies produced by the epidermis constitute a novel cutaneous defence barrier that protects us from pathogen infections. Moreover, IgG and IgA derived from HaCaT cells induced by different microbes showed cross-recognition characteristics with other pathogens (Jiang et al. 2015) (Fig. 12.1).

12.6 SIA-IgG Is Located in the Cytoplasm and Interacts with the Cytoskeleton-Associated Protein in Keratinocytes

In addition to acting as a secreted protein and exhibiting natural antibody activity, IgG produced by epidermal cells, Zhou et al. found that the SIA-IgG was mainly distributed in the cytoplasm of keratinocytes and showed a filamentous structure in epidermal cells, colocalising with cytokeratin 1 and plectin. Plectin is a cytoskeletal cross protein and maintains the assembly of microfilaments, intermediate filaments, and microtubules. Subsequently, the co-immunoprecipitation results further revealed that SIA-IgG could interact with desmoplakin, desmoglein, and plectin. These findings suggest that the function of SIA-IgG may be involved in assembling microfilaments, intermediate filaments, and microtubules. The specific mechanism underlying this interaction requires further investigation.

12.7 Possible Clinical Significance of Epidermal Cell-Derived Igs

The concept of non B cell-derived Ig and its function have gradually been recognised (Zhang et al. 2010; Jiang et al. 2006; Zheng et al. 2009). Considering the involvement of Ig in various skin diseases, we speculate that Ig produced by epidermal cells participates in the occurrence of a variety of skin diseases. For example, Pemphigus Vulgaris (PV) is a potentially life-threatening autoimmune blistering disease that affects skin and mucous membranes (Tucker et al. 2014), with desmoplakin-specific autoantibodies being the main pathogenic factors. Before the advent of systemic corticosteroids and immunosuppressants, the prognosis of pem-

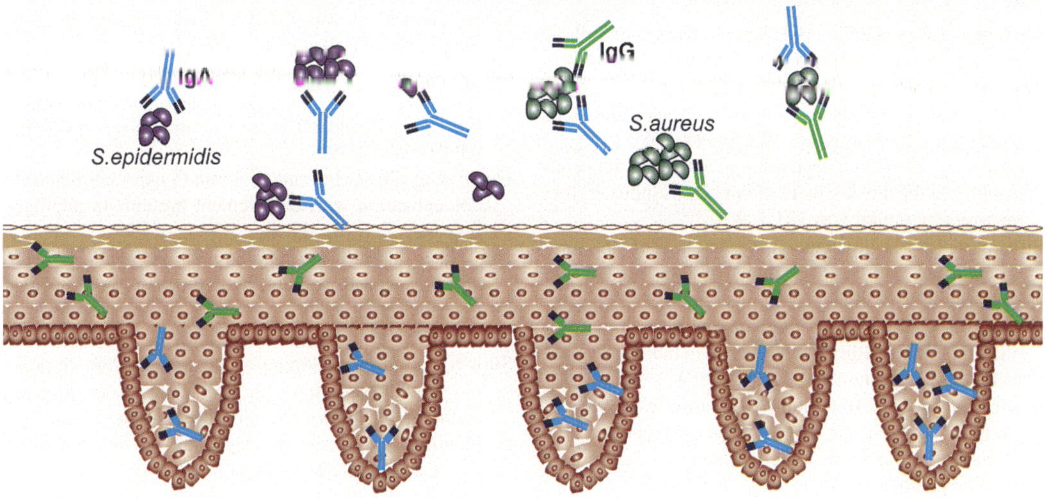

Fig. 12.1 IgG or IgA secreted by keratinocytes as natural antibodies to recognise microbial antigens. Specific pathogens were identified in all skin samples, including Staphylococcus epidermidis, Staphylococcus aureus, Candida albicans, and Escherichia coli, among others. The epidermal-derived immunoglobulins of keratinocytes include IgG and IgA, both of which have the ability to bind microorganisms to pathogenic or commensal bacteria. Of these, epidermis-derived IgG strongly binds to Staphylococcus aureus, a major cause of skin infections, while epidermis-derived IgA binds to both Staphylococcus aureus and the commensal bacterium Staphylococcus epidermidis

phigus was fatal within 5 years with a mortality of 50% after just 2 years due to secondary complications, including infections and malnutrition. However, when patients were treated simultaneously with systemic glucocorticoids and immunosuppressive agents, most of them remitted within a few weeks. Moreover, the CD20-targeted B cell depletion results in short-term disease remission in 95% of pemphigus patients, but still, 81% of them relapse and suffer fatal infection. Short-lived plasmablasts were indicated as the source of autoantibodies in PV because reduced serum autoantibody titres were observed after CD20-positive cells depletion (Silos et al. 1996).

The failure of B cell deletion in PV therapy suggests the existence of another source of Ig apart from B cells. Since IgG produced by epidermal cells can bind to desmosomes, it is possible that the autoantibodies in the epidermis against the pathogen, such as those in PV, may come from keratinocytes or other non B cells.

12.8 Progress and Prospects

The skin is the first barrier for the human to defend against invasive pathogens, playing an essential role in both innate and adaptive immune responses. IgG and IgA in epidermal cells were involved in anti-infectious, which suggests that the epidermal cell-derived Ig serves as an effective defender with those from B cells, providing a novel insight into skin immunity.

IgG was also found in the cytoplasm of keratinocytes. The SIA-IgG showed a filamentous structure in epidermal cells, colocalising with cytokeratin1 and plectin. These findings suggest that SIA-IgG function may participate in the assembly of microfilaments, intermediate filaments, and microtubules, potentially involving the migration and regeneration of epidermal cells and contributing to the progression of skin cancer. This provides novel insights into the fields of cell biology and tumour biology in the context of the skin.

References

Biernaskie J et al (2009) SKPs derive from hair follicle precursors and exhibit properties of adult dermal stem cells. Cell Stem Cell 5(6):610–623

Coruh G et al (1980) Serum proteins in human squamous epithelium. Br J Dermatol 102(5):497–505

Dong Y et al (2017) The deposition of anti-DNA IgG contributes to the development of cutaneous lupus erythematosus. Immunol Lett 191:1–9

Hu F et al (2012) Spontaneous production of immunoglobulin M in human epithelial cancer cells. PLoS One 7(12): p. e51423

Huang J et al (2014) Rearrangement and expression of the immunoglobulin mu-chain gene in human myeloid cells. Cell Mol Immunol 11(1):94–104

Huang X et al (2023) A Self-Propagating c-Met-SOX2 Axis Drives Cancer-Derived IgG Signaling That Promotes Lung Cancer Cell Stemness. Cancer Res 83(11): p. 1866–82

Jiang R et al (2006) A functional variant in the transcriptional regulatory region of gene LOC344967 cosegregates with disease phenotype in familial nasopharyngeal carcinoma. Cancer Res 66(2):693–700

Jiang D et al (2015) IgG and IgA with potential microbial-binding activity are expressed by normal human skin epidermal cells. Int J Mol Sci 16(2):2574–2590

Jordanovski D et al (2013) The hypoxia-inducible transcription factor ZNF395 is controlled by IkB kinase-signaling and activates genes involved in the innate immune response and cancer. PLoS One 8(9):e74911

Koster M et al (2006) TAp63alpha induces AP-2gamma as an early event in epidermal morphogenesis. Dev Biol 289(1):253–261

Liu Y et al (2015) Binding of the monoclonal antibody RP215 to immunoglobulin G in metastatic lung adenocarcinomas is correlated with poor prognosis. Histopathology 67(5):645–653

Oren-Suissa M et al (2007) Cell fusion during development. Trends Cell Biol 17(11):537–546

Qiu X et al (2003) Human epithelial cancers secrete immunoglobulin g with unidentified specificity to promote growth and survival of tumor cells. Cancer Res 63(19):6488–6495

Qiu Y et al (2012) Immunoglobulin G expression and its colocalization with complement proteins in papillary thyroid cancer. Mod Pathol 25(1):36–45

Qiu X et al (2013) Immunoglobulin gamma heavy chain gene with somatic hypermutation is frequently expressed in acute myeloid leukemia. Leukemia 27(1):92–99

Silos S et al (1996) Cloning of the gene for human pemphigus vulgaris antigen (desmoglein 3), a desmosomal cadherin. Characterization of the promoter region and identification of a keratinocyte-specific cis-element. J Biol Chem 271(29):17504–17511

Tang J et al (2018) Lung squamous cell carcinoma cells express non-canonically glycosylated IgG that activates integrin-FAK signaling. Cancer Lett 430:148–159

Tucker D et al (2014) Plakophilin-1 protects keratinocytes from pemphigus vulgaris IgG by forming calcium-independent desmosomes. J Invest Dermatol 134(4):1033–1043

Zhang S et al (2010) Immunoglobulin gene locus events in epithelial cells of lactating mouse mammary glands. Cell Mol Life Sci 67(6):985–994

Zheng J et al (2009) Immunoglobulin gene transcripts have distinct VHDJH recombination characteristics in human epithelial cancer cells. J Biol Chem 284(20):13610–13619

Non B Cell-Derived Immunoglobulins in Lung Epithelial Cells and Lung Cancer

13

Jingshu Tang, Erya Gao, Xinmei Huang, Yang Liu, and Wenwei Shao

Abstract

As the locus for air exchange, lung tissue is perpetually exposed to a significant quantity of foreign pathogens. Consequently, lung has developed a refined and intricate immune system. Beyond their physical and chemical barrier roles, lung epithelial cells can contribute to immune defence through the expression of Toll-like receptors (TLRs) and other pattern recognition receptors, along with the secretion of cytokines. Emerging evidence demonstrates that lung epithelial cells can generate and secrete immunoglobulins (Igs), including IgM, IgA, or IgG, thus performing antibody function. Moreover, malignantly transformed lung epithelial cells have been discovered to produce high levels of Ig, predominantly IgG, which do not fulfill the role of antibodies, but instead carries out tumour-promoting activity. Structural analysis has indicated that the biological activity of IgG produced by lung cancer cells differs from that of Igs produced by normal lung epithelial cells due to the unique glycosylation modification. Specifically, the sialylated IgG (SIA-IgG), characterised by a non-traditional N-glycosylation modification at the Asn162 site of Igγ CH1, is highly expressed in tumour stem cells. It has been demonstrated that SIA-IgG relies on this unique sialylation modification to promote tumorigenesis, metastasis, and immune evasion. Current results have proven that the Ig produced by lung epithelial cells has multifaceted biological activities, including immune defence functions under physiological conditions, while acquiring tumour-promoting activity during malignant transformation. These insights possess potential for the diagnosis and treatment of lung cancer as novel biomarkers and targets.

J. Tang (✉)
State Key Laboratory of Bioactive Substances and Functions of Natural Medicines, Institute of Materia Medica, Chinese Academy of Medical Sciences & Peking Union Medical College, Beijing, China
e-mail: tangjingshu@imm.ac.cn

E. Gao
People's Medical Publishing House Co., Ltd, Beijing, China

X. Huang
Department of Respiratory and Critical Care Medicine, Nanjing Drum Tower Hospital, The Affiliated Hospital of Nanjing University Medical School, Nanjing, China
e-mail: huangxm2017@pku.edu.cn

Y. Liu
Neuro-Oncology Branch, Center for Cancer Research, National Cancer Institute, Bethesda, MD, USA

W. Shao
Academy of Medical Engineering and Translational Medicine, Tianjin University, Tianjin, China
e-mail: wenwei.shao@tju.edu.cn

X. Qiu et al. (eds.), *Non B cell-Derived Immunoglobulins*, Advances in Experimental Medicine and Biology 1445, https://doi.org/10.1007/978-981-97-0511-5_13

Keywords

SIA-IgG · NSCLC · Integrin-FAK signalling · c-MET · Antibody therapy

13.1 General Considerations

The surface of the lung epithelium is the site of gas exchange in the body. Assuming that the average respiratory rate of a human is 15 times/min and the average tidal volume is 500 mL, the lung may be exposed to more than 10, 000 L of air every day. As a result, the lung will be exposed to a variety of foreign substances, including dust, pollen, bacteria, fungi, and viruses (Zarogoulidis et al. 2014). To effectively resist the invasion of different pathogens, the lung must coordinate with other cells. Growing studies have proven that, in the process of host defence, lung epithelial cells form a physical barrier through tight connection, recruit immune cells, produce cytokines, and regulate immune and inflammatory responses. Additionally, lung epithelial cells can also recognise pathogens through their own pattern recognition receptors (PRRs) and secrete anti-bacterial proteins to directly kill pathogens, which is crucial for host defence, repair, and maintenance of homeostasis in the innate lung immunity process (Hewitt and Lloyd 2021).

Immunoglobulin (Ig) is an essential molecule in humoral immunity. It has been long believed that Ig can only be produced by B cells. However, a large amount of evidence from our and other groups demonstrates that, apart from B cells, other cell lineages, including lung epithelial cells (Qiu et al. 2003), neurons (Huang et al. 2008), cardiomyocytes (Zhu et al. 2017), and germ cells (Huang et al. 2009), can express Ig (non-B-Ig). Non-B-Ig can fulfil the traditional antibody function, but it can also perform diverse cell biological functions, including promoting cell proliferation, adhesion, migration, and survival. Recently, IgG has been found to be overexpressed in malignant cells, particularly epithelial tumour cells, including breast cancer, lung cancer, colon cancer, oesophageal cancer, ovarian cancer, prostate cancer, and bladder cancer. Particularly, tumour stem cells overexpress a group of IgG with unique glycosylation, which promotes tumour progression, tumour stem cell self-renewal, metastasis, and tumour immune evasion (Qiu et al. 2003; Babbage et al. 2006; Chen and Gu 2007; Yang et al. 2013; Sheng et al. 2016; Zhang et al. 2012a; Lee and Ge 2009).

13.2 Lung Epithelial Cells Can Produce Igs

To determine if lung epithelial cells can produce Igs, Gao et al. sorted the lung epithelial cells of μMT mice by fluorescence-activated cell sorting (FACS) using EpCAM as an epithelial cell marker. The results demonstrated that lung epithelial cells can produce and secrete different types of Igs, which is further confirmed by the RNA-Seq and 10× Genomics single-cell sequencing assays showing the existence of Ig transcripts in these cells. More evidence came from the analysis of the lung tissue of mouse single-cell sequencing dataset (GSE124872) which is established by sequencing the whole lung tissue cells of C57BL/6J mice using Dropseq sequencing. Analysis of the results, with plasma cells as a positive control, revealed that Ig was expressed in all epithelial types of lung tissues, acting as a cell type-specific marker. Goblet cells primarily expressed Ighm; type II alveolar epithelial cells (AT2 cells), ciliated cells (ciliated cells), and clara cells (club cells) mainly expressed Ighm and Igha; plasma cells from lung tissue expressed Ighm, Igha, Ighg2b, and Ighg2c. Compared to plasma cells in lung tissue, epithelial cells only expressed Igκ, while plasma cells expressed both Igκ and Igλ. These data demonstrated that lung epithelial cells undergo Ig gene rearrangement and express Ig transcripts.

13.3 Lung Epithelial Cells Can Produce Specific Antibodies

Antibodies are ubiquitously present in various bodily fluids and can be classified as natural antibodies and antibodies produced following induction (provisionally referred to as induced antibodies or immune antibodies). Natural antibodies are generated spontaneously and can offer resistance against a range of pathogenic microorganisms. Immune antibodies are produced in response to specific antigens and have greater specificity and a stronger immune response to certain antigens. Under physiological conditions, the skin and mucosa are rich in natural antibodies, and as the primary line of defence, these natural antibodies play a significant role in local immunity. Thus far, according to classical immunology theory, it is believed that the natural antibodies present in the skin and mucosa are spontaneously produced by B cells/plasma cells in the lamina propria or dermis. However, our study has discovered that Ig can be produced by epithelial cells, such as liver epithelial cells (Shao et al. 2016) and epidermal cells (Jiang et al. 2015), and exhibits natural antibody activity. Yet it remains unclear whether the Ig produced by epithelial cells possesses the activity of immune antibodies.

Immune antibodies can be divided into two types based on whether the process requires the assistance of $CD4^+$ helper T cells (Th) after antigen stimulation. One type is referred to as T-dependent antigen (TD-Ag) specific antibody, and the other is known as T-independent antigen (TI-Ag) specific antibody (Kitamura 2021). Our recent findings have shown that B cell-deficient μMT mice could not produce TI antigen-specific antibodies but could produce TD antigen-specific antibodies, suggesting that the production of TI antigen-specific antibodies, but not TD antigen-specific ones, depends on B cells. Furthermore, we also observed high titres of TD antigen-specific IgG and IgA in bronchoalveolar lavage fluid of μMT mice, indicating that Ig produced by lung epithelial cells might possess immune antibody activity (Gao et al. 2021).

13.3.1 Specific Antibodies Against TD Antigen Can Be Produced Locally in the Lungs of B Cell-Deficient Mice

The μMT mice and CD19 knockout ($CD19^{-/-}$) mice are both B cell-deficient. To address whether non-B cells can produce specific antibodies, Gao et al. discovered that the first immunisation with 4-hydroxy-3-nitrophenylacetyl-Keyhole Limpet Hemocyanin (NP-KLH) took the μMT mice or $CD19^{-/-}$ mice about 2–3 weeks to produce low-level antibodies whereas the same procedure only took wild-type mice 1 week to make high-titre antibodies. Unexpectedly, both μMT and $CD19^{-/-}$ mice responded quickly to the secondary immunisation in producing high-level antibodies. The finding suggests that non-B cells can produce specific antibodies, whilst B cells are necessary for the rapid production of specific antibodies in response to the initial antigen challenge (Gao et al. 2021).

Subsequently, Gao et al. further investigated whether specific antibodies could be produced locally in the lung tissue. They examined the specific antibody levels in lungs, alveolar lavage fluid, and supernatant from cultured lung tissue, and found that high titres of NP-KLH-specific IgG and IgA appeared in lung lysate, bronchoalveolar lavage fluid, and the lung tissue after immunising the μMT mice twice with NP-KLH, which suggested that local cells in the lung can produce immune antibodies (Gao et al. 2021).

13.3.2 Lung Epithelial Cells Can Express Specific Antibodies Against TD Antigen

To explore whether lung epithelial cells can produce specific antibodies in response to TD antigen, Gao et al. sorted the lung epithelial cells from NP-KLH immunised mice and cultured them in vitro. The results showed that lung epithelial cells could produce and secrete specific antibodies. Furthermore, they crossed the $Ig\kappa^{fl/fl}$ mice with AAV-CC10-Cre transgenic mice to knock down the Igκ in Clara cells, then found that the level of NP-KLH-specific antibodies significantly decreased. These results indicated that a large part of the specific antibodies in lung tissue are derived from lung epithelial cells.

To further investigate the timing and titre of antibodies produced by lung epithelial cells after direct stimulation with NP-KLH, Gao et al.

established a lung unilateral immune model, which involved immunising one lung while leaving the other lung non-immunised. Systemic immunisation of mice was used as a control. The results showed a high level of specific antibodies in the cultured supernatant of immune side lung epithelial cells in both wild-type and μMT mice, which were significantly higher than the non-immunised counterparts and the systemically immunised group. In addition, we also analysed the possibility of lung epithelial cells transdifferentiating into antigen-presenting cells by RNA-Seq. The results showed a significant expression upregulation of not only multiple Ig transcripts but also the MHCII molecules and costimulatory molecules (CD80 and CD86) associated with antigen presentation, in epithelial cells of the immune-side lung in unilateral immune mice, suggesting that lung epithelial cells are potential antigen-presenting cells in the lung (Gao et al. 2021).

The discovery of lung epithelial cells can produce immune antibodies against TD antigens challenges the classical theory of immunology and it suggests that the classical theory regarding antibody source needs to be revised.

13.4 IgG Expression in Lung Cancer

Lung cancer is the leading cause of cancer-related deaths worldwide (Siegel et al. 2023). Non-small cell lung cancer (NSCLC) primarily consists of lung squamous cell carcinoma (LSCC) and adenocarcinoma (LADC), which together account for approximately 80% of the cases. Despite aggressive frontline treatment, the 5-year survival rate remains poor for most patients (Field et al. 2013; Reck et al. 2013; Rosell et al. 2013). The high mortality of this disease is attributable to its local invasion and metastasis to distant organs, and clinical treatment of NSCLC remains challenging due to this high incidence of tumour metastasis. Although PD-1/PD-L1 immune checkpoint therapy can benefit patients with NSCLC, it has a limit up to 20% of the patient population. Given that the incidence of NSCLC ranks first in the world, it is imperative to find new therapeutic targets. The highly sialylated IgG expressed in NSCLC is a potential novel therapeutic target for treating NSCLC.

13.4.1 SIA-IgG Discovered in Non-Small Cell Lung Cancer (NSCLC)

In the early study, Qiu et al. (2003) and Jiang et al. (2014) used commercialised anti-human IgG antibodies to perform immunohistochemical staining on various lung cancer tissues, and found that IgG was expressed in multiple lung cancer tissues and normal lung tissues. However, the anti-IgG antibody cannot distinguish IgG produced by lung cancer cells from that by normal lung epithelial cells. Monoclonal antibody RP215 has been identified that specifically recognised the IgG expressed by epithelial cancer cells (lung cancer, breast cancer, colon cancer, etc.), but little or no IgG by normal epithelial cells as well as IgG produced by other lineage cells, including IgG produced by B cells (Lee et al. 1992, 2008). However, the antigen epitopes specifically recognised by RP215 have not been revealed until recently, when it was found that the IgG recognised by RP215 has an atypical N-glycosylation modification at position Asn162 on Igγ CH1, and its glycan end shows high sialylation modification (called sialylated IgG, SIA-IgG). Both the CH1 sequence on Igγ and high sialylation modification on this new glycan formed the epitope recognised by RP215 (Tang et al. 2018).

Liu et al. have analysed the expression of RP215-recognised IgG in different types of lung cancers by immunohistochemical staining. SIA-IgG was found to be widely expressed in NSCLC but not in small-cell lung cancer (Liu et al. 2015). Expression profile analysis of SIA-IgG showed that the expression level of SIA-IgG varied in various types of NSCLCs. SIA-IgG was expressed at a high frequency (89%) and high level in almost all LSCC cells but not in tumour stroma and adjacent normal tissues. In contrast, SIA-IgG was expressed at a low frequency (40%) and low levels in lung adenocarcinoma.

Interestingly, significantly high SIA-IgG expression was mainly observed in frontier cancer cells, scattered invasive cancer cells, and cancer nest margin cells. In contrast, other cancer cells, such as large cell lung cancer cells and alveolar cell carcinoma cells, showed weak expression. Similarly, Tang et al. examined five NSCLC cell lines, including lung adenocarcinoma cell lines A549, Calu-3, and NCI-H1299, as well as lung squamous cell carcinoma cell lines NCI-H520 and SK-MES-1. They found that SIA-IgG was also expressed in the five lung cancer cell lines, with much higher expression levels in two lung squamous cell lines than in three lung adenocarcinoma cell lines (Tang et al. 2018).

Analysis of the correlation between SIA-IgG and survival of lung cancer patients revealed that the high expression of SIA-IgG in lung adenocarcinoma (Liu et al. 2015) and lung squamous cell carcinoma (Tang et al. 2018) indicated a poor prognosis. These findings suggest that the SIA-IgG expression level and distribution are essential for clinical diagnosis and prognosis evaluation of lung cancer, further indicating that SIA-IgG is involved in the occurrence and development of lung cancer.

13.4.2 SIA-IgG Promotes Proliferation, Migration, and Invasion of Lung Cancer Cells

In 2003, Qiu et al. reported that non-B cancer cells could produce IgG and further found that blocking or knocking down IgG could inhibit cancer cell growth and induce apoptosis of cancer cells, suggesting that cancer cell-derived IgG has growth factor-like activity (Qiu et al. 2003). Liu et al. (2015) and Jiang et al. (2014) found that knocking down IgG in SK-MES-1 and A549, respectively, resulted in the inhibition of proliferation in lung cancer cells and promotion of apoptosis in cancer cells. Similarly, Tang et al. found that IgG silencing in NCI-H520 led to a significant decrease in colony formation of cancer cells, whereas overexpression of IgG in NCI-H520 significantly promoted the colony-forming ability and subcutaneous tumorigenicity (Tang et al. 2018). In addition, Liu et al. sorted A549 cells with either high (SIA-IgG^{high}) or low (SIA-IgG^{low}) expression of SIA-IgG by FACS and found that tumour cells with high expression of SIA-IgG had stronger proliferation and colony-forming abilities. To analyse if SIA-IgG promotes migration and invasion of lung cancer cells, Liu et al. examined clinical biopsy samples from lung cancer patients and found that some cancer cells showed intense positive staining for SIA-IgG. Most of these cells were single invasive and frontal cells, which have strong migration and invasion abilities and are closely related to lung cancer metastasis. Moreover, the expression of SIA-IgG in lymph node metastases of cancer patients was significantly higher than that in primary tumours. The above findings suggest that SIA-IgG may be closely related to tumour metastasis and invasion.

Biological phenotypic experiments further confirmed that IgG can promote cell migration. Using siRNA to knock down IgG or adding IgG antibodies into the cell culture system can inhibit the migration and invasion of lung cancer cells. Flow cytometry sorting was used to enrich tumour cells with either high or low expression of SIA-IgG. The result showed that, in SIA-IgG^{high} cells, E-cadherin expression was down-regulated, and the matrix metalloproteinase MMP9 was significantly upregulated; the epithelial-mesenchymal transformation (EMT)-related transcription factors Slug, Twist, and ZEB1/2 were upregulated considerably, while vimentin and MMP2 remained unchanged. The results suggested that SIA-IgG is involved in EMT.

13.4.3 SIA-IgG Is Involved in Maintaining Cancer Stem Cells Properties of Lung Cancer Cells

Our previous study revealed that SIA-IgG is highly expressed in stem-like cancer cells and bronchial basal stem cells, contributing to tumour initiation and metastasis in breast cancer. Thus, using NSCLC as a model, Huang et al., further found that SIA-IgG was upregulated in lung cancer stem cells (CSCs) and positively correlated with CSCs markers CD44v6, SOX2, and OCT4. To further characterise SIA-IgG^{high} cells, they compared the gene expression profiles of SIA-IgG^{high} and SIA-$IgGl^{ow}$ NSCLC cells by RNA-seq analysis. Genes upregulated in SIA-IgG^{high} cells were associated with known CSC signatures. In addition, gene set enrichment analysis (GSEA) of The Cancer Genome Atlas (TCGA) lung cancer samples showed significant enrichment of EMT gene modules in samples with high expression of IGHG1, which is relevant to the acquisition and maintenance of stem cell-like characteristics. The Kaplan-Meier analysis showed that higher IgG expression levels in lung cancer patients are associated with a significantly worse overall survival (Huang et al. 2023). Together, these results suggest that SIA-IgG is upregulated in lung CSCs and may be involved in the maintenance of CSCs stemness.

To further determine whether SIA-IgG plays a critical role in CSC properties and tumorigenesis, Huang et al. sorted both the SIA-IgG^{high} and SIA-IgG^{low} lung cancer cells using FACS. As expected, SIA-IgG^{high} cells showed significantly stronger tumoursphere and clone-forming abilities compared to SIA-IgG^{low} cells. In contrast, silencing SIA-IgG reduced the size and number of tumour spheroids, which could be rescued by SIA-IgG purified from RP215-affinity chromatography. More importantly, they performed the limiting dilution assays with FACS-sorted SIA-IgG^{high} and SIA-IgG^{low} cells and observed that SIA-IgG^{high} cells significantly promoted the tumour-forming potential (Liao et al. 2015). Collectively, these results indicate that SIA-IgG plays a vital role in self-renewal, differentiation potential, and tumorigenicity of CSCs.

13.5 SIA-IgG Promoting Mechanism to Cancer Cells

13.5.1 SIA-IgG Activates the Downstream FAK Signalling Pathway by Binding to Integrin α6β4

Tang et al. identified integrins as potential membrane receptors for SIA-IgG using Immunoprecipitation-Mass Spectrometry (IP-MS) and label-free quantitative proteomic analysis (Tang et al. 2018). Stewart et al. analysed the expression profile of integrin β4 in lung cancer and found that integrin β4 significantly elevated in NSCLC but not in small cell lung cancer (Stewart et al. 2016). Further analysis revealed that integrin β4 expression in lung squamous cell carcinoma (85%) was significantly higher than in lung adenocarcinoma (25%), and it was significantly associated with venous invasion and poor prognosis. Moreover, SIA-IgG and integrin β4 exhibited similar expression profiles. In addition, several studies have shown the essential role of integrin β4 in tumour cell migration, invasion, and maintenance of stemness (Zheng et al. 2013; Fisher et al. 2016; Yoshioka et al. 2013). Tang et al. confirmed the accuracy of proteomics results by co-immunoprecipitation (CO-IP) and found that SIA-IgG and integrin β4 co-expressed and co-localised at tissue and cell levels by immunohistochemistry, flow sorting, and immunofluorescence (Tang et al. 2018). These findings provided evidence for the interaction between SIA-IgG and integrin β4 under physiological conditions.

As an extracellular matrix protein receptor, the integrin family does not possess intrinsic tyrosine kinase activity but primarily recruits and activates non-receptor tyrosine kinases for signal transduction (Miranti and Brugge 2002). When integrin β4 binds to its ligand, focal adhesion kinase (FAK) can be recruited via its cytoplasmic domain, resulting in the phosphorylation of the

Tyr397 site of FAK. Subsequently, the FAK attaches to the SH2 domain of Src and promotes the phosphorylation of the Tyr416 site of Src. As a result, the phosphorylation of the Tyr925 site of FAK can be achieved by feedback regulation following the phosphorylation of Src. These sequential events ultimately trigger downstream Ras-MAPK or PI3K-AKT cascade reactions. Knockdown of IgG or exogenous addition of antibody RP215 in lung cancer cells led to a downregulation of phosphorylation levels of FAK pathway-related molecules downstream of integrin. Conversely, overexpression of IgG or exogenous administration of tumour-derived SIA-IgG reversed the changes in these signalling pathway-related molecules, suggesting that SIA-IgG promotes activation of the FAK signalling pathway (Fig. 13.1).

13.5.2 SIA-IgG Activates c-MET Signalling and Promotes CSCs Stemness Through the Asn162 N-Glycan Modification

Reports suggest that Integrin α6β4 not only recruits and activates non-receptor tyrosine kinases for signal transduction, but also interacts with receptor tyrosine kinases (RTKs) to activate the corresponding RTK and its downstream signalling pathways in malignant tumours (Folgiero et al. 2007; Trusolino et al. 2001; Guo et al. 2006). IP-MS results indicated that c-MET and epidermal growth factor receptor (EGFR), members of the RTK family, were also detected in the components bound to SIA-IgG, suggesting that SIA-IgG can also interact with the RTK family. Next, phosphorylation chips for RTK were used

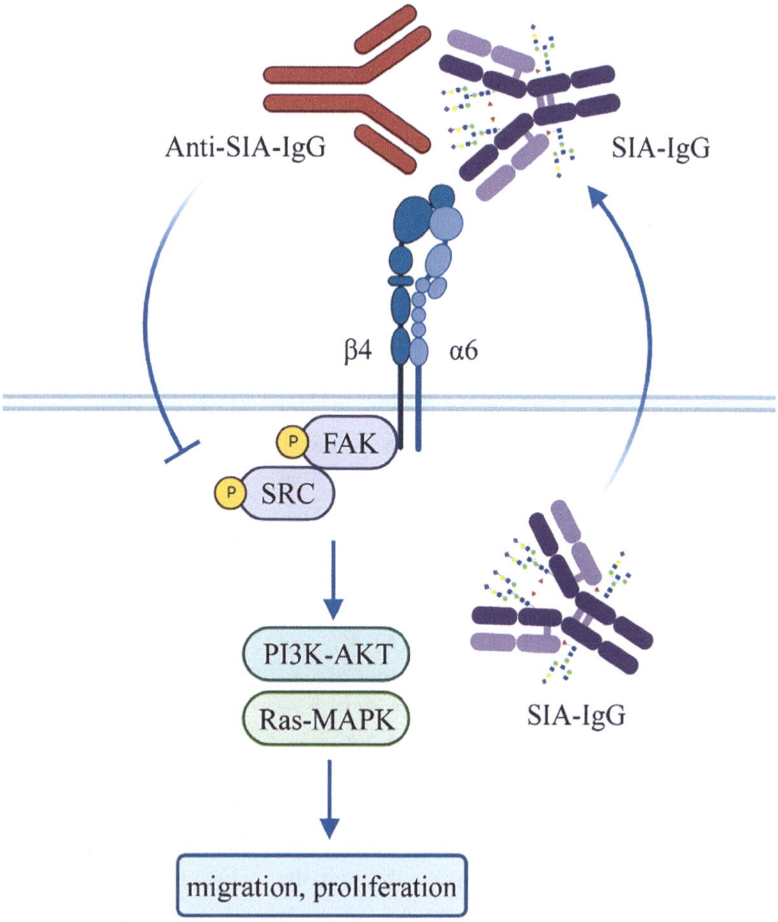

Fig. 13.1 A schematic graph depicting the role of SIA-IgG in lung squamous cell carcinoma

to screen potential SIA-IgG-regulated molecules in the RTK family. The results showed that the phosphorylation level of c-Met at Tyr1234/1235 site in lung squamous cell carcinoma and lung adenocarcinoma cells significantly decreased following IgG knockdown. In contrast, the phosphorylation levels of EGFR and HER2 remained unchanged. Moreover, overexpression of IgG increased the phosphorylation level of c-MET. All these results suggest that SIA-IgG is involved in activating the c-MET pathway (Huang et al. 2023). To further characterise the functions of N-glycan modification Asn162 in the c-MET signalling activation and stemness maintenance, Huang et al. stably overexpressed wild-type IgG heavy chain (WT) and constructs with mutations in the CH1 or CH2 region (CH1mu/CH2mu) in lung cancer cells. Interestingly, they observed that overexpression of WT or CH2mu SIA-IgG, but not the CH1mu construct, enhanced c-MET activation and stemness maintenance of CSCs. Consistently, SIA-IgG, but not neuraminidase-digested IgG, significantly reversed the inhibition of tumour spheroids by SIA-IgG silencing (Huang et al. 2023).

13.5.3 SOX2 and OCT4 Coordinately Transactivate the Expression of IgG in CSCs

Huang et al. further explored the mechanisms underpinning the regulation of IgG transcript expression in CSCs. Interestingly, a genome-wide CRISPR activation screen revealed that OCT4 and SOX2 co-regulated SIA-IgG expression. To confirm this, they further analysed the promoter region of the human IgHV gene using the JASPAR database and identified the existence of octamer element (5′-ATGCAAAT-3′) and SOX2 binding sites in the IgHV promoter of cancer cells. Moreover, knockdown of SOX2 or OCT4 repressed the expression of both IgG transcript and protein. Consistently, Huang et al. observed a specific interaction between SOX2 and OCT4, and further confirmed that SOX2 and OCT4 transactivated the *Igγ* gene through the specific binding sites in CSCs using chromatin immunoprecipitation (CHIP), luciferase assay, and electrophoretic mobility experiment assays (EMSA). Importantly, they revealed that SOX2 and OCT4 exhibited a tendency to bind to the promoter of IgHV in CSCs compared to non-CSC cells, indicating that SOX2 and OCT4 might mediate the upregulation of IgG expression in CSCs (Huang et al. 2023).

13.5.4 SIA-IgG/c-MET/SOX2-OCT4 Axis Upregulates the Expression of SOX2 and SIA-IgG Through Feedback

Huang et al. discovered that SIA-IgG activates the c-MET signalling and upregulates the expression of SOX2, while treatment with PHA-665752 or RP215 completely reversed the activation of c-MET and significantly inhibited SOX2 expression. Consequently, they further investigated if SIA-IgG is regulated by SIA-IgG/c-MET/SOX2-OCT4 axis through positive feedback. They found that RP215 and PHA-665752 inhibited the protein and mRNA expression levels of SIA-IgG in a dose-dependent manner in lung cancer cells, consistent with the previous report showing that SOX2 is positively regulated by c-MET signalling. Additionally, treatment with SIA-IgG activated c-MET signalling and dramatically increased the protein and mRNA expression levels of SIA-IgG. In summary, Huang et al. explored the functional role and mechanism of SIA-IgG in CSC, revealing that SIA-IgG is significantly upregulated in CSCs and plays a crucial role in maintaining CSCs properties and promoting tumorigenesis by activating c-MET signalling. Moreover, SOX2 and OCT4, the stemness-related transcription factors, coordinate IgG expression to confer self-renewal capacity in CSCs, and the expression of SOX2 and SIA-IgG is triggered by activation of the SIA-IgG-c-MET-SOX2/OCT4 axis in CSCs (Huang et al. 2023) (Fig. 13.2).

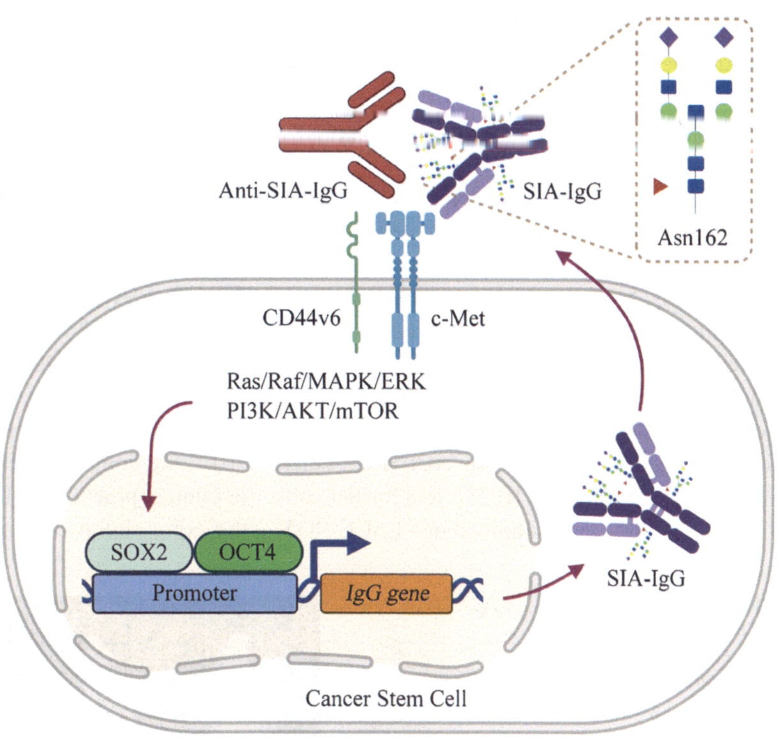

Fig. 13.2 SIA-IgG activates c-MET signalling and promotes CSCs stemness through the Asn162 N-glycan modification. SOX2 and OCT4 collaboratively transactivate the expression of IgG in CSCs, and the SIA-IgG/c-MET/SOX2-OCT4 axis upregulates the expression of SOX2 and SIA-IgG through feedback

13.6 The Application Potential of SIA-IgG in the Diagnosis and Treatment of Lung Cancer

13.6.1 SIA-IgG Serves as a Novel Therapeutic Target for NSCLC

Globally, deaths from lung cancer exceed those from any other type of cancer, with NSCLC accounting for approximately 80% of cases. Despite various first-line treatments, the 5-year survival rate for most patients remains exceedingly poor. Targeted therapies such as the EGFR-TK inhibitors gefitinib and erlotinib, and the EML4-ALK fusion gene inhibitor crizotinib, have shown relatively positive clinical effects for lung adenocarcinoma. However, these drugs prove ineffective in treating LSCC, and effective targeted therapies for LSCC are currently lacking (Niu et al. 2022). Immunohistochemistry and immunofluorescence results have demonstrated that SIA-IgG is primarily localised on the cell membrane of LSCC cells. Consequently, membrane proteins are currently regarded as the most effective targets for tumour treatment. Additionally, SIA-IgG plays a crucial role in the development and metastasis of LSCC, making it a potential novel target for treating this type of cancer.

Tang et al. found that the antibody RP215 effectively inhibits both the integrin-FAK pathway activation and colony formation of two LSCC cell lines in a dose-dependent manner in vitro. The patient-derived xenograft (PDX) model omits the in vitro culture step and preserves the histological and molecular biological characteristics of the primary tumour cells. Therefore, the pharmacodynamic results obtained by the PDX model will be more accurate and closer to the clinical results. The PDX model has been used in biological and translational medical research in many kinds of tumours, such as tumour heterogeneity, pre-clinical drug evaluation, drug screening and so on (DeRose et al. 2011; Zhang et al. 2012b). Tang et al. further established a PDX model and found that the antibody RP215 exhibited an excellent therapeutic effect in three PDX models of LSCC, suggesting that the antibody RP215 has potential as an anti-tumour therapeutic agent (Fig. 13.3) (Tang et al. 2018).

Recently, Huang et al. further explored whether SIA-IgG could serve as a promising target for the prevention and treatment of CSCs. They first evaluated the anti-tumour effects of RP215, the SIA-IgG-specific monoclonal antibody (mAb). Soft agar colony formation assay and sphere formation assay showed that cells treated with RP215 formed significantly fewer spheres and colonies. Moreover, the RP215 treatment significantly decreased the colony formation and migration capacity of NCI-H520 cells. The anti-tumour effect of RP215 was further substantiated by the results attained in a LSCC PDX tumour model, demonstrating that the tumour growth was significantly slower in mice receiving intravenous administration of RP215 than in mice treated with mIgG. They also observed that RP215 treatment inhibited c-MET activation in PDX tumours (Huang et al. 2023).

13.6.2 SIA-IgG Can Be Used as a Molecular Marker for the Diagnosis of LSCC

The CK5/6 (Kargi et al. 2007) and P40 (Ao et al. 2014) are commonly used as molecular markers in clinical practice for LSCC diagnosis. Notably, the expression pattern of SIA-IgG is comparable

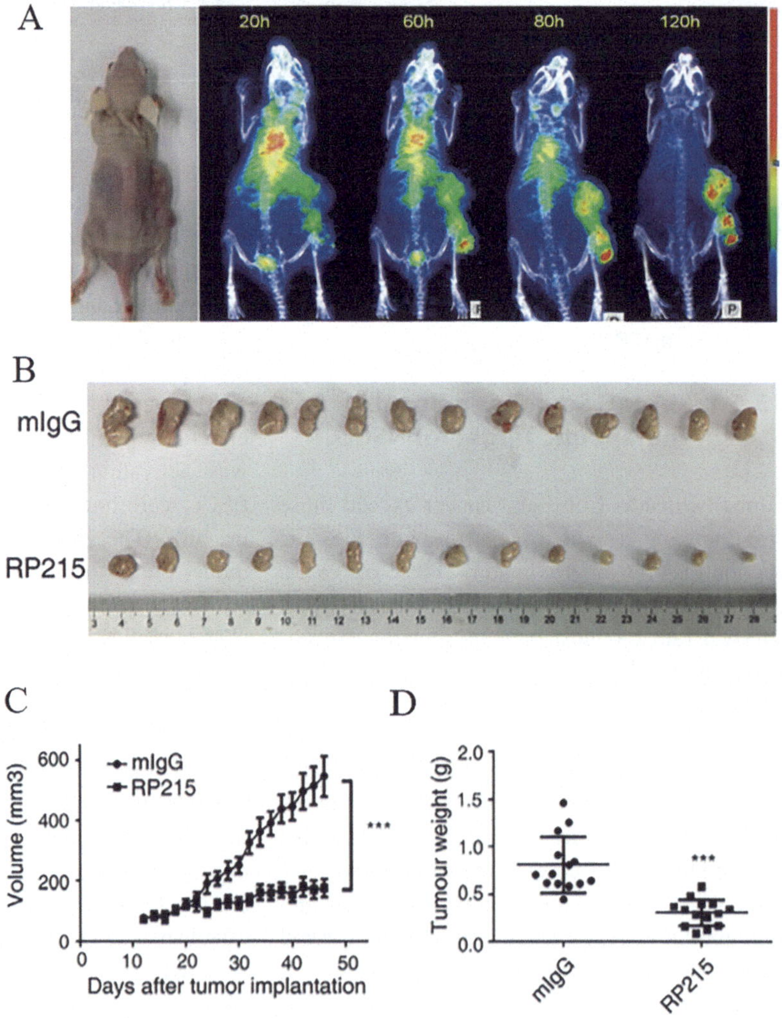

Fig. 13.3 RP215 has therapeutic effects on LSCC. SCID mice bearing PDX tumours were injected with 5 mg/kg of RP215 or mIgG twice a week intravenously (i.v.). The growth curve (**a**) Detection of antibody targeting in vivo after injection of RP215 antibody. (**b**) Tumor images of the RP215 group and mIgG group at the end of the experiment. (**c**) and weight plots (**d**) of harvested tumours are illustrated, indicating the suppressive effect of RP215 on tumour growth

with those of CK5/6 and P40. Furthermore, statistical analysis reveals a significant correlation between SIA-CIgG and CK5/6. Importantly, SIA-IgG is more sensitive than CK5/6 and P40 in LSCC diagnosis. Liu et al. have shown that combining the diagnosis of SIA-IgG, CK5/6, and P40 in LSCC significantly improves the detection rate, using the receiver operating characteristic (ROC) curve to comprehensively evaluate the sensitivity and specificity of SIA-IgG, CK5/6, and P40. The above result suggests that SIA-IgG is a valuable addition to the lung cancer diagnostic system.

It is noteworthy that the treatment with the antibody RP215 in two PDX models improves the weight loss in mice caused by malignant tumours (Tang et al. 2018), indicating that the antibody RP215 has potential application value in reducing malignant heterogeneity in patients with LSCC.

13.7 Conclusions and Prospects

The lung is a vital defence organ, and the lung epithelium is the first line of defence against pathogens. It is a significant discovery in immunology that lung epithelial cells can produce specific antibodies. This discovery challenges the traditional understanding of lung local immunity and humoral immune mechanisms and promotes the in-depth development of mucosal immunity and acquired immune response. Specifically, this discovery will reshape our understanding of the pathogenesis of lung-related diseases and potentially provide a new entry point for their diagnosis and treatment. Particularly, SIA-IgG is expressed in NSCLC and is significantly associated with poor prognosis of LSCC and lung adenocarcinoma, and its high expression in LSCC makes it a potential diagnostic marker. SIA-IgG promotes the occurrence and metastasis of NSCLC and participates in cell stemness maintenance. Mechanistic studies have revealed that SIA-IgG binds to integrin α6β4 and c-MET, thereby initiating downstream PI3K/AKT or MAPK cascades by mediating downstream FAK and Src activation. The monoclonal antibody RP215 binds to SIA-IgG secreted by CSC cells, inhibiting the activation of FAK and the c-MET pathways and tumorigenesis. Therefore, SIA-IgG can be used as a potential target for treating NSCLC.

References

Ao MH et al (2014) The utility of a novel triple marker (combination of TTF1, napsin A, and p40) in the subclassification of non-small cell lung cancer. Hum Pathol 45(5):926–934

Babbage G et al (2006) Immunoglobulin heavy chain locus events and expression of activation-induced cytidine deaminase in epithelial breast cancer cell lines. Cancer Res 66(8):3996–4000

Chen Z, Gu J (2007) Immunoglobulin G expression in carcinomas and cancer cell lines. FASEB J 21(11):2931–2938

DeRose YS et al (2011) Tumor grafts derived from women with breast cancer authentically reflect tumor pathology, growth, metastasis and disease outcomes. Nat Med 17(11):1514–1520

Field JK et al (2013) Prospects for population screening and diagnosis of lung cancer. Lancet 382(9893):732–741

Fisher ML et al (2016) Transglutaminase interaction with alpha6/beta4-integrin stimulates YAP1-dependent deltaNp63alpha stabilization and leads to enhanced cancer stem cell survival and tumor formation. Cancer Res 76(24):7265–7276

Folgiero V et al (2007) The alpha6beta4 integrin can regulate ErbB-3 expression: implications for alpha6beta4 signaling and function. Cancer Res 67(4):1645–1652

Gao E et al (2021) Lung epithelial cells can produce antibodies participating in adaptive humoral immune responses. BioRxiv Preprint. https://doi.org/10.1101/2021.05.13.443498

Guo W et al (2006) Beta 4 integrin amplifies ErbB2 signaling to promote mammary tumorigenesis. Cell 126(3):489–502

Hewitt RJ, Lloyd CM (2021) Regulation of immune responses by the airway epithelial cell landscape. Nat Rev Immunol 21(6):347–362

Huang J et al (2008) Expression of immunoglobulin gene with classical V-(D)-J rearrangement in mouse brain neurons. Int J Biochem Cell Biol 40(8):1604–1615

Huang J et al (2009) Expression of immunoglobulin gene with classical V-(D)-J rearrangement in mouse testis and epididymis. J Histochem Cytochem 57(4):339–349

Huang X et al (2023) A self-propagating c-met-SOX2 axis drives cancer-derived IgG signaling that promotes lung cancer cell stemness. Cancer Res 83(11):1866–1882

Jiang C et al (2014) Immunoglobulin G expression in lung cancer and its effects on metastasis. PLoS One 9(5):e97359

Jiang D et al (2015) IgG and IgA with potential microbial-binding activity are expressed by normal human skin epidermal cells. Int J Mol Sci 16(2):2574–2590

Kargi A, Gurel D, Tuna B (2007) The diagnostic value of TTF-1, CK 5/6, and p63 immunostaining in classification of lung carcinomas. Appl Immunohistochem Mol Morphol 15(4):415–420

Kitamura D (2021) Mechanisms for the regulation of memory B-cell recall responses in mice. Int Immunol 33(12):791–796

Lee G, Ge B (2009) Cancer cell expressions of immunoglobulin heavy chains with unique carbohydrate-associated biomarker. Cancer Biomark 5(4): 177–188

Lee CY et al (1992) Studies of a tumor-associated antigen, COX-1, recognized by a monoclonal antibody. Cancer Immunol Immunother 35(1):19–26

Lee G et al (2008) Molecular identity of a pan cancer marker, CA215. Cancer Biol Ther 7(12):2007–2014

Liao Q et al (2015) Aberrant high expression of immunoglobulin G in epithelial stem/progenitor-like cells contributes to tumor initiation and metastasis. Oncotarget 6(37):40081–40094

Liu Y et al (2015) Binding of the monoclonal antibody RP215 to immunoglobulin G in metastatic lung adenocarcinomas is correlated with poor prognosis. Histopathology 67(5):645–653

Miranti CK, Brugge JS (2002) Sensing the environment: a historical perspective on integrin signal transduction. Nat Cell Biol 4(4):E83–E90

Niu Z et al (2022) Signaling pathways and targeted therapies in lung squamous cell carcinoma: mechanisms and clinical trials. Signal Transduct Target Ther 7(1):353

Qiu X et al (2003) Human epithelial cancers secrete immunoglobulin g with unidentified specificity to promote growth and survival of tumor cells. Cancer Res 63(19):6488–6495

Reck M et al (2013) Management of non-small-cell lung cancer: recent developments. Lancet 382(9893):709–719

Rosell R, Bivona TG, Karachaliou N (2013) Genetics and biomarkers in personalisation of lung cancer treatment. Lancet 382(9893):720–731

Shao W et al (2016) Epithelial cells are a source of natural IgM that contribute to innate immune responses. Int J Biochem Cell Biol 73:19–29

Sheng Z et al (2016) Involvement of cancer-derived IgG in the proliferation, migration and invasion of bladder cancer cells. Oncol Lett 12(6):5113–5121

Siegel RL et al (2023) Colorectal cancer statistics, 2023. CA Cancer J Clin 73(3):233–254

Stewart RL et al (2016) Elevated integrin alpha6beta4 expression is associated with venous invasion and decreased overall survival in non-small cell lung cancer. Hum Pathol 54:174–183

Tang J et al (2018) Lung squamous cell carcinoma cells express non-canonically glycosylated IgG that activates integrin-FAK signaling. Cancer Lett 430:148–159

Trusolino L, Bertotti A, Comoglio PM (2001) A signaling adapter function for alpha6beta4 integrin in the control of HGF-dependent invasive growth. Cell 107(5):643–654

Yang B et al (2013) Correlation of immunoglobulin G expression and histological subtype and stage in breast cancer. PLoS One 8(3):e58706

Yoshioka T et al (2013) Beta4 integrin signaling induces expansion of prostate tumor progenitors. J Clin Invest 123(2):682–699

Zarogoulidis P et al (2014) Defense mechanisms of the respiratory system and aerosol production systems. Med Chem 10(2):123–136

Zhang L et al (2012a) Expression of immunoglobulin G in esophageal squamous cell carcinomas and its association with tumor grade and Ki67. Hum Pathol 43(3):423–434

Zhang J et al (2012b) Translating the therapeutic potential of AZD4547 in FGFR1-amplified non-small cell lung cancer through the use of patient-derived tumor xenograft models. Clin Cancer Res 18(24):6658–6667

Zheng Y et al (2013) A rare population of CD24(+) ITGB4(+)Notch(hi) cells drives tumor propagation in NSCLC and requires Notch3 for self-renewal. Cancer Cell 24(1):59–74

Zhu Z et al (2017) Immunoglobulin M, a novel molecule of myocardial cells of mice. Int J Biochem Cell Biol 88:172–180

14 Expression and Function of Mammary Epithelial Cell-Derived Immunoglobulins

Qinyuan Liao, Dongyang Jiang, Shuai Zhang, and Xiaoyan Qiu

Abstract

Over the past 20 years, increasing evidence has demonstrated that immunoglobulins (Igs) can be widely generated from non B cells, including normal and malignant mammary epithelial cells. In normal breast tissue, the expression of IgG and IgA has been identified in epithelial cells of mammary glands during pregnancy and lactation, which can be secreted into milk, and might participate in neonatal immunity. On the other hand, non B-IgG is highly expressed in breast cancer cells, correlating with the poor prognosis of patients with breast cancer. Importantly, a specific group of IgG, bearing a unique N-linked glycan on the Asn162 site and aberrant sialylation modification at the end of the novel glycan (referred to as sialylated IgG (SIA-IgG)), has been found in breast cancer stem/progenitor-like cells. SIA-IgG can significantly promote the capacity of migration, invasiveness, and metastasis, as well as enhance self-renewal and tumorigenicity in vitro and in vivo. These findings suggest that breast epithelial cells can produce Igs with different biological activities under physiological and pathological conditions. During lactation, these Igs could be the main source of milk Igs to protect newborns from pathogenic infections, while under pathological conditions, they display oncogenic activity and promote the occurrence and progression of breast cancer.

Keywords

Immunoglobulins · Mammary epithelial cell · Breast cancer · SIA-IgG · Cancer stem cell · Cancer metastasis

14.1 General Considerations

Mammals produce milk through the mammary gland, an exocrine gland located in the breast. The breast is divided into 15–25 lobes by connective tissues, and each lobe is divided into several lobules. The lobule is the basic functional unit of the compound tubulovesicular mammary gland,

Q. Liao (✉)
Department of Immunology, Guilin Medical University, Guilin, Guangxi, China

D. Jiang
Department of Cardiology, Pan-vascular Research Institute, Shanghai Tenth People's Hospital, Tongji University School of Medicine, Shanghai, China
e-mail: jiangdy@tongji.edu.cn

S. Zhang
Strategic Marketing Department of Central Marketing, GeneralElectric Healthcare Co., Shanghai, China

X. Qiu
Department of Immunology, School of Basic Medical Sciences, Peking University, Beijing, China
e-mail: qiuxy@bjmu.edu.cn

X. Qiu et al. (eds.), *Non B cell-Derived Immunoglobulins*, Advances in Experimental Medicine and Biology 1445, https://doi.org/10.1007/978-981-97-0511-5_14

consisting of intralobular ducts and acini. The major cell types of mammary gland are the inner luminal epithelial cells and the myoepithelial cells in the basal compartment. Polarised secretory epithelial cells are responsible for the synthesis and secretion of milk, and myoepithelial cells are necessary for the contraction of mammary gland ducts (McManaman and Neville 2003). Notably, myoepithelial cells possess stem/progenitor cell properties and give rise to basal or luminal progeny under certain conditions, such as pregnancy (Prater et al. 2014; Gieniec and Davis 2022).

The development of the mammary gland starts at puberty in females, and its structure varies with different physiological status. In adults, the mammary glands are usually in a quiescent phase, with underdeveloped breast acini and ductal systems, as well as abundant connective tissues, especially adipose tissues. During the secretory phase of each menstrual cycle, the mammary acini and ducts proliferate slightly. During pregnancy, the mammary gland becomes active. Oestrogen and progesterone promote the proliferation of the mammary epithelial cells and the development of acini. In late pregnancy and lactation, prolactin stimulates the growth and differentiation of breast epithelial cell and milk secretion (Inman et al. 2015).

14.2 Abundant Igs with Undetermined Sources in Colostrum and Milk

During lactation, milk is synthesised by luminal epithelial cells and secreted into the ductal lumen. Besides providing nutrition, colostrum and milk offer passive immunity to the offspring by maternal Igs. There are large amounts of Igs, including IgA, IgG, and IgM, in milk and especially colostrum. The composition of Igs in milk varies among different animal species. Human milk primarily contains IgA, whereas ruminant milk, such as that of cows, sheep, horses, and so on, predominantly contains IgG (Hurley and Theil 2011).

Maternal Igs exert diverse functions. The acquisition of maternal Igs from breast milk protects the infants from enteric pathogens, making breast milk a vital defence against infections (Niewiesk 2014; Newburg 2005). The transmission of maternal IgG and IgA can also coordinate to establish proper immune function early in life by regulating mucosal T-cell responses (Koch et al. 2016). Moreover, milk Igs show high activities of natural catalytic antibody (abzymes) by binding pockets capable of catalysing specific chemical reactions, functioning as phosphokinase, phosphatase, nuclease, lipase, protease, and so on (Nevinsky et al. 2000; Kompaneets et al. 2020).

Current theories suggest that Igs in colostrum and milk are derived from either circulating blood and/or produced by plasma cells in the mammary glands, then these Igs transfer into milk through the neonatal Fc receptors (FcRn) for IgG (Cianga et al. 2003; Cui et al. 2014), or the poly-Ig receptors (pIgRs) for IgA (Rincheval-Arnold et al. 2002; De Groot et al. 2000). However, the evidence regarding the source and transportation of milk Igs is insufficient. The contribution of Igs from blood and plasma cells alone appears to be insufficient to account for the levels of Igs in milk. The reasons include: (1) Although plasma cells can produce Igs, plasma cell infiltration is rarely observed in healthy mammary gland during lactation, (2) Considering the physicochemical similarity of milk IgA and intestinal IgA, it was hypothesised that milk IgA is produced by IgA plasma cells in Peyer's patches and transferred to milk via blood circulation (Usami et al. 2021). However, the hypothesis is questionable since the colostrum IgA level (about 32 mg/mL) (Mickleson and Moriarty 1982) is much higher than serum IgA (about 4.73 mg/mL) level (Soto-Ramirez et al. 2012) in humans, and (3) The catalytic activity of milk abzymes is higher than that of abzymes in blood, even in the blood of patients with autoimmune diseases, suggesting different sources of milk abzymes and blood abzymes (Nevinsky et al. 2000). Therefore, it is of great significance to characterise the generation of milk Igs.

14.3 Healthy Mammary Epithelial Cells Express Igs

As early as 1996, Qiu discovered that IgG-like protein exists in breast cancer cells and proliferative epithelial cells. Later, by a series of evidences proved that IgG like Protein was IgG produced by breast cancer and proliferative breast epithelial cells. Inspired by this discovery, Zhang *et al.* examined whether epithelial cells in the lactating mammary gland of mice could also produce Ig molecules (Zhang et al. 2010) (Fig. 14.1).

Firstly, they carried out immunohistochemical staining and observed the high expression of IgG protein in mammary epithelial cells of pregnant (day 17) and lactating (day 1) Balb/c mice, but not in virgin mammary glands, as anticipated. However, immunohistochemistry cannot distinguish whether these Igs are transported from extracellular of the mammary epithelial cells or produced by the epithelial cells themselves. Hence, they investigated Ig transcripts. Surprisingly, IgG1 transcripts were quite abundant in the mammary gland of lactating mice, even comparable to IgG1 in the spleen, as determined by Northern blot. Since only a few B cells or plasma cells were observed in the mammary gland, these cells could not account for the large amount of IgG1 transcripts. In situ hybridisation further confirmed that these IgG1 transcripts were in mammary epithelial cells of lactating mice, indicating the expression of Igs by these mammary epithelial cells (Zhang et al. 2010).

Given it is required of gene rearrangement in the production of functional Ig, Ig mRNA was analysed by PCR in $CK18^{+}CD19^{-}$ mammary epithelial cells sorted by flow cytometry. Transcripts of both the IgG heavy chain (rearranged VHDJH-CHγ1) and the rearranged Igκ light chain were detected in the mammary gland epithelial cells of lactating BALB/c, Kunming, and C57BL/6 mice (Zhang et al. 2010). These data confirmed the expression of Igs by mammary epithelial cells.

Interestingly, mammary epithelial cell-derived Ig, like other non B cell-derived Igs (Qiu et al. 2003; Geng et al. 2007), displays distinct characteristics in sequences compared to classical Igs from B cells. Igκ from mammary epithelial cells shows restricted recombinant patterns even in different strains and individuals of mice (Zhang et al. 2010). These specific usages of Igκ patterns are also found in V segments used by liver epithelial cell-derived Igκ, which contribute to innate immune responses as natural antibodies (Hu et al. 2012). Thus, milk Igs are likely produced by mammary epithelial cells to mediate the adoptive transfer to offspring.

In humans, IgG protein expression has also been observed in normal mammary epithelial cells, particularly with high levels of expression

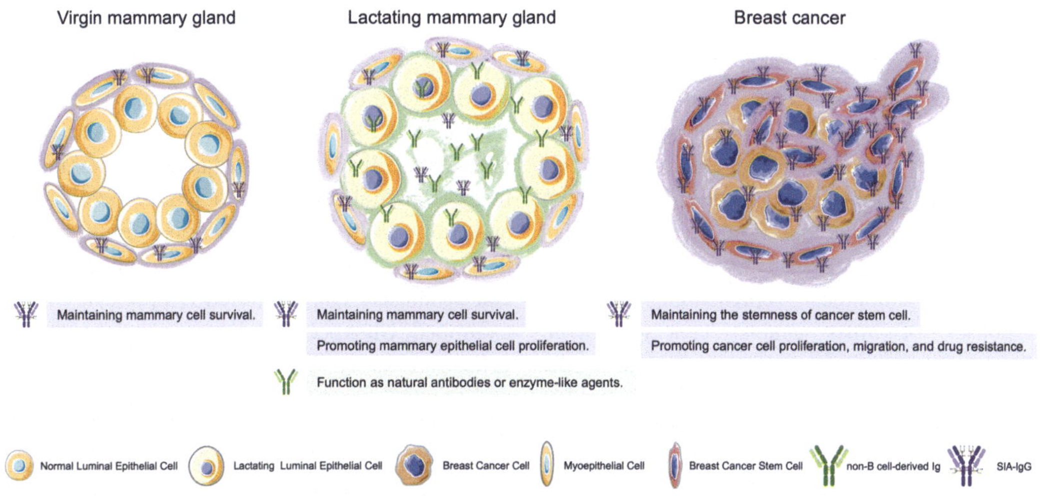

Fig. 14.1 Schematic illustration of the biological activities of mammary epithelial cell-derived Igs

in myoepithelial cells (Liao et al. 2015). Moreover, mammary myoepithelial cells can function as stem/progenitor cells. It will be instructive to explore whether this IgG is required for the stem/progenitor cell properties of mammary myoepithelial cells, as it can reveal new mechanisms for the development of the mammary gland during pregnancy and lactation.

14.4 Elevated Igs in Patients with Breast Cancer

Breast cancer is one of the most common cancers in women worldwide, accounting for 23% of the total cancer cases and 14% of the cancer deaths. Approximately one in eight to ten women will develop breast cancer during their lifetime. Early-stage breast cancer is considered potentially curable, whereas treatment for metastatic breast cancer aims to prolong the survival of patients and maintain the quality of life (Harbeck and Gnant 2017; Siegel et al. 2012; Ferlay et al. 2015). Despite the great efforts in basic research in the past, the mechanisms underlying the pathogenesis, invasion, and metastasis of breast cancer remain unclear, and patients with breast cancer have not benefitted substantially.

So far, the roles of tumour-reactive Igs in cancer have been recognised. There is wealth evidence describing the existence of Igs in serum or Ig deposition in malignant tissues of patients with cancer (Tomer et al. 1998; Fernandez Madrid 2005). Elevated levels of serum IgG, IgA, or IgM are frequently observed in patients with cancers of epithelial origin, including breast cancer. Similarly, IgG deposition in malignant mammary tissues is more pronounced than in benign lesions or normal tissues (Qiu et al. 2003; Ma et al. 2013). At first, these Igs were interpreted as protective antibodies representing humoral responses of hosts against cancer growth. However, increasing evidence has led to the opposite conclusion that high levels of Ig predict a poor prognosis. Expression of IgG and Igκ in breast cancer tissues positively correlates with AJCC stages of breast carcinoma (Liao et al. 2015; Yang et al. 2013). Tumour tissues of more malignant invasive breast cancer have greater IgG expression than breast carcinoma in situ (Yang et al. 2013). The percentage of IgG-positive cells in tumours is also correlated with the occurrence of axillary lymph node metastasis (Liao et al. 2015; Yang et al. 2013). Intriguingly, therapeutic plasma exchange involving the removal of IgG from plasma also implied the involvement of these IgG in cancer progression. This therapy, used from 1970s to 1990s, showed effectiveness in treating cancer patients. In this therapy, 200 mL peripheral blood was drawn, treated by protein A affinity chromatography column to remove or reduce IgG, and re-transfused back to the patient (Messerschmidt et al. 1988). As a result, the tumour cells showed growth retardness or even necrosis (Messerschmidt et al. 1988). Although the mechanism has not been identified, it indicated that IgG from cancer patients might not be protective but rather promote the development of cancer. These findings raise scientific questions regarding the origin of these tumour-reactive Igs and their roles in the genesis, progression, and metastasis of cancer.

14.5 Breast Cancer Cell-Produced Igs Promote the Progression of Cancer

14.5.1 Breast Cancer Cells Produce Igs

As early as 1989, Qiu noted strong positive staining of IgG in breast cancer cells quite serendipitously. IgG-like protein existed in nearly all malignant cells, but not in adjacent normal mammary epithelial cells or mesenchymal cells (Qiu and Yang 1996). As the classical immunology theory posits that only B cells and plasma cells can synthesise Igs, infiltrated lymphocytes were examined. Unexpectedly, lymphocytes or plasma cells were rarely found in the IgG-positive area, and the IgG staining did not correlate with the infiltration of lymphocytes and plasma cells. Furthermore, in the lymph nodes of these patients, plasma cells did not increase and the germinal centres displayed no significant changes.

Therefore, the heightened level of IgG-like protein appeared to be independent of immune cells. Importantly, Ig gene rearrangement, a key event in Ig expression, was observed using Southern blot and RT-PCR (Qiu and Yang 1996). Another intriguing observation was the presence of monoclonal gammopathy not only in patients with neoplastic proliferation of B cells and/or plasma cells but also in those with malignant tumours of epithelial origin (Hori et al. 2000; Gregersen et al. 2001). Taken together, these findings led to a bold hypothesis that breast cancer cells can produce Ig molecules independently, apart from B cells or plasma cells.

In recent decades, the investigation of non-B cell-derived Igs has been significantly facilitated by technological advancements. Although the earliest study and subsequent reports revealed positive staining of IgG in breast cancer tissues (Qiu et al. 2003; Qiu and Yang 1996), the structure of the IgG-like protein was unresolved. To identify the nature of this protein, Qiu et al. isolated cancer cells from breast cancer tissues, and identified the IgG-like protein by Western blot using anti-human IgG. The results revealed that this protein showed bands of molecular sizes equivalent to the typical heavy chain and light chain of IgG, but no other bands, suggesting that the IgG-like protein in breast cancer cells was indeed IgG (Qiu et al. 2003). At the transcriptional level, IgG transcripts were markedly elevated in breast cancer cells compared to normal mammary epithelial cells, as determined by in situ hybridisation (Qiu et al. 2003).

Ig gene rearrangement is indispensable for the functional expression of Ig, and according to classical theory, it occurs only in B cells, not in other cell lineages. To clarify the source of the IgG, Qiu et al. investigated whether Ig gene rearrangement occurred in these non-B cancer cells. Genomic DNA was extracted from cancer cells, and Ig gene rearrangement was identified by Southern blot and RT-PCR. Notably, transcripts of rearranged functional IgG and IgM heavy chains, as well as Igκ light chain, were detected in primary breast cancer cells isolated by laser capture microdissection (LCM) from breast invasive ductal carcinoma samples (Qiu et al. 2003; Wang et al. 2022; Zheng et al. 2009). Similarly, Ig gene rearrangement was also found in several cancer cell lines, which eliminated the interference of plasma cells and B cells from breast cancer tissues. Recombinant IgG, IgM, and IgA transcripts were identified in MDA-MB-231 (Liao et al. 2015; Babbage et al. 2006), MCF-7 (Liao et al. 2015; Zheng et al. 2007), SKBR3 (Babbage et al. 2006), T47D (Babbage et al. 2006), and ZR75-1 (Babbage et al. 2006), alongside transcripts of Ig constant regions (Fig. 14.2) (Zheng et al. 2007; Kimoto 1998). Expression of activation-induced cytidine deaminase (AID), which mediates the somatic hypermutation, class switching recombination, and locus suicide recombination of Ig, was also detected in MDA-MB-231, MCF-7, SKBR3, T47D, and ZR75-1 (Babbage et al. 2006; Zheng et al. 2007). Therefore, Ig genes do undergo rearrangement to produce functional Ig molecules in breast cancer cells.

The sequence of Ig, determining the physicochemical property and the binding capacity to various antigens, is critical for its biological activity. VHDJH and VκJκ rearrangement patterns were further determined in primary breast cancer cells isolated by LCM, as well as in breast cancer cell lines. Consistent with other non B cell-derived Igs, breast cancer cell-derived Igs displayed distinct features compared with B cell-derived Igs: (1) Unlike B cell-derived Igs that exhibited great diversity, the breast cancer cell-derived Ig sequences displayed restricted VHDJH and VκJκ patterns, with VH3, VH4, and VH5 gene family being the primary V segments for heavy chains (Qiu et al. 2003; Liao et al. 2015; Zheng et al. 2009; Babbage et al. 2006) and Vκ4-1/Jκ3 being the dominant usage of κ light chain recombination (Wang et al. 2022) (Fig. 14.2), suggesting the unique biological activity of breast cancer cell-derived Igs, and (2) Breast cancer cell-derived Igs did not appear to originate from classical class switching, as the VHDJH recombination of IgG, IgM, and IgA from the same tumour tissue or the same cells could be completely different (Zheng et al. 2009; Babbage et al. 2006). No identical VHDJH recombination was shared between IgM and IgG

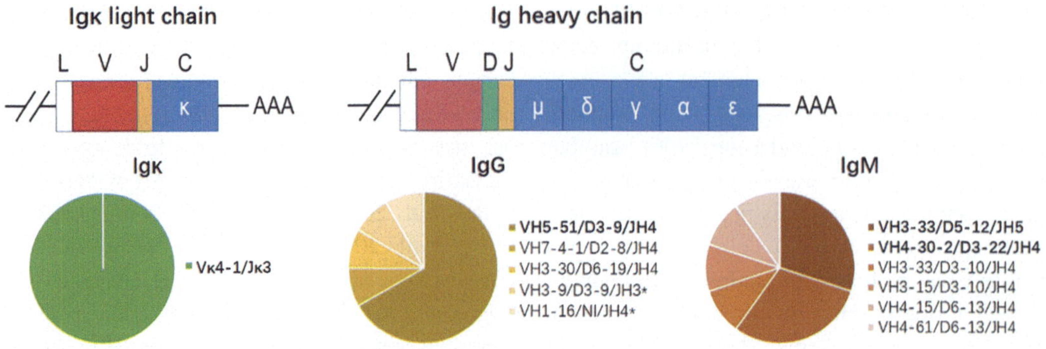

Fig. 14.2 Predominant V(D)J recombination patterns in primary breast cancer cells. Breast cancer cells were isolated by laser capture microdissection (LCM) from breast invasive ductal carcinoma samples, and rearrangement of Igκ, IgG, and IgM were detected. * indicated nonfunctional VHDJH sequences with stop codons (Wang et al. 2022; Zheng et al. 2009)

from the same cancer samples. Moreover, although the feature of the same cell line might have distinctions in different labs, the recombination patterns of VHγDγJHγ and VHμDμJHμ differed in ZR75-1, and the recombination patterns of VHγDγJHγ and VHαDαJHα also differed in SKBR3 from the same group (Babbage et al. 2006). The detailed mechanism requires further exploration.

14.5.1.1 Breast Cancer Stem-Like Cells Exhibit High Expression of a Unique Sialylated IgG (SIA-IgG)

Investigations into non-B cell-derived Ig have been limited due to the lack of specific antibodies. Although IgG, recognised by commercial anti-IgG antibodies, is broadly distributed in breast cancer tissues, these antibodies are produced using circulating IgG (mainly B cell-derived IgG) as the antigen and cannot distinguish B cell-derived IgG from cancer cell-derived IgG. Hence, a specific antibody for non B cell-derived IgG is crucial to probe the unique structure and biological activity of breast cancer cell-derived IgG.

RP215 is a monoclonal antibody against a specific form of IgG known as sialylated IgG (SIA-IgG), which carries a unique N-linked sialylated glycan on the Asn162 site of the CH1 domain. This novel glycan contrasts with the previously known N-glycan on Asn297 located in the CH2 domain of IgG, suggesting the unique biological activity of SIA-IgG (Tang et al. 2018). SIA-IgG predominantly originates from non B cells, which account for only a portion of IgG in circulation, rather than the majority. Immunohistochemical staining revealed high expression of SIA-IgG in cancer cells of epithelial origin, including breast cancer cells, particularly metastatic cancers, but not in lymphomas. Notably, SIA-IgG expression positively correlated with CD44v6, a major marker of tumour invasion (Liao et al. 2015). The findings underscore the association of tumour-related Igs with poor cancer prognosis.

Functional analysis has demonstrated that SIA-IgG is essential for stem/progenitor-like cells during tumour initiation and progression (Fig. 14.1). Unlike IgG recognised by commercial anti-IgG antibodies, SIA-IgG is abundantly concentrated in a small population of basal/myoepithelial-like cancer cells and the cancer cells at the invasive tumour front. These cells with high expression of SIA-IgG (SIA-IgG^{high} cells) were co-localised with $CD44^{+}/CD24^{-/low}$ breast cancer stem cells in human tissues. In vitro studies have revealed that SIA-IgG^{high} MDA-MB-231 cells exhibit significant properties of stem/progenitor-like cells, including high capacities for proliferation, division, migration, and invasion. Furthermore, the self-renewal and drug-resistance abilities of SIA-IgG^{high} cells were significantly higher than those of SIA-IgG^{low} cells. The findings were further substantiated by an RNAi assay using siRNA specifically targeting IgG in breast cancer cells. Additionally, blocking SIA-IgG with the specific antibody effectively suppressed the growth, migration, and invasion of breast cancer cells. Importantly, in vivo analysis further demonstrated the critical role of SIA-IgG in tumorigenicity and cancer metastasis in breast cancer (Liao et al. 2015; Zhang et al. 2022).

Concerning the underlying mechanism, SIA-IgG is involved in regulating the expression of MMP2 and MMP9, which mediate the degradation of the extracellular matrix to promote cancer metastasis. Moreover, SIA-IgG was located on the cell membrane, filamentous junction between cancer cells, and pseudopodia-like structure in MDA-MB-231 and MCF-7 cells, maintaining the anchoring of cancer cells to the extracellular matrix and promoting cell proliferation and migration (Liao et al. 2015).

14.5.1.2 Clinical Significance of Breast Cancer Cell-Derived Igs

14.5.1.2.1 SIA-IgG as a Novel Predictor of Poor Prognosis in Breast Cancer

The prognostic indicators widely used for breast cancer in the clinic include progesterone receptor (PR), PS-2, NM23, human epidermal growth factor receptor-2 (Her-2), and P-glycoprotein (P170), with PR, PS-2, and NM23 predicting good prognosis, whereas Her-2 and P170 represent poor prognosis. The expression of SIA-IgG is higher in metastatic breast cancer cells than in in situ cancer cells. Notably, SIA-IgG expression negatively correlates with the expression of PR, PS-2, and NM23, whilst positively correlating with Her-2 and P170 (Liao et al. 2015), providing a reasonable explanation for the association between IgG levels and poor prognosis in cancer patients. However, none of these markers mentioned above can independently predict poor prognosis in breast cancer. Nevertheless, SIA-IgG holds significant potential in predicting poor prognosis and metastasis in breast cancer.

14.5.1.2.2 SIA-IgG as a Promising Therapeutic Target for Breast Cancer

Cancer stem cells are the seedbeds of tumours, determining the occurrence and metastasis of cancer. Targeting cancer stem cells has been considered a prospective avenue in cancer therapy. As stated above, SIA-IgG is a novel marker of stem/progenitor-like cells and is critical for the self-renewal of stem cells. HER-2 positive breast cancer ($ER^-PR^-Her\text{-}2^+$) and triple-negative breast cancer ($ER^-PR^-Her\text{-}2^-$) are more aggressive than hormone receptor-positive breast cancer. Interestingly, SIA-IgG positively correlates with Her-2 and negatively correlates with ER and PR (Liao et al. 2015). Furthermore, Zhang et al. demonstrated that blocking SIA-IgG significantly attenuated the growth of xenografts using MDA-MB-231, a triple-negative breast cancer cell line (Zhang et al. 2022). The unique sialylated N-glycan on the Asn162 site highlights the applicability and specificity of SIA-IgG as a drug target. Therefore, SIA-IgG might be an attractive therapeutic target for the treatment of HER2-positive and triple-negative breast cancers.

14.6 Prospect

The mammary gland is a highly dynamic tissue and is constantly remodelled throughout a woman's lifetime. Growing evidence describes the expression of Ig, mainly IgG, in normal and malignant mammary epithelial cells. Mammary epithelial cell-derived Igs may have significant physiological and pathological implications, which require further investigations.

Under physiological conditions, the proliferation of mammary epithelial cells is a critical biological process during pregnancy, depending on the pluripotency of myoepithelial cells. Normal myoepithelial cells express SIA-IgG with unknown functions. Given the significant role of tumour cell-produced SIA-IgG in sustaining tumour cell proliferation, it is plausible that the functionality of SIA-IgG present in mammary myoepithelium might be associated with mammary gland development during pregnancy and lactation (Fig. 14.1). However, further investigations are warranted to substantiate this hypothesis.

Furthermore, it is an attractive scientific topic to investigate the expression, secretion, and function of Ig proteins, especially IgA, in luminal epithelial cells of the mammary gland during lactation. These non B cell-derived maternal Igs may contribute to the innate immune system as natural antibodies or improve the functions of the digestive system of infants as protease- or lipase-like abzymes, as mentioned above.

In breast cancer cells, the underlying mechanisms of the effects of breast cancer cell-derived Igs have not yet been fully elucidated, which is another active area of breast cancer research field. Moreover, future clinical research is of great significance to explore whether SIA-IgG will be another independent predictor or a valuable combined index with Her-2 for prognosis prediction in breast cancer. Related studies will provide a new perspective on the initiation and development of breast cancer.

References

Babbage G et al (2006) Immunoglobulin heavy chain locus events and expression of activation-induced cytidine deaminase in epithelial breast cancer cell lines. Cancer Res 66(8):3996–4000

Cianga P et al (2003) The MHC class I related Fc receptor, FcRn, is expressed in the epithelial cells of the human mammary gland. Hum Immunol 64(12):1152–1159

Cui D et al (2014) Bovine FcRn-mediated human immunoglobulin G transfer across the milk-blood barrier in transgenic mice. PLoS One 9(12):e115972

De Groot N et al (2000) Increased immunoglobulin A levels in milk by over-expressing the murine polymeric immunoglobulin receptor gene in the mammary gland epithelial cells of transgenic mice. Immunology 101(2):218–224

Ferlay J et al (2015) Cancer incidence and mortality worldwide: sources, methods and major patterns in GLOBOCAN 2012. Int J Cancer 136(5):E359–E386

Fernandez Madrid F (2005) Autoantibodies in breast cancer sera: candidate biomarkers and reporters of tumorigenesis. Cancer Lett 230(2):187–198

Geng LY et al (2007) Expression of SNC73, a transcript of the immunoglobulin alpha-1 gene, in human epithelial carcinomas. World J Gastroenterol 13(16):2305–2311

Gieniec KA, Davis FM (2022) Mammary basal cells: stars of the show. Biochim Biophys Acta Mol Cell Res 1869(1):119159

Gregersen H et al (2001) The impact of M-component type and immunoglobulin concentration on the risk of malignant transformation in patients with monoclonal gammopathy of undetermined significance. Haematologica 86(11):1172–1179

Harbeck N, Gnant M (2017) Breast cancer. Lancet 389(10074):1134–1150

Hori H et al (2000) Signet ring cell carcinoma of the colon with monoclonal gammopathy of undetermined significance. Nihon Shokakibyo Gakkai Zasshi 97(11):1373–1377

Hu F et al (2012) Spontaneous production of immunoglobulin M in human epithelial cancer cells. PLoS One 7(12):e51423

Hurley WL, Theil PK (2011) Perspectives on immunoglobulins in colostrum and milk. Nutrients 3(4):442–474

Inman JL et al (2015) Mammary gland development: cell fate specification, stem cells and the microenvironment. Development 142(6):1028–1042

Kimoto Y (1998) Expression of heavy-chain constant region of immunoglobulin and T-cell receptor gene transcripts in human non-hematopoietic tumor cell lines. Genes Chromosomes Cancer 22(1):83–86

Koch MA et al (2016) Maternal IgG and IgA antibodies dampen mucosal T helper cell responses in early life. Cell 165(4):827–841

Kompaneets IY et al (2020) IgGs from human milk hydrolyze microRNAs. Molecules 25(10):2366

Liao Q et al (2015) Aberrant high expression of immunoglobulin G in epithelial stem/progenitor-like cells contributes to tumor initiation and metastasis. Oncotarget 6(37):40081–40094

Ma C et al (2013) Immunoglobulin G expression and its potential role in primary and metastatic breast cancers. Curr Mol Med 13(3):429–437

McManaman JL, Neville MC (2003) Mammary physiology and milk secretion. Adv Drug Deliv Rev 55(5):629–641

Messerschmidt GL et al (1988) Protein A immunoadsorption in the treatment of malignant disease. J Clin Oncol 6(2):203–212

Mickleson KN, Moriarty KM (1982) Immunoglobulin levels in human colostrum and milk. J Pediatr Gastroenterol Nutr 1(3):381–384

Nevinsky GA, Kanyshkova TG, Buneva VN (2000) Natural catalytic antibodies (abzymes) in normalcy and pathology. Biochemistry (Mosc) 65(11):1245–1255

Newburg DS (2005) Innate immunity and human milk. J Nutr 135(5):1308–1312

Niewiesk S (2014) Maternal antibodies: clinical significance, mechanism of interference with immune responses, and possible vaccination strategies. Front Immunol 5:446

Prater MD et al (2014) Mammary stem cells have myoepithelial cell properties. Nat Cell Biol 16(10):942–950, 1–7

Qiu X, Yang G (1996) Charecteristics and gene structure of Ig like molecules in malignant cancer cells. Chinese J Immunol 05:295

Qiu X et al (2003) Human epithelial cancers secrete immunoglobulin g with unidentified specificity to promote growth and survival of tumor cells. Cancer Res 63(19):6488–6495

Rincheval-Arnold A et al (2002) Up-regulation of polymeric immunoglobulin receptor mRNA in mammary epithelial cells by IFN-gamma. Mol Cell Endocrinol 194(1–2):95–105

Siegel R, Naishadham D, Jemal A (2012) Cancer statistics, 2012. CA Cancer J Clin 62(1):10–29

Soto-Ramirez N et al (2012) Maternal immune markers in serum during gestation and in breast milk and the risk of asthma-like symptoms at ages 6 and 12 months: a longitudinal study. Allergy Asthma Clin Immunol 8(1):11

Tang J et al (2018) Lung squamous cell carcinoma cells express non-canonically glycosylated IgG that activates integrin-FAK signaling. Cancer Lett 430:148–159

Tomer Y, Sherer Y, Shoenfeld Y (1998) Autoantibodies, autoimmunity and cancer (review). Oncol Rep 5(3):753–761

Usami K et al (2021) The gut microbiota induces Peyer's-patch-dependent secretion of maternal IgA into milk. Cell Rep 36(10):109655

Wang Q et al (2022) A widely expressed free immunoglobulin kappa chain with a unique Vkappa4-1/Jkappa3 pattern promotes colon cancer invasion and metastasis by activating the integrin beta1/FAK pathway. Cancer Lett 540:215720

Yang B et al (2013) Correlation of immunoglobulin G expression and histological subtype and stage in breast cancer. PLoS One 8(3):e58706

Zhang S et al (2010) Immunoglobulin gene locus events in epithelial cells of lactating mouse mammary glands. Cell Mol Life Sci 67(6):985–994

Zhang M et al (2022) SIA-IgG confers poor prognosis and represents a novel therapeutic target in breast cancer. Bioengineered 13(4):10072–10087

Zheng H et al (2007) Expression and secretion of immunoglobulin alpha heavy chain with diverse VDJ recombinations by human epithelial cancer cells. Mol Immunol 44(9):2221–2227

Zheng J et al (2009) Immunoglobulin gene transcripts have distinct VHDJH recombination characteristics in human epithelial cancer cells. J Biol Chem 284(20):13610–13619

Expression, Function, and Significance of Non B Cell-Derived Immunoglobulin in Haematological System

15

Lina Wu, Miaoran Xia, Chong Wang, Huige Yan, Xiaoting Gong, and C. Cameron Yin

Abstract

Acute myeloid leukaemia (AML) is a collection of genetically diverse diseases characterised by abnormal proliferation of immature haematopoietic cells and disruption of normal haematopoiesis. Myeloid cells and lymphocytes originate from different haematopoietic precursors within the bone marrow. It has been traditionally assumed that myeloid cells cannot produce immunoglobulin (Ig), a marker of B cells and plasma cells. However, in recent years, all five Ig classes have been detected in CD34+ haematopoietic stem cells, mature monocytes and neutrophils, differentiated macrophages and tumour-associated macrophages, acute myeloid leukaemia cell lines, as well as myeloblasts of AML. The rearranged V(D)J sequences exhibit unique restricted or biased V gene usage and evidence of somatic mutation. Furthermore, AML-derived Igs could promote cell proliferation, induce apoptosis, and enhance migration. Elevated levels of Ig expression predict inferior clinical outcomes. These findings indicate that AML-derived Ig plays a role in AML pathogenesis and progression, and could serve as a novel biomarker for risk stratification, disease monitoring, and targeted therapy. In this chapter, we provide a comprehensive review of recent literature on the expression, function, and significance of non B cell-derived Ig in the haematological system, with a focus on AML.

L. Wu (✉)
Central Laboratory, Key Laboratory of Carcinogenesis and Translational Research (Ministry of Education/Beijing), Peking University Cancer Hospital & Institute, Beijing, China
e-mail: lnwu@bjmu.edu.cn

M. Xia
Department of Immunology, School of Basic Medical Sciences, Capital Medical University, Beijing, China
e-mail: mxia@ccmu.edu.cn

C. Wang
Department of Analytical Development, Cell Therapy Technology Product Engine, Oncology Therapeutic Area Unit, Takeda Pharmaceutical Company, Cambridge, MA, USA
e-mail: chong.wang@takeda.com

H. Yan
Department of Orthopaedic Surgery, Sir Run Run Shaw Hospital, Zhejiang University School of Medicine, Hangzhou, China
e-mail: yanhg@zju.edu.cn

X. Gong
Department of Clinical Laboratory, Tongde Hospital of Zhejiang Province, Hangzhou, Zhejiang, China

C. C. Yin
Department of Hematopathology, The University of Texas MD Anderson Cancer Center, Houston, TX, USA
e-mail: cyin@mdanderson.org

X. Qiu et al. (eds.), *Non B cell-Derived Immunoglobulins*, Advances in Experimental Medicine and Biology 1445, https://doi.org/10.1007/978-981-97-0511-5_15

Keywords

Immunoglobulins · Myeloid cells · Haematopoietic stem cells · Acute myeloid leukaemia

15.1 General Considerations

Haematopoiesis, the primary function of the haematopoietic system, is the process of differentiating from haematopoietic stem cells (HSCs) to mature, functional cell types of the blood lineages (Orkin 1995; Hofer and Rodewald 2018; Dharampuriya et al. 2017). The two primary branches of the haematopoietic lineage are myeloid and lymphoid, and the myeloid lineage is distinguished by the common myeloid progenitor (CMP) and all of its descendent cell types, including red blood cells, platelets, monocytes, neutrophils, etc. The lymphoid lineage is characterised by the common lymphoid progenitor (CLP) and all of its resulting cell types, including B and T cells, NK cells, and lymphoid dendritic cells (Spits and Cupedo 2012).

The conventional immunological theory posits that immunoglobulins (Igs) are solely produced by lymphoid B cells and plasma cells. The Igs exist in two forms, with one form expressed on the membrane serving as B cell receptors (BCR) and another form secreted as soluble molecules being known as antibodies (Cyster and Allen 2019).

15.2 Expression and Characteristics of Non B Cell-Derived Immunoglobulins in Haematologic Cells

Myeloid cells and lymphocytes originate from different haematopoietic precursor cells within the bone marrow, leading to the assumption that myeloid cells cannot produce Ig. However, as early as 1980, Kemp et al. found the expression of Ig mu genes (IgCμ) transcripts not only in T and B lymphocytes but also in T cells and myeloid cell lineage (Kemp et al. 1980). Over the past decade, a series of studies have confirmed that Igs are widely expressed in $CD34^+$ umbilical cord blood (UCB) haematopoietic stem/progenitor cells (HSPCs) (Liu et al. 2016), mature non-neoplastic monocytes and neutrophils from both patients with non-haematopoietic neoplasms and healthy individuals (Huang et al. 2014; Wang et al. 2015), ex vivo differentiated macrophages and macrophages from the tumour microenvironment (TME) (Fuchs et al. 2018), as well as myeloblasts of acute myeloid leukaemia (AML) (Wang et al. 2015; Fuchs et al. 2018; Qiu et al. 2013; Wu et al. 2020; Xia et al. 2022). The diagram of expression and Ig repertoire of immunoglobulins in the haematological system are shown in Fig. 15.1.

In this chapter, we provide a comprehensive review of recent literature on the expression, function, and clinical significance of non B cell-derived Ig in the haematological system.

15.2.1 Non B Cell-Derived Ig in $CD34^+$ Haematopoietic Stem/Progenitor Cells (HSPC) and Myeloid Progenitors

According to the classical theory of immunology, the Ig gene should be present at the stage of pre-B cell rather than in HSCs according to the developmental order. However, Liu et al. found that multiple types of Ig genes, especially IgM, the earliest Ig class expressed during evolution, were expressed in umbilical cord blood-derived HSPCs by analysing Gene Expression Omnibus (GEO) microarray database. They further confirmed IgM expression in HSPCs by RT-PCR sequencing and immunofluorescence (Liu et al. 2016). Though this is inconsistent with the traditional theory, it aligns with previous reports that Ig genes could be expressed in various non-B cells, including epithelial cells and myeloid cells (Huang et al. 2014; Wang et al. 2015; Fuchs et al. 2018; Qiu et al. 2003).

Recently, Fuchs et al. detected the rearranged V region genes of IgG, IgM, and IgA in myeloid progenitors using $CD34^+$ haematopoietic progen-

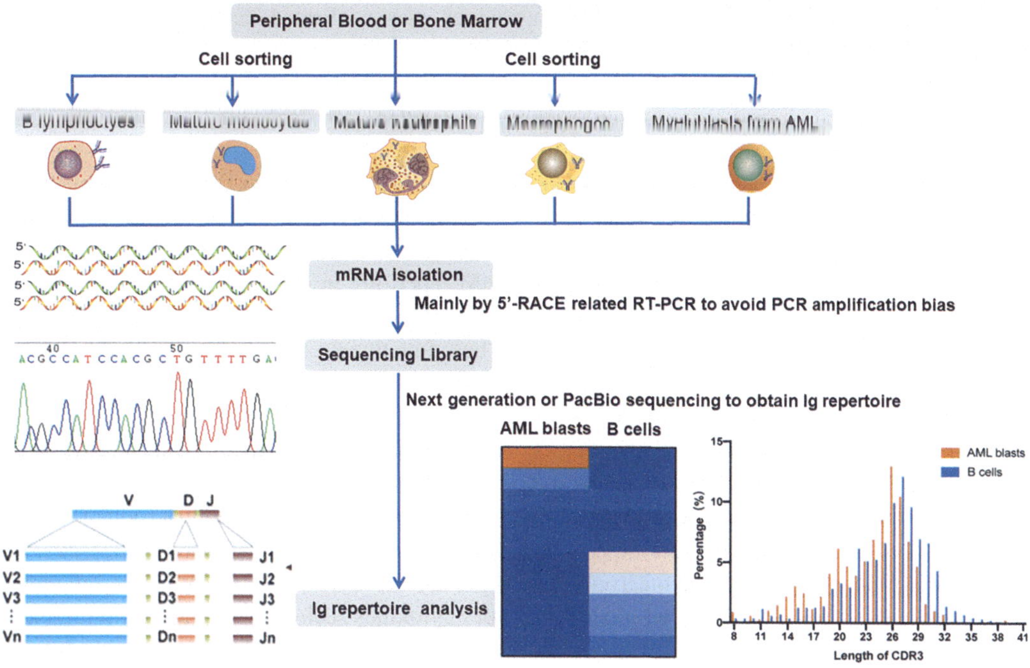

Fig. 15.1 The diagram of expression and immune repertoire of immunoglobulins in the haematological system. Igs are widely expressed in B lymphocytes, mature monocytes, mature neutrophils, macrophages, and myeloblast from AML patients by using fluorescence-activated cell sorting from peripheral blood or bone marrow in healthy persons or patients. Next, the mRNAs of these Igs are isolated. Then mainly through 5'-RACE-related PCR and next-generation or PacBio sequencing, the Ig repertoire sequences are obtained and analysed

itor cells stimulated with Granulocyte-macrophage colony-stimulating factor (GM-CSF) in vitro. The diversity of Igs in B cells results from V(D)J unrestricted recombination, somatic hypermutation (SHM), and class switching. Liu et al. demonstrated that HSPC-derived Ig displayed restricted/biased VHDJH rearrangement patterns. Among the seven VH families (VH1-VH7), only the VH3 family, such as VH3-15, VH3-21, VH3-23, and VH3-30, were observed in the HSPCs. Fuchs et al. found that Ig of $CD34^+$ haematopoietic progenitor cells also displayed a monoclonal expression pattern. CDR3 length characteristics revealed the expression of single or few rearranged VH clonotypes. Among them, the VH3 in IgA, VH2 in IgG, and VH6 in IgM were frequently used (Fuchs et al. 2018). These results further suggest that Ig expression in HSPCs is not random and may be functional. However, VHμDJHμ (heavy chain of IgM) in HSPCs showed no evidence of somatic mutation (Liu et al. 2016). Since IgM is essential for B cell survival and function, it is likely that HSPC-derived IgM also is involved in the survival and function of HSPCs, but this remains to be verified.

15.2.2 Expression of Non B Cell-Derived Igs in Differentiated Macrophages

Macrophages are myeloid immune cells residing in all tissues and function as pillars of the innate immune system (Janeway Jr and Medzhitov 2002; Wynn and Vannella 2016; Wang et al. 2019). Macrophages are a crucial component of innate immunity, recognising and eliminating pathogens or dead cells by their phagocytosis and producing inflammatory mediators, including cytokines such as IL-1β, IL-6, and TNF-α, and reactive oxygen species (Locati et al. 2020). Macrophages have characteristics such as foreign pathogen surveillance, quick migration, and the ability to patrol almost all tissues, enabling them to be the primary line of defence against infection (Wu and Hirschi 2020).

It has long been believed that macrophages neither produce Igs nor directly, functionally relate to Igs. However, as shown above, a series of studies have provided evidence for the existence and function of Igs outside the lymphoid lineage. Fuchs et al., for the first time, demonstrated the constitutive expression of both Ig heavy chains (IgM, IgG, and IgA) and Ig light chains (Igκ) in ex vivo differentiated macrophages and macrophages from the tumour microenvironment (tumour-associated macrophages (TAMs)) (Fuchs et al. 2018). They analysed the Ig VHDJH rearrangement and CDR3 mRNA expression profiles and found that the expression and rearrangement of Ig isotypes in macrophages were confirmed by the complexity and individual variations in Ig repertoires. The comparison of Ig repertoires between macrophages and normal B cells, which used the first- and second-generation sequencing technology, revealed that the Ig repertoire in TAMs is restricted, which cannot be found in the B cell heavy chain repertoires. The most common VHDJH heavy chain sequences present in TAM from a human oesophagus carcinoma patient and a melanoma patient were VH4-61/D2-2/JH3 and VH4-31/D3-3/JH4, respectively. Furthermore, a complete Ig gene encoding a functional IgG heavy and a corresponding Igκ light chain was detected in a single TAM. These results indicate that TAM is capable of producing functional Igs, which may be the result of an antigen-directed immune response of macrophages against tumours. Moreover, Fuchs et al. found that Igs were produced in macrophages of the knockout mice that lacked B and T cells. The function and significance of these rearranged Igs produced by macrophages remain to be discovered, but it is conceivable that macrophage-derived Ig is involved in the anti-tumour immune response.

In addition, Gong et al. identified the expression of Igμ heavy chain (IgM) in mice macrophages (Gong et al. 2021). They found that macrophage-expressed IgM mainly located in intracellular compartments rather than on the cell surface or secreted like B cell-expressed IgM, which was confirmed using both normal mice and B cell-deficient mice, as well as macrophage cell line RAW264.7. Interestingly, the μ chain in macrophages existed alone without the light chain, including κ and λ chain, suggesting that IgM derived from mouse macrophages might exert bioactivity as a free heavy chain by itself. Next-generation sequencing revealed that the IgM repertoire in macrophage showed restricted recombination and individual conservativeness. The functional rearrangement of VHDJH accounted for 90% of the repertoire, indicating that macrophage-derived intracellular IgM might have a unique structure and function. Furthermore, functional analysis showed that IgM knocking down significantly promoted macrophage migration and FAK/Src-Akt axis activation. Under the lipopolysaccharide (LPS) stimulation, inflammatory cytokines such as MCP1 and IL-6 increased after IgM knockdown. The mechanistic investigation revealed that IgM regulated LPS-induced inflammatory response and unfolded protein response (UPR) activation through interacting with binding immunoglobulin protein (Bip) in macrophages, thereby participating in endoplasmic reticulum stress (Gong et al. 2021).

15.2.3 Expression of Non B Cell-Derived Igs in Mature Monocytes and Neutrophils

Huang et al. examined IgM expression in human myeloid cells and detected VHμDJHμ transcripts in both AML cell lines and primary myeloblasts from patients with AML. Interestingly, unlike IgG, IgM was found to be expressed in monocytes and neutrophils from patients with non-haematopoietic system malignant diseases (Huang et al. 2014). Similarly, Igκ was also detected in monocytes and neutrophils from patients with non-haematopoietic neoplasms (Wang et al. 2015). In order to further investigate the expression of Ig derived from normal myeloid cells, Yan et al. first used flow cytometry to separate mature monocytes and neutrophils from health individuals, and detected the expression frequencies of 5 Ig heavy chains (IgH) and 2 Ig light chains (IgL) in mature neutrophils and monocytes at the transcriptional level. This result reveals for the first time that both mature monocytes and neutrophils have Ig heavy and light chain transcripts, but the frequency of Ig expression varies among different categories and types of individuals.

Yan et al. further sorted mature neutrophils and monocytes from 17 individuals and used a combination of 5′-RACE-related RT-PCR and Pacbio sequencing scheme. By adding a universal primer connector at the 5′ end of the variable region, the bias issue caused by PCR amplification was avoided. Then, through third-generation high-throughput sequencing, different Ig variable region sequence characteristics were analysed. They obtained a total of 640,000 Ig sequences from mature neutrophils and monocytes, and for the first time obtained comprehensive variable region sequence characteristics of Ig from myeloid cells, confirming the functional V(D)J rearrangement of five categories of heavy chains and two types of light chains in these mature myeloid cells.

Different from high diversity of Ig heavy and light chain variable regions derived from B cells, IgH derived from myeloid cells exhibits a biased VHDJH rearrangement pattern with oligoclonal features, but the VHDJH rearrangement pattern used by myeloid cells varies among different individuals. However, Igκ and Igλ from the myeloid cells exhibit higher consistent VLJL rearrangement between different individuals, such as IGKV4-1/IGKJ1 of κ sequences; IGLV2-14/IGLJ2 and IGLV2-14/IGLJ3 of λ sequences, are commonly used by mature neutrophils and monocytes (unpublished data). However, the physiological significance and expression regulation mechanism of Ig derived from the medullary system need to be addressed.

15.2.4 Expression of Non B Cell-Derived Igs in Myeloblasts from Acute Myeloid Leukaemia (AML)

Before the first report on Ig expression in AML myeloblasts by Qiu et al. (Qiu et al. 2013), there were a few reports showing low frequencies of Ig gene expression in bone marrow or peripheral blood samples from AML patients (Greinix et al. 1995). However, in these studies, it remained unknown whether Ig was produced by myeloblasts or was secreted by residual B cells and non-specifically bound to myeloblasts. Here, we summarised recent literature on the expression of non B cell-derived Ig in the myeloblasts from AML.

As early as 1999, Qiu et al. found that Ig rearrangement sequences were amplified by PCR in a myeloid leukaemia cell line HL-60, suggesting that myeloid cells can express Igs (Qiu et al. 1999). In 2013, Qiu et al. further explored IgG expression in several AML cell lines, including HEL, NB4, HL-60, KG-1, OCI-AML3, and THP-1, and found high levels of IgG expression and secretion in all six AML cell lines. Furthermore, they investigated IgG gene rearrangement and transcription in primary myeloblasts from patients with AML and detected the VHDJH transcripts in $CD33^+CD19^-CD138^-$ myeloblasts at high frequencies and levels, but not in $CD33^+$ non-neoplastic monocytes or neutrophils from patients with non-haematopoietic neoplasms or healthy individuals. These results confirm that IgG is produced by myeloblasts rather than residual B cells in the sample (Qiu et al. 2013). Besides IgG, Huang et al. examined IgM expression in human myeloid cells and detected VHμDJHμ transcripts in both AML cell lines and primary myeloblasts from patients with AML (Huang et al. 2014). Consequently, whether Ig light chain is also expressed and functional in haematopoietic cells was addressed. Wang et al. demonstrated that rearranged Igκ light chain was frequently expressed in AML cell lines and sorted primary $CD33^+CD19^-CD138^-$ myeloblasts from patients with AML at both RNA and protein levels (Wang et al. 2015). Similarly, Igλ expression was also detected in sorted myeloblasts from patients with AML by RT-PCR (Wang et al. 2015; Xia et al. 2022). Moreover, the frequency of Igκ expression was higher than that of IgG and IgM in myeloblasts sorted by FACS from patients with AML (Wu et al. 2020). Using next-generation sequencing (NGS), Xia et al. confirmed the transcripts of all classes and types of Igs were detected in the sorted myeloblasts of patients with AML (Xia et al. 2022). The significance of Ig molecules present in the haematopoietic system is uncertain; however, one explanation may be that the expression of IgH and IgL in both AML cell lines and myeloid cells in patients may be induced by certain tumour-related factors.

Unlike B cell-derived Ig, which shows a polyclonal pattern, myeloid cell-derived IgG VHDJH rearrangements display a unique biased monoclonal or oligoclonal IgG repertoire. AML cell line-derived IgG VHDJH rearrangements illustrate the same CDR3 sequence (VH3-48/D4-7/JH4) across different AML cell lines. VHDJH rearrangements from primary myeloblasts demonstrate diversity in each patient but with biased usage of VH3 and VH4 families (Qiu et al. 2013). In addition, IgM VHμDJHμ gene rearrangements exhibit a restricted V usage and repertoire in AML cell lines, primary myeloblasts, as well as monocytes and neutrophils from patients with non-haematopoietic neoplasms. V segments, such as VH3-23 and VH3-48, are more often used in AML cells, in contrast to VH5-51, VH3-33 and VH6-1 are used more often in epithelial cancer cells (Zheng et al. 2009). Although AML-derived VHμDJHμ rearrangements show an oligoclonal or monoclonal pattern in each patient with AML, they do not possess conservative sequences. In addition, both AML-derived and monocytes/neutrophils-derived Igκ display a monoclonal or oligoclonal rearrangement pattern, with several shared IGKV and IGKJ gene usages but at different frequencies, in that IGKV1-5*03 and IGKV1-NL1*01 are more frequently used in myeloblasts, whereas IGKV3-20*01 and IGKV1-27*01 are more frequently used in mature myeloid cells.

Previous studies evaluated the Igs rearrangements mainly using the first-generation sequencing which is prone to bias because of technical limitations. Xia et al. conducted a comprehensive analysis of Ig repertoire in AML using next-generation sequencing (NGS). As for Ig light chain, normal B cells-derived Igκ used dominant Vκ segments, including IGKV3-20, IGKV3-11, and IGKV3-15, showing biased IGKV usage (Juul et al. 1997). AML-derived Igκ frequently presents three rearrangement patterns, IGKV3-20*01/IGKJ1*01, IGKV2D-28*01/IGKJ1*01, and IGKV4-1*01/IGKJ1*01 in different patients, also demonstrating biased Igκ rearrangements. As for Ig heavy chain, IgH rearrangements exhibit great diversity in normal B cells, enabling recognition of various antigens (Briney et al. 2019). However, AML-derived IgH shows much less diversity and is individual-specific. Some IGHV fragments preferentially used in AML, such as IGHV3-23, IGHV3-48, and IGHV1-69, are partially similar to IGHV fragments in CLL/SLL, such as IGHV1-69, IGHV3-07, IGHV3-23, and IGHV4-34 (Ghia et al. 2005). The differences in V(D)J usage among different types of myeloid cells might be lineage-specific and likely the result of selection under different differentiated or neoplastic conditions.

Interestingly, AML-derived IgG VHDJH transcripts also exhibit somatic hypermutation (SHM), with the hotspots of mutation predominantly situated in the CDR regions (Qiu et al. 2013). Moreover, AML-derived IgM gene sequences display SHM of the V gene segments, whilst VHμDJHμ gene sequences in monocytes and neutrophils from healthy individuals demonstrate greater diversity and fewer SHMs (Huang et al. 2014). Furthermore, nearly all AML-derived IGKV show SHM, whilst only rare mutations are detected in monocytes/neutrophils-derived IGKV. Surprisingly, AML-derived Igκ differs from B cell-derived Igκ in high mutation rates, especially at the unique serine (S) site mutations. Serine is one of the most commonly phosphorylation sites by kinases and also an O-linked glycosylation site. The change to Asn (S30N and S32N) in AML-derived IGκ could constitute classic (N-X-T/S) or non-classic (T/S-X-N) N-glycosylation (Xia et al. 2022). Protein glycosylation plays a vital role in maintaining Ig function, such as N-glycosylation at Asn297 can promote IgG to bind FcγR. Therefore, the alteration of serine may influence the phosphorylation or glycosylation of Ig, leading to the modification of Igκ functions. The SHM pattern of AML-derived IgG, IgM, and Igκ may be associated with leukemogenesis and/or AML progression, and may hold potential significance in the diagnosis and therapy of AML.

15.3 Bioactivity of Non B Cell-Derived Immunoglobulin in the Haematological System

As demonstrated above, Ig is expressed by AML cells. However, the precise effect of AML-expressed Ig, particularly its role in cancer initiation and progression, remains unclear. It has been reported that the blockade of tumour-derived IgG increased programmed cell death and inhibited the growth of epithelial cancer cells in vitro and in immunodeficient nude mice (Qiu et al. 2003). Furthermore, the monoclonal antibody RP215, which specifically recognises cancer-derived IgG, led to a reduction in adhesion, invasion, migration, and self-renewal, and increased apoptosis of cancer cells (Liao et al. 2015). These data suggest IgG expressed in AML cells plays a role in tumour progression, metastasis, and cancer stem cell maintenance, and may serve as a potential therapeutic target.

Qiu et al. showed that anti-human IgG reduced cell viability and induced apoptosis in AML cell lines in culture, suggesting a role for IgG in cell survival (Qiu et al. 2013). Similarly, anti-human IgM inhibited cell proliferation in a dose-dependent manner but did not induce apoptosis in AML cell lines, suggesting a role for IgM in cell proliferation but not survival (Huang et al. 2014). To address the functional significance of Igκ expression, Wang et al. constructed an expression vector containing IgKV1-5*03/IGKJ3*01 and transfected it into HL-60 and NB4 cell lines. They found that, unlike AML-derived IgG or IgM, overexpression of Igκ did not affect the proliferation or survival of AML cells. Instead, it significantly promoted migration and chemotaxis of the cells in the two AML cell lines assessed. Consistently, knocking down Igκ expression resulted in a decrease in migration of cells of these two AML cell lines. These data suggest that AML-derived Igκ plays a role in the migration of AML cells rather than proliferation (Wang et al. 2015). In summary, these data support a role for AML-derived Igs in AML pathogenesis and progression.

15.4 Clinical Significance of Non B Cell-Derived Immunoglobulin in AML

AML is the most common acute leukaemia in adults, and the current 2-year and 5-year overall survival (OS) rates are 32.0% and 24.0%, respectively, according to the National Cancer Institute (NCI) in the US (SEER 2018). It is a heterogeneous disease characterised by an abnormal overproduction of undifferentiated myeloid cells, resulting in haematopoiesis impairment and bone marrow failure. Cytogenetic changes, such as chromosomal abnormalities and genetic mutations with prognostic value, have been identified in over 90% of patients with AML, and have been used to establish prognostic stratification. Patients with AML can be stratified into favourable, intermediate, or adverse prognostic risk groups based on their cytogenetic profiles. Nearly 60 gene mutations, such as C-KIT, FLT3-ITD, CEBPA, IDH1/2, NPM1, and RUNX, have helped refine risk stratification and precise therapy (De Kouchkovsky and Abdul-Hay 2016; Dohner et al. 2017). More and more novel biomarkers for AML are being discovered, not only for guiding risk stratification but also for monitoring minimal/measurable residual disease and the development of target therapy.

Since Igs are widely expressed in myeloid immature or mature cells in AML, the clinical significance of the Ig gene in AML is worth discussing. Recently, Wu et al. assessed the expression of different classes of Ig in peripheral blood and/or bone marrow samples from 132 patients with AML by flow cytometry. Using ROC curve analysis, they identified two distinct groups of patients with AML with different Ig expression. They subsequently explored the relationship between levels of Ig expression and known prognostic factors as well as clinical outcome. Blasts from patients with high IgG, Igκ, or Igλ expression more often showed monocytic differentiation and dysplasia and had a higher frequency of CD14 and CD64 expression. Moreover, a high level of IgG expression was associated with TET2 mutations, whilst a high level of Igλ expression was associated with KRAS and TET2 mutations (Wu et al. 2020). RAS is a family of oncogenes encoding the GTP-binding protein p21, which plays an important role in cell proliferation (Van Meter et al. 2007; Jin et al. 2018). TET2 is a methylcytosine dioxygenase and catalyses the conversion of 5-methylcytosine to 5-hydroxymethylcytosine, leading to a loss of DNA methylation (Gaidzik et al. 2012; Solary et al. 2014). It has been shown that TET2 and NRAS mutations can collaborate to promote cell proliferation and enhance haematopoietic stem cell self-renewal through the activation of the JAK/STAT pathway (Jin et al. 2018). TET2 mutation may result in demethylation, which in turn induces Ig gene activation and expression. On the other hand, the higher rates of KRAS/NRAS and TET2 mutations in the Ig high-expression group may, at least in part, contribute to the role of Ig in promoting cell proliferation. Wu et al. further assessed whether the levels of Ig expression are related to the clinical outcome of patients with AML. AML patients with blasts showing high levels of Ig expression had significantly poorer overall survival than those with low levels of Ig expression (Wu et al. 2020). A higher transcriptional level of AML-derived IgG expression correlated with a significantly shorter disease-free survival (Xia et al. 2022). These results demonstrate a potential role for Ig expression as a useful marker for prognostic stratification.

15.5 Progress and Prospects

In summary, non B cell-derived Igs are rearranged, transcribed, and expressed in HSPCs, macrophages, mature myeloid cells, TAM, myeloblast in AML, etc., with unique characteristics and functions. High levels of Ig expression are correlated with a poor clinical outcome in patients with AML. Therefore, non B cell-derived Ig may become a novel biomarker for prognosis prediction, monitoring of residual disease, and developing targeted therapies.

Traditionally, rearrangement and production of Igs have been viewed as a unique feature of lymphoid lineage B cells and plasma cells. However, some experimental evidence from the past has been challenged with the development of technologies such as single-cell sequencing, which can identify gene transcripts in a single cell (Kulkarni et al. 2019). Through single-cell sequencing technology, Shi et al. demonstrated that multiple V(D)J rearrangements exist in hundreds to thousands of single B cells, and different Ig classes in the same B cell exhibit distinct V(D)J recombination patterns (Shi et al. 2019). These data are seemingly inconsistent with the principle of "one B cell-one V(D)J copy" and "different classes of Igs from the same B cell share the same V(D)J pattern". It is likely that the usages of non B cell-derived IgM and IgA were different in the same individual, suggesting that perhaps the classical Ig class switching does not exist in non B cell-derived Igs. Fuchs et al., using the single-cell sequencing method, found that functional Ig was expressed in a single macrophage isolated from the tumour microenvironment, providing strong evidence that Igs are indeed produced by myeloid cells, from neither residual B cells nor any non-specific membranous binding of extracellular Ig. Furthermore, myeloid cell-derived Ig rearrangements display a unique biased monoclonal or oligoclonal IgG repertoire. How are these biased sequences selected during the process of VDJ rearrangement? The ongoing comprehensive sequencing analysis of the Ig repertoire will shed light on a deeper understanding of the molecular genetic mechanisms of myeloid-derived Ig regulation and function in the haematopoietic system.

Igs have a common structure consisting of two identical heavy and two identical light chains. In Wang's study, the expression frequency of Igκ in myeloid-derived cells was higher than that of IgG and IgM. Moreover, Igκ was also found expressed in monocytes and neutrophils from patients with non-haematopoietic system malignant diseases (Wang et al. 2015). In contrast, IgG was only expressed in AML cells but not in mature myeloid cells (Qiu et al. 2013). It is known that Ig light chains are part of the Ig tetramer, however, they can also exist in free forms, which may play a role in various pathological conditions, such as inflammation, amyloidosis, light chain deposition disease, and autoimmune diseases (Connors et al. 2007; Nakano et al. 2011; Chaulagain and Comenzo 2013; van der Heijden et al. 2006). It is possible that myeloid-derived Igκ may also be present in a free form, or be coupled with other subclasses of Ig heavy chains.

On the evidence of the literatures that the AML-derived Igs is involved in the proliferation, apoptosis, and migration of leukaemic cells (Huang et al. 2014; Wang et al. 2015; Qiu et al. 2013), high-level Ig expression is associated with a poor clinical outcome (Wu et al. 2020), these data suggest that AML-derived Ig is a novel molecule with unique characteristics and functions that may play a role in leukemogenesis and serve as a marker for disease surveillance, risk stratification, and targeted therapy.

However, the mechanism of non B cell-derived IgG in the haematopoietic system remains unclear. Jiang et al. reported that IgG may play a key role in lung cancer growth and metastasis by regulating the metastatic gene MTA1 (Jiang et al. 2014). Tang et al. demonstrated that non B cell-derived IgG interacts with integrin α6β4 and activates the downstream oncogenic FAK and Src signalling pathways in lung squamous cell carcinoma, thereby promoting cancer initiation, progression, and metastasis (Tang et al. 2018). Qin et al. found that non B cell-derived IgG promoted the development of prostate cancer through the SOX2/SIA-IgG pathway (Qin et al. 2020). Our ongoing elucidation of Ig downstream signal transduction pathways will be significant for understanding the mechanism of Ig association with leukemogenesis, disease surveillance, risk stratification, and targeted therapy.

References

Briney B et al (2019) Commonality despite exceptional diversity in the baseline human antibody repertoire. Nature 566(7744):393–397

Chaulagain CP, Comenzo RL (2013) New insights and modern treatment of AL amyloidosis. Curr Hematol Malig Rep 8(4):291–298

Connors LH et al (2007) Heterogeneity in primary structure, post-translational modifications, and germline gene usage of nine full-length amyloidogenic kappa1 immunoglobulin light chains. Biochemistry 46(49):14259–14271

Cyster JG, Allen CDC (2019) B cell responses: cell interaction dynamics and decisions. Cell 177(3):524–540

De Kouchkovsky I, Abdul-Hay M (2016) Acute myeloid leukemia: a comprehensive review and 2016 update. Blood Cancer J 6(7):e441

Dharampuriya PR et al (2017) Tracking the origin, development, and differentiation of hematopoietic stem cells. Curr Opin Cell Biol 49:108–115

Dohner H et al (2017) Diagnosis and management of AML in adults: 2017 ELN recommendations from an international expert panel. Blood 129(4):424–447

Fuchs T et al (2018) Expression of combinatorial immunoglobulins in macrophages in the tumor microenvironment. PLoS One 13(9):e0204108

Gaidzik VI et al (2012) TET2 mutations in acute myeloid leukemia (AML): results from a comprehensive genetic and clinical analysis of the AML study group. J Clin Oncol 30(12):1350–1357

Ghia P et al (2005) Geographic patterns and pathogenetic implications of IGHV gene usage in chronic lymphocytic leukemia: the lesson of the IGHV3-21 gene. Blood 105(4):1678–1685

Gong X et al (2021) Macrophage-derived immunoglobulin M inhibits inflammatory responses via modulating endoplasmic reticulum stress. Cell 10(11):2812

Greinix HT et al (1995) Heavy-chain immunoglobulin gene rearrangement and cytoplasmic immunoglobulin expression in acute monocytic leukemia following primary germ cell tumor. Ann Hematol 70(2):109–112

Hofer T, Rodewald HR (2018) Differentiation-based model of hematopoietic stem cell functions and lineage pathways. Blood 132(11):1106–1113

Huang J et al (2014) Rearrangement and expression of the immunoglobulin mu-chain gene in human myeloid cells. Cell Mol Immunol 11(1):94–104

Janeway CA Jr, Medzhitov R (2002) Innate immune recognition. Annu Rev Immunol 20:197–216

Jiang C et al (2014) Immunoglobulin G expression in lung cancer and its effects on metastasis. PLoS One 9(5):e97359

Jin X et al (2018) Oncogenic N-Ras and Tet2 haploinsufficiency collaborate to dysregulate hematopoietic stem and progenitor cells. Blood Adv 2(11):1259–1271

Juul L et al (1997) The normally expressed kappa immunoglobulin light chain gene repertoire and somatic mutations studied by single-sided specific polymerase chain reaction (PCR); frequent occurrence of features often assigned to autoimmunity. Clin Exp Immunol 109(1):194–203

Kemp DJ et al (1980) Expression of the immunoglobulin C mu gene in mouse T and B lymphoid and myeloid cell lines. Proc Natl Acad Sci U S A 77(5):2876–2880

Kulkarni A et al (2019) Beyond bulk: a review of single cell transcriptomics methodologies and applications. Curr Opin Biotechnol 58:129–136

Liao Q et al (2015) Aberrant high expression of immunoglobulin G in epithelial stem/progenitor-like cells contributes to tumor initiation and metastasis. Oncotarget 6(37):40081–40094

Liu J et al (2016) Immunoglobulin gene expression in umbilical cord blood-derived CD34(+) hematopoietic stem/progenitor cells. Gene 575(1):108–117

Locati M, Curtale G, Mantovani A (2020) Diversity, mechanisms, and significance of macrophage plasticity. Annu Rev Pathol 15:123–147

Nakano T et al (2011) Free immunoglobulin light chain: its biology and implications in diseases. Clin Chim Acta 412(11–12):843–849

Orkin SH (1995) Hematopoiesis: how does it happen? Curr Opin Cell Biol 7(6):870–877

Qin C et al (2020) Cancer-driven IgG promotes the development of prostate cancer though the SOX2-CIgG pathway. Prostate 80(13):1134–1144

Qiu X et al (1999) Expression of Ig like protein and sequence analysis of IgH V-D-J in HL-60 cells. Chinese J Hematol 20(6):3

Qiu X et al (2003) Human epithelial cancers secrete immunoglobulin g with unidentified specificity to promote growth and survival of tumor cells. Cancer Res 63(19):6488–6495

Qiu X et al (2013) Immunoglobulin gamma heavy chain gene with somatic hypermutation is frequently expressed in acute myeloid leukemia. Leukemia 27(1):92–99

SEER (2018) Surveillance, epidemiology, and end results (SEER) program Cancer stat facts: Leukemia—acute myeloid leukemia (AML). https://seer.cancer.gov/statfacts/html/amyl.html

Shi Z et al (2019) More than one antibody of individual B cells revealed by single-cell immune profiling. Cell Discov 5:64

Solary E et al (2014) The Ten-Eleven Translocation-2 (TET2) gene in hematopoiesis and hematopoietic diseases. Leukemia 28(3):485–496

Spits H, Cupedo T (2012) Innate lymphoid cells: emerging insights in development, lineage relationships, and function. Annu Rev Immunol 30:647–675

Tang J et al (2018) Lung squamous cell carcinoma cells express non-canonically glycosylated IgG that activates integrin-FAK signaling. Cancer Lett 430:148–159

van der Heijden M, Kraneveld A, Redegeld F (2006) Free immunoglobulin light chains as target in the treatment of chronic inflammatory diseases. Eur J Pharmacol 533(1–3):319–326

Van Meter ME et al (2007) K-RasG12D expression induces hyperproliferation and aberrant signaling in primary hematopoietic stem/progenitor cells. Blood 109(9):3945–3952

Wang C et al (2015) IGK with conserved IGKappaV/IGKappaJ repertoire is expressed in acute myeloid leukemia and promotes leukemic cell migration. Oncotarget 6(36):39062–39072

Wang S et al (2019) Metabolic reprogramming of macrophages during infections and cancer. Cancer Lett 452:14–22

Wu Y, Hirschi KK (2020) Tissue-resident macrophage development and function. Front Cell Dev Biol 8:617879

Wu L et al (2020) High levels of immunoglobulin expression predict shorter overall survival in patients with acute myeloid leukemia. Eur J Haematol 105(4):449–459

Wynn TA, Vannella KM (2016) Macrophages in tissue repair, regeneration, and fibrosis. Immunity 44(3):450–462

Xia M et al (2022) Next-generation sequencing revealed a distinct immunoglobulin repertoire with specific mutation hotspots in acute myeloid leukemia. Biology (Basel) 11(2):161

Zheng J et al (2009) Immunoglobulin gene transcripts have distinct VHDJH recombination characteristics in human epithelial cancer cells. J Biol Chem 284(20):13610–13619

SPRINGER NATURE

GPSR Compliance

The European Union's (EU) General Product Safety Regulation (GPSR) is a set of rules that requires consumer products to be safe and our obligations to ensure this.

If you have any concerns about our products, you can contact us on ProductSafety@springernature.com

In case Publisher is established outside the EU, the EU authorized representative is:

Springer Nature Customer Service Center GmbH
Europaplatz 3
69115 Heidelberg, Germany

The manufacturer's authorised representative in the EU is Springer Nature Customer Service Centre GmbH, Europaplatz 3, 69115 Heidelberg, Germany. If you have any concerns regarding our products, please contact ProductSafety@springernature.com

Printed and bound by CPI Group (UK) Ltd, Croydon, CR0 4YY
01/10/2026
02225862-0019